ADVANCES IN

Applied Microbiology

Edited by

SAUL L. NEIDLEMAN
Emeryville, California

ALLEN I. LASKIN
Somerset, New Jersey

VOLUME 35

Academic Press, Inc.
Harcourt Brace Jovanovich, Publishers
San Diego   New York   Boston
London   Sydney   Tokyo   Toronto

This book is printed on acid-free paper. ∞

COPYRIGHT © 1990 BY ACADEMIC PRESS, INC.
All Rights Reserved.
No part of this publication may be reproduced or transmitted in any form or by any means, electronic or mechanical, including photocopy, recording, or any information storage and retrieval system, without permission in writing from the publisher.

ACADEMIC PRESS, INC.
San Diego, California 92101

United Kingdom Edition published by
ACADEMIC PRESS LIMITED
24-28 Oval Road, London NW1 7DX

LIBRARY OF CONGRESS CATALOG CARD NUMBER: 59-13823

ISBN 0-12-002635-X (alk. paper)

PRINTED IN THE UNITED STATES OF AMERICA
90 91 92 93   9 8 7 6 5 4 3 2 1

CONTENTS

Preface.. ix

Production of Bacterial Thermostable α-Amylase by Solid-State Fermentation: A Potential Tool for Achieving Economy in Enzyme Production and Starch Hydrolysis

B. K. Lonsane and M. V. Ramesh

I. Introduction	1
II. Possible Modes for Economy	3
III. Solid-State Fermentations Involving Bacterial Strains	7
IV. Specific Advantages of the Solid-State Fermentation Technique in α-Amylase Production	8
V. Current Status of Industrial Production of Bacterial α-Amylase	13
VI. Bacterial α-Amylase Production Using Solid-State Fermentation: Present Status	18
VII. Epilogue	46
References	47

Methods for Studying Bacterial Gene Transfer in Soil by Conjugation and Transduction

G. Stotzky, Monica A. Devanas, and Lawrence R. Zeph

I. Introduction	58
II. Literature Review of Conjugation and Transduction in Soil	73
III. Terrestrial Microcosms	105
IV. Selection, Preparation, Maintenance, and Storage of Soils	111
V. Methods for Studying Conjugation	115
VI. Methods for Studying Transduction	132
VII. Identification, Characterization, and Confirmation of Recombinants	139
VIII. Quality Assurance and Quality Control	152
IX. Appendix	156
References	159

Microbial Levan

Youn W. Han

I. Introduction	171
II. Occurrence	172

III. Biosynthesis	174
IV. Chemical Structure and Properties	178
V. Analytical Methods	184
VI. Production of Levan	186
VII. Utilization of Levan	188
VIII. Summary	191
References	191

Review and Evaluation of the Effects of Xenobiotic Chemicals on Microorganisms in Soil

R. J. HICKS, G. STOTZKY, AND P. VAN VORIS

I. Introduction	197
II. Soil as a Microbial Habitat	198
III. Microorganisms and Their Activities in Soil	203
IV. Xenobiotics	209
V. Environmental Factors That Influence Interactions between Microorganisms and Xenobiotics	216
VI. Effects of Xenobiotics on Microorganisms in Soil	219
VII. Methods of Assessing the Effects of Xenobiotics on Microorganisms	234
VIII. Conclusions and Recommendations	245
References	249

Disclosure Requirements for Biological Materials in Patent Law

SHUNG-CHANG JONG AND JEANNETTE M. BIRMINGHAM

I. Principles of the Patent System	256
II. History of Depositing Microbial Cultures in Connection with Patent Applications	257
III. Legal Decisions Affecting Deposits for United States Patents	258
IV. Deposit of a Living Organism in a Culture Collection for the Purpose of Patent Disclosure	261
V. United States Regulations Concerning the Deposit of Biological Materials	266
VI. Budapest Treaty	268
VII. Other International Agreements Relevant to Biological Inventions	274
VIII. Patent Law in European Patent Convention Countries	279
IX. Patent Law in European Countries outside the European Patent Convention	279
X. Patent Law in Non-European Countries	280
XI. Summary of Deposit of Biological Materials	281
XII. Patent Protection for New Plant Varieties and New Animals	284

XIII. Summary of International Protection for Microorganisms, Plants, and Animals	289
XIV. Conclusion	292
References	292
Index	295
Contents of Previous Volumes	309

PREFACE

Change is sometimes short-lived. Dr. Allen Laskin has rejoined *Advances in Applied Microbiology* as co-editor after a one-year hiatus. He could not stay away from the tumult of editorship. We welcome his return.

The two of us will continue to invite contributions on timely topics for the *Advances*. As always, we welcome proposals for manuscripts from the readers of these volumes.

The chapters in this volume cover a broad spectrum of topics, from soil microbiology to patent law. We trust that you will find them useful in your research and other professional endeavors.

Saul L. Neidleman

Production of Bacterial Thermostable α-Amylase by Solid-State Fermentation: A Potential Tool for Achieving Economy in Enzyme Production and Starch Hydrolysis

B. K. LONSANE AND M. V. RAMESH

Fermentation Technology and Bioengineering Discipline
Central Food Technological Research Institute
Mysore 570 013, India

I. Introduction
II. Possible Modes for Economy
III. Solid-State Fermentations Involving Bacterial Strains
IV. Specific Advantages of the Solid-State Fermentation Technique in α-Amylase Production
V. Current Status of Industrial Production of Bacterial α-Amylase
VI. Bacterial α-Amylase Production Using Solid-State Fermentation: Present Status
    A. Historical Highlights
    B. Bacterial Cultures
    C. Genetic Improvements of Cultures
    D. Solid Substrates
    E. Supplementary Nutrients
    F. Moisture Level
    G. pH
    H. Culture Vessels
    I. Autoclaving
    J. Nature and Ratio of Inoculum
    K. Incubation Temperature and Time
    L. Aeration
    M. Growth Characteristics
    N. Enzyme Yields
    O. Recovery of Enzyme
    P. Clarification and Purification
    Q. Enzyme Characteristics
    R. Economic Considerations
    S. Research and Development Needs
VII. Epilogue
    References

I. Introduction

Bacterial thermostable α-amylase (1,4-α-D-glucan glucanohydrolase, EC 3.2.1.1), one of the enzymes used in saccharification of starch, randomly attacks α-D-(1 → 4)-glucosidic linkages in starch thereby leading to the production of limit dextrins (Kerr, 1950; Fullbrook, 1984). It makes the starch amenable to the action of amyloglucosidase (1,4-α-D-glucan glucohydrolase, EC 3.2.1.3) for saccharification to reducing sugars. α-Amylase is one of the industrially important enzymes produced in higher quantities and represents ~12% of the sales value of the world market for enzymes (Aunstrup et al., 1979). In Japan alone, 600 tons of bacterial α-amylase powder was produced annually as early as 1963–1964 by submerged fermentation or SmF (Arima, 1964).

In addition to its use in starch-processing industries, the ability of α-amylase to cause midchain random degradation of starch (Fogarty, 1983) has been of vital importance to several other industries. Notable among these are the brewing, textile, and paper industries (Wiseman, 1975). It facilitates the use of highly concentrated barley syrup in brewing, whereas it hydrolyzes starch to a product of suitable viscosity for coating and surface sizing of paper. In textile industries it is used for desizing purposes. In all these cases, the enzymic hydrolysis of starch is a critical step that strongly influences further unit operations. For example, it influences the wort sugar composition and consequently the quality of fermented beer (Slaughter, 1985; Weig et al., 1969; Quittenton, 1969; Stewart and Russel, 1985). Similarly, the product yield, filtration characteristics of the hydrolysate, and overall efficiency of the process are influenced in the manufacture of glucose syrups, dextrose, high-fructose syrup, and ethanol (Madsen and Norman, 1973; Barfoed, 1967; Coker and Venkatasubramanian, 1985; Howling, 1984).

The recent worldwide explosion of interest in the use of renewable raw materials for manufacture of organic chemical feedstocks, sweeteners, proteins, energy, and other industrially important products has opened new horizons in biotechnology. Consequently, several countries have launched major biotechnological plans and policies with substantial financial input. One of the promising renewable raw materials in many countries is starch and starchy substrates such as corn, cassava, sweet potato, sweet sorghum, potato, and other plants rich in starch. Many products such as glucose syrups, dextrose, high-fructose syrup, maltose syrups, cyclodextrins, and ethanol are currently produced from these substrates. Ethanol is also considered one of the key chemical feedstocks (Nyiri and Charles, 1977) as well as an energy source of the future, and major ethanol programs based on starchy materials are being initiated by most countries. Consequently, the production of reducing sugars from starch has acquired prime importance as a first step in many industrial processes.

The Board of Science and Technology for International Development, Office of International Affairs and the National Research Council in

Washington, D.C. (1982) have also identified the use of starchy materials for energy production as one of the main priorities in biotechnology research for international development. However, improvements in the saccharification process and a reduction in capital investment and production costs are specified by them as necessary for achieving acceptable economics. With a view to meeting these requirements, this chapter analyzes various possible modes for reducing the cost of starch hydrolysis. It also establishes that the liquefaction and partial saccharification of starch by bacterial thermostable α-amylase is one of the most cost-intensive unit operations, chiefly because of the cost of the enzyme. In addition, it identifies the potential of the solid-state fermentation (SSF) technique for reducing the cost of producing bacterial thermostable α-amylase and reviews the literature on its production using the SSF process.

II. Possible Modes for Economy

Analyses of various unit operations involved in starch hydrolysis, their present status, and the potential for achieving further economies through advancements in scientific and technological aspects are discussed in the following paragraphs.

1. Vacuum concentration of the hydrolysate to a specific total solid content is involved in the industrial process for the production of glucose syrups and dextrose. It is necessary to evaporate a minimum of about an equal weight of water as compared to the final weight of the product in the case of glucose syrups. The use of an increased starch concentration in the slurry during hydrolysis can therefore reduce the expenses of vacuum concentration. However, the physicochemical properties of starch impose limitations in the use of higher starch concentrations as a result of gelatinization of the starch and undesirable viscosity development. The starch concentration used by the industry is 30–40%. There is no potential for reduction in cost in this aspect, as any further increase in starch concentration will lead to operational problems during hydrolysis of the starch. The possibility that the hydrolysate may contain some ungelatinized or unliquefied starch may also pose problems in downstream processing.

2. Although starch can be hydrolyzed by acid, acid–enzyme, and enzyme–enzyme techniques, the latter is preferred because of various advantages in hydrolysis, downstream processing operations, process economy, and product quality. The important advantages of the saccharification of starch by the enzyme–enzyme mode include higher

yield, higher purity, easier crystallization, better process control, lower cost of production, lower requirement for ion exchange capacity, lower capital and recurring expenses, significant reduction in energy requirement, elimination of heavy depreciations on expensive corrosion-resistant equipment, possible potential for production of new products, need for using milder process operations, and the formation of fewer by-products (Barfoed, 1967; Madsen and Norman, 1973; Fullbrook, 1984). The remainder of our discussion is therefore confined to the enzyme–enzyme technique.

3. Many of the larger enzyme users, such as starch-processing industries, are cognizant of the fact that a significant proportion of their consumable costs are due to the expenditure on enzymes (Fullbrook, 1984). It is estimated that ~12% of the raw-material cost, other than the expense of starch, is due to the expense of bacterial thermostable α-amylase used in the saccharification of the starch (Ramesh and Lonsane, 1987a).

4. It is essential to carry out the liquefaction and partial saccharification of starch by α-amylase at high temperature because of the initiation of gelatinization of starch around 65°–70°C and its completion at 130°–140°C or even at 150°C, although 105°–110°C is considered sufficient for most starches (Kerr, 1950; Norman, 1981). This is effected in industry either at 85°–90°C with the use of enzyme in two doses or at 105°–115°C in a jet cooker with single-enzyme dose. The former method requires more enzyme as compared to the latter and also involves cooking of the slurry at 120°–150°C after the action of the first dose of the enzyme to achieve complete gelatinization. The use of precisely controlled jet cookers with shorter cooking time, which involve high energy cost, offers little opportunity for cost reduction.

5. The recent advances in enzymic hydrolysis of native or raw starch at room temperature, without any cooking, are of considerable industrial importance as they could lead to a substantial reduction in energy cost (Ueda and Koba, 1980; Matsuoka et al., 1982; Ueda et al., 1984). The hydrolysis of uncooked glutinous rice starch to glucose by glucoamylase produced by black mold was found to be complete though accomplished at a slow rate (Ueda, 1957). However, the hydrolysis rate could be improved 3-fold with the synergistic action of α-amylase in the process (Ueda et al., 1984). Other enzymes that have synergistic action include isoamylase from Pseudomonas sp., pullulanase from Aerobacter sp., and β-amylase (Ueda et al., 1974; Ueda and Ohba, 1976; Ueda and Marshall, 1980). The use of a mixture of enzymes from hog pancreas and Aspergillus oryzae also leads to complete hydrolysis of uncooked starch (Balls and Schwimmer, 1945). It is interesting to note that the

glucoamylases with stronger debranching activity are highly active in raw-starch hydrolysis (Ueda and Kano, 1975). It is also established that the action of the enzymes in hydrolysis of raw starch is related to their ability to be adsorbed onto starch granules. In spite of studies on an industrial scale (Matsumoto et al., 1982), industrial exploitation of this technique has not yet developed, probably as a result of the lack of supporting technical information. It should be emphasized that this technique offers tremendous economic advantages over the presently employed methods for saccharification of starch.

6. Bacterial α-amylases are also considered to be present in two forms: liquefying α-amylase and saccharifying α-amylase. These are distinguished based on their action on starch. Saccharifying α-amylase has the capability to increase reducing-sugar formation by $\sim$2-fold over that formed by liquefying α-amylase (Fogarty and Kelly, 1979). However, saccharifying α-amylase does not function, as does glucoamylase, by exclusively producing glucose from the nonreducing end (Keay, 1970). Consequently, the commercial use of this enzyme is limited because the rate of glucose production is much slower than that of dextrins. Moreover, it does not possess liquefying properties (Fogarty and Kelly, 1979) and its action is also much lower than that of glucoamylase. It is interesting to note that saccharifying α-amylase exhibits amylase, cyclodextrinase, and maltase activities, and all these are due to one molecular species (Keay, 1970).

7. Pullulanase, a debranching enzyme produced by Klebsiella aerogenes, splits α-1,6-glucosidic linkages in pullulan and amylopectin (Peppler and Reed, 1987) and thus can accelerate the rate of hydrolysis of α-1,6 linkages in liquefied starch to glucose when used in combination with glucoamylase. The latter is also capable of hydrolyzing α-1,6 linkages but at a much slower rate (Fogarty and Kelly, 1979). Consequently, pullulanases have become commercially available in recent years (Peppler and Reed, 1987). Isoamylase, another debranching enzyme produced by Pseudomonas sp., Cytophaga sp., and yeasts, is also capable of hydrolyzing all of the α-1,6 interchain linkages in amylopectin (Fogarty and Kelly, 1979). However, it is not capable of acting on pullulan. Moreover, β-limit dextrins are not completely debranched and a minimum of three α-D-glucose residues are required by this enzyme in the B or C chains of the substrate (Fogarty and Kelly, 1979). Hence, it is not preferred over pullulanase because of this strong specificity. It is doubtful that the economic benefits from the use of these debranching enzymes will be substantial.

8. Successful techniques are available for efficient immobilization of bacterial thermostable α-amylase (Ramesh and Singh, 1981; Linko et

al., 1975; Walton and Eastman, 1973) and amyloglucosidase (Smily, 1971; Lee *et al.*, 1975, 1976). The use of immobilized enzymes could lead to a reduction in the quantity of enzymes required as compared to that needed with free enzymes. However, the use of immobilized enzymes in starch hydrolysis poses technical and operational problems and, hence, these are presently used in free form by the industry. Consequently, the contribution of enzyme cost to the overall product cost is of comparatively high order. There is no opportunity at present to reduce it by repeated use of the enzymes, as would have been possible with immobilized enzymes.

9. Improvement in the yield of product and the consequent cost reduction are known to make a major difference in the overall economics of the process (Bull *et al.*, 1982). The technology used today for the production of bacterial α-amylase by SmF is of the highest possible efficiency because of improvements achieved during extensive work carried out over the last two decades. All the possible angles— including the selection of strains, optimization of parameters, factors affecting the biosynthesis of enzyme, stabilization of the enzyme, characterization of the enzyme, kinetic studies, and genetic improvements (Chandra *et al.*, 1980; Aiba *et al.*, 1983; Zhang *et al.*, 1983; Saito and Yamamoto, 1975; Chiang *et al.*, 1979; Rosendal *et al.*, 1979; Akamatsu and Sekiguchi; 1984) were worked on to develop the present technologies. Unless highly improved yields are achieved with greater reduction in cost of production, there does not seem to be much room for additional economies in the processes presently employed on an industrial scale for the production of the enzyme.

10. It was stressed by Arima (1964) that the most important factor for the production of enzyme at lower cost is to have a method of culture that gives a high rate of enzyme production. The SSF technique fulfils this requirement and hence has been receiving increased attention in recent years. The application of the SSF technique for the production of enzymes involved in the saccharification of starch may therefore offer the desired opportunity for cost reduction.

11. Between the two enzymes used in the saccharification of starch, a keen competition among enzyme manufacturers to provide the most cost-effective product is documented only for amyloglucosidase (Fullbrook, 1984). The technology used is also of higher efficiency (Fogarty and Kelly, 1979; Chang and Terry, 1973; Smith and Frankiewicz, 1976; Barton *et al.*, 1972; Hayashida, 1975). The SSF technique is used by most of the industrial concerns for the production of amyloglucosidase (Ramakrishna *et al.*, 1982; Ghildyal *et al.*, 1985), and thus it enjoys the benefits and advantages of the SSF technique. The price of amyloglu-

cosidase has dropped dramatically since 1984, primarily as a result of lowered production cost due to yield improvement and other changes in the process technology (Scott et al., 1987). It is stated to be one of the cheapest enzymes (Aunstrup, 1979). Further discussion is therefore confined to bacterial thermostable α-amylase.

12. The bacterial thermostable α-amylase is currently produced throughout the world by employing the SmF technique (Aunstrup et al., 1979; Böing, 1982), although liquid surface fermentation (LSF) was also used in an earlier period (Wallerstein, 1939; Boidin and Effront, 1917a,b; Schultz et al., 1931). The advantages and benefits of the SSF technique will thus be available when bacterial α-amylase is produced with this technique. It deserves attention and research and development efforts for possible industrial exploitation.

III. Solid-State Fermentations Involving Bacterial Strains

The industrial exploitation of the SSF technique is mainly confined to processes involving fungi (Chahal, 1983). It is generally believed that these techniques are not suitable for bacterial cultivation (Lonsane et al., 1985) mainly because of the bacterial requirement for higher water activity (Hahn-Hägerdal, 1986; Troller and Christian, 1978; Scott, 1957). However, successful bacterial growth using the SSF technique is known in many natural fermentations. Examples include composting (Ralph, 1976; Aidoo et al., 1982), several types of food fermentation, and bacterial attack on underground pipes (Cragnoline and Tuovinen, 1984) and pipe-jointing rings (Cundell and Mulcock, 1976). Composting is reported to involve >2000 bacterial species of many different genera (Biddlestone and Gray, 1985). Bacterial counts of 10^8–10^9 per gram of moist compost indicate their vast numbers and extensive growth under SSF conditions. The normal metabolic activities of bacteria in composting are not altered, since they are even able to form endospores and complete their life cycle. Moreover, their ability to withstand considerable heat and desiccation, both common phenomena in the SSF technique, was also reported (Biddlestone and Gray, 1985).

Some industrial fermentation processes involving bacterial strains in the SSF technique are also being successfully exploited. Examples include several types of food fermentation (Stamer, 1983; Lehrian and Patterson, 1983; Castelein and Verachtert, 1983; Peppler, 1983; Liepe, 1983; Beuchat, 1983). The industrial production of itahikinatto involves the fermentation of soybeans in the SSF technique by *Bacillus subtilis* or *Bacillus natto* (Hesseltine, 1981). The secondary microflora development in cheese involves *Lactobacillus* species and *Propionibacterium shermanii* (Scott, 1981; Davis, 1965). Work was carried out to determine the feasibility of mass growth of bacteria on different solid media in closed vessels (Porter and Nack, 1960). The technique involving growth of bacteria on semipermeable membranes was reported to be more effective than SmF, while other solid media were found to compare unfavorably. Bacterial proteases are reported to be produced using *B. subtilis* in the SSF technique (Yasunobu and McConn, 1970). In fact, some reports (Beckord et al., 1945; Qadeer et al., 1980; Feniksova et al., 1960; Park and Rivera, 1982; Tobey and Yousten, 1976; Lulla and Subrahmanyan, 1954) and patents (Oji Paper Co., Ltd., 1961; Minagawa and Hamaishi, 1962; Christensen, 1944) are available on the production of bacterial α-amylase under SSF conditions. Thus it may be possible to harvest the benefits of SSF techniques in the industrial production of bacterial α-amylase.

IV. Specific Advantages of the Solid-State Fermentation Technique in α-Amylase Production

A large number of advantages are specifically obtainable when bacterial thermostable α-amylase is produced in an SSF process. These are enumerated here.

1. In recent years, it has been argued that a major obstacle in developing biotechnological processes is the large quantity of water that usually has to be disposed of during product upgrading in SmF processes (Hahn-Hägerdal, 1986). Thus, any other method of reduction in the amount of fermentation broth that has to be processed will have far-reaching effects on economics, and the SSF technique offers this opportunity. This advantage of SSF may be illustrated by the example of amyloglucosidase production (Ramakrishna et al., 1982). *Aspergillus niger* produces 600 units of enzyme per milliliter in SmF as compared to 6000 units/ml extract obtainable from 1 g dry moldy bran using the SSF technique. Thus it would be necessary to handle 10 times the volume with SmF as compared to the SSF process. Similarly, it is necessary to handle 10,000 liters of extract when 9.1 kg gibberellic acid per batch is produced with the SSF technique as compared to 58,000 liters in the case of SmF (Kumar and Lonsane, 1987c). These advantages are available because of the comparatively higher concentration of the product in the SSF technique. Despite great advances that have been made in increasing the yield of product in industrial SmF processes, the concentration of products in cell-free fermentation broths is still very low. Consequently, more research on the SSF technique has been undertaken

as a spin-off effect (Hahn-Hägerdal, 1986). In fact, SSF is considered as a ripe area for further investigation from a biotechnological point of view (Steinkraus, 1984).

2. Because of the lower concentration of products in fermentation broths, the volume of effluent to be disposed of by efficient waste treatment will also be substantially higher with SmF. Waste treatment techniques are costly and involve capital and recurring expenses without any return to the industry. This, in turn, increases the cost of manufacture of products under SmF conditions. The involvement of ~10-fold less liquid handling in SSF processes (Ramakrishna et al., 1982) will therefore correspondingly reduce these expenses.

3. The cost of removal of water by separation or concentration is a costly unit operation. It was reported that cell harvesting using centrifugation and microfiltration involved between 48 and 76% of the total production cost of the product (Datar, 1986). The ratio of product recovery to fermentation costs in enzyme production was also found to be 2.0 (Rosen and Datar, 1983). Moreover, the primary separation steps, which are known to be volume-dependent, were recently pinpointed as acutely problematic and poorly understood unit operations (Datar, 1986).

4. Enzymes and most other biological products are unstable unless handled within limited pH and temperature ranges. This instability imposes a constraint on the choice of unit operations that can be used. For example, gas phase processing is seldom possible (Bull et al., 1982). Solid–liquid separation normally involves vacuum or pressure filtration or centrifugation, which are characterized by heat input to processed liquor and sedimented solids. The low relative densities and compressible nature of biological solids also pose serious problems in solid–liquid separation (Bull et al., 1982). The instability of the product also dictates product stream handling at low temperature and with reasonable speed (Bull et al., 1982). Consequently, the need exists to employ a comparatively higher capacity infrastructure for faster recovery operations, which leads to idling of the infrastructure. The comparatively higher concentration of metabolites in the SSF technique obviates these difficulties to a considerable extent.

5. The recovery of enzyme from wheat bran (WB) koji to an extent of ~85% is possible by using two to three volumes of water (Ramakrishna et al., 1982; Arima, 1964). By employing countercurrent extraction technique, it is possible to get 1 liter of extract from 1 kg of dry moldy bran (Ramakrishna et al., 1982). The concentration of the enzyme in the extract thus obtained is sufficiently high to allow the next processing step, in which the enzyme is either precipitated by addition of salt or

concentrated under vacuum. When the enzyme is produced by SmF, the concentration of enzyme in the fermentation broth is much less. Consequently, concentration of the cell-free fermentation broth is needed before subjecting it to precipitation. Some destruction of the enzyme is inevitable, even when enzyme is concentrated at low temperature under vacuum.

6. The difficulty in maintaining sterility in large stirred-tank reactors is well known. Incidentally, this is responsible for the avoidance by the antibiotic industry of submerged fermentors >200 m^3 in volume (Bull et al., 1982). Contamination results in substantial financial losses. Most of the media used in SmF, including those for production of bacterial α-amylase, are favorable for the growth of many contaminant microorganisms. However, the use of the SSF technique may, to a large extent, obviate this problem. It is our observation that the contamination problem is largely controlled by the SSF approach (Lonsane et al., 1985), even though much concern is expressed by some workers about the vulnerability of the SSF technique to contamination. We have observed that contamination in SSF processes occurs rarely, even when the cooling of the sterilized SSF medium and the mixing of inoculum with the medium are carried out manually in an unprotected room. This may be due to low water activity of the system, which does not allow many contaminating microorganisms to grow. Moreover, the use of exceptionally high percentages of inoculum in SSF processes also aids in overcoming the contamination problem.

7. The growth characteristics of microorganisms on solid substrates are well worked out and involve the penetration of the solid particles by mycelial cells or adsorption of the cells onto the solid substrate (Lonsane et al., 1985; Chahal, 1983). It is essential that the microorganisms adhere to the substrate particle for uptake of nutrients. Enzyme production is considerably reduced in some SSF processes that involve agitation of the fermenting mass, probably because of breakage of the attachment of cells to the solid substrate particles (Mudgett, 1986). This effect was reported to be very prominent in A. niger and other coenocytic genera such as Rhizopus (Arima, 1964). All these observations indicate that the ability of the microorganism to grow in mycelial form and to adhere to the substrate particle is probably the most important characteristic for their successful exploitation in the SSF technique. Species of the genus Bacillus and especially the strains used for α-amylase production are known to form filamentous cells (Gibson and Gordon, 1974). Bacterial transport through porous solids (Shales and Kumarasingham, 1987), the influence of ionic strength, pH, and protein layer on the interactions between bacterial cells and glass surfaces

(Abbott et al., 1983), other physicochemical interactions (Rutter and Vincent, 1984; Zo Bell and Anderson, 1936), mechanism of bacterial adhesion to solids (Costerton et al., 1985; Rutter, 1984), and other related factors and phenomena (Shales, 1985; Wardell et al., 1983) have been thoroughly studied and well documented. Bacterial adhesion to solids is made irreversible by the production of exopolysaccharides by the cells during initial growth (Wardell et al., 1983; Marshall, 1984). The capability of B. subtilis to adhere to solids has been reported (Daniels, 1967; Krishnamurti and Soman, 1951). Thus, the successful exploitation of bacteria in SSF processes is also scientifically possible.

8. The production of bacterial thermostable α-amylase in the SmF technique requires, in general, a fermentation time of 96–120 hours for achieving maximum production (Saito and Yamamoto, 1975). On the other hand, it takes only 24–48 hours in the SSF technique (Beckord et al., 1945; Ramesh and Lonsane, 1987b; Qadeer et al., 1980). Thus, with the reduced batch time, the use of the SSF technique for bacterial α-amylase production will lead to considerable reduction in capital and recurring expenditures.

9. The influence of some specific parameters such as limited aeration and increased CO_2 concentration on improving the yield of bacterial α-amylase under SmF conditions is well known (Markkanen and Bailey, 1975; Gandhi and Kjaergaard, 1975). These parameters normally prevail in the SSF technique and thus may lead to higher yields without any extra effort.

10. The SSF technique offers an interesting advantage in the drying of the fermented solid substrate with its enzyme content (Beckord et al., 1945) and the use of dried substrate in saccharification of starch. The fermented substrate could be dried rapidly in a variety of ways including dehydration at room temperature in an evacuated chamber, heating in a forced-air oven up to 41°C, or a combination of both (Weller et al., 1984). The dried fermented substrate can be stored for about a year without any appreciable loss in enzyme activity if it is treated with 5–10% ethanol to inactivate the cells before drying (Srikanta et al., 1987). Thus, the use of the SSF technique eliminates recovery and purification operations, thereby leading to significant economic advantages.

11. Bacterial culture variability is a formidable problem (Lockwood, 1952) in the production of α-amylase under SmF conditions. It was reported that culture variants outgrow the original culture throughout the fermentation tank in few hours. On the other hand, the variant does not influence all trays in the case of still culture with LSF (Lockwood,

1952). This advantage will also be applicable in the case of bacterial α-amylase production using SSF.

12. The advantages of the combined action of two or more α-amylases from different sources as compared to any α-amylase alone is well recognized for achieving more rapid and more complete digestion of starch and, thus, the mixture approach may have considerable merit. Christensen (1944) has also developed a process for increasing the diastatic power of barley malt or soybeans by cultivating α-amylase-producing fungi on these substrates without subjecting the grains to sterilization. Thus the diastatic power of these substrates is augmented by the enzyme produced by fungi on these substrates by SSF. This approach is not achievable in SmF involving bacterial cultures, since sterilization of the liquid medium is a must for preventing contamination. However, such an approach is possible in the SSF technique involving bacterial strains.

13. It has been reported that enzyme extracts contain comparatively fewer chemical impurities when produced on solid media as compared to liquid media (Aunstrup, 1974). Such enzyme extracts from SSF processes can be concentrated to a higher extent than the usual 2- to 4-fold concentration possible in the case of liquid medium under SmF fermentation, when they become too viscous for further handling (Meyrath and Volavšek, 1975). In addition, smaller volumes of salts or solvents are required for precipitation and further purification is facilitated (Meyrath and Volavsek, 1975).

14. In the production of bacterial α-amylase under LSF and SmF conditions, improved yields were reported resulting from the inclusion of extracts from proteinaceous as well as starchy materials of vegetable and animal origin in the production media. These include extracts from soybean, soybean cake, linseed cake, cottonseed meal, potato, rice bran, wheat bran, defatted peanut meal, cornmeal, rolled oats, and juices from other plants (Arima, 1964; Boidin and Effront, 1924; Wallerstein, 1939; Tilden and Hudson, 1942; Mahmoud et al., 1983; Yabuchi and Okita, 1985; Babbar et al., 1960). Preparation of the extract from these is a costly process as it involves steam cooking, filtration of fibrous residue from the cooked extract, and sterilization of the final medium involving the extract. In fact, these substances can be used in solid form in SSF and thus will result in considerable economy.

15. It is well known that the ability of B. subtilis to produce α-amylase in the SmF techniques is somewhat unstable (Casida, 1964). Moreover, it is difficult to detect culture degeneration that may occur during the course of fermentation, thereby leading to the loss of the whole fermentation batch. In LSF, the degeneration of B. subtilis is

reported to be less serious than in SmF culture (Casida, 1964). This may also hold true for the SSF process.

Thus the production of bacterial α-amylase in an SSF process has promise. It offers many technological and economical advantages that are not available with SmF.

V. Current Status of Industrial Production of Bacterial α-Amylase

α-Amylase is produced by a variety of plants, animals, and microorganisms (Underkofler, 1976). Although considerable quantities of some enzymes are currently produced commercially from animal and plant sources, enzymes from microbial sources play an important role in the technical and economic aspects of enzyme production on an industrial scale (Vincent and Priestley, 1975). The economic reasons cited include the possibility of bulk production, ease of extraction, predictable output, and product uniformity (Vincent and Priestley, 1975). In contrast to the limited availability of animal tissue, and dependence on available acreage, labor, and vagaries of the weather in the case of plant material, some of the major technical advantages of using microorganisms for production of enzymes are the unlimited capacity for expansion and the ability to select and bring about mutation of microbial cultures to obtain enzymes of desired characteristics (Underkofler, 1976).

Although yeasts, fungi, bacteria, actinomycetes, and algae are able to produce α-amylase (Windish and Mhatre, 1965), only the enzyme from fungal and bacterial sources has industrial applicability. Fungal α-amylase is used in the baking, food, and pharmaceutical industries to enhance the α-amylase levels of diastatically deficient raw materials or as a digestive aid (Wiseman, 1975). The fungal α-amylases are produced using either SmF or SSF techniques (Cassida, 1964; Smith, 1969). They have temperature optima for enzyme activity at 50°–60°C and hence are not useful in starch hydrolysis.

For complete gelatinization of starch, the thermostability of α-amylase at a minimum temperature of 80°–90°C in a 30% starch slurry is of vital importance. Another important characteristic is a broad pH range for enzyme activity, which is highly desirable for use of the enzyme in various processes and applications. These two characteristics are of profound importance in the manufacture of glucose syrups, dextrose, reducing sugars, and high-fructose syrup. For example, the use of high temperature during liquefaction of starch also leads to the precipitation of proteins, present in starch as an impurity, and the consequent lower expenses in downstream processing. Bacterial α-amylase with pH optima on the acidic side are known to improve the yield of glucose from starch as a result of lower formation of maltulose precursors during liquefaction (Luenser, 1983). Alkaline amylases are also reported to be less sensitive to chelating agents like EDTA than α-amylase from *Bacillus amyloliquefaciens* (Boyer et al., 1979).

Bacterial α-amylases produced by *Bacillus* species are thermostable even up to 110°C (Madsen et al., 1973). The commercially produced enzymes have, however, pH optima for enzyme activity at neutral or slightly acidic pH ranges (Norman, 1981). In spite of some reports on bacterial α-amylases with broad pH range (Saito, 1973; Medda and Chandra, 1980), or with pH optima on the alkaline side (Horikoshi and Akiba, 1982; Boyer et al., 1979; Medda and Chandra, 1980), or with two or three peaks of pH optima in acidic, neutral, and alkaline ranges (Kelly and Fogarty, 1976), these have not been adopted commercially, most likely because of their lower thermostability, low yield of the enzyme, and insufficient information. The bacterial α-amylase from *B. amyloliquefaciens* or *B. licheniformis* are widely accepted by the starch-processing industry for the hydrolysis of starch. These enzymes have superior characteristics and thus dominate the market. It should be emphasized that the application of α-amylase in different industrial processes demands specific characteristics and these are more or less satisfied by the bacterial thermostable α-amylases. They are produced by the fermentation industry employing the SmF culture technique, using technical know-how of the highest possible efficiency (Aunstrup et al., 1979).

The submerged culture process involves cultivation of the selected culture in fermentation vessels with a diameter–height ratio of about 1:3. Consequently, the depth of the fermentation medium in the vessel is also greater as compared to the shallow layer of medium employed in the LSF process. The nutrients are dissolved in a large body of water. Substrates that are insoluble or immiscible in water can also be used but at concentrations ranging from 1 to 10%. Consequently, the fermentation medium is free-flowing in nature as compared to the absence of free water in the SSF technique. Various physical and chemical parameters are maintained either manually or automatically at predetermined optimal values for achieving highest possible yields of the product. The SmF process involves many steps and unit operations (Aunstrup et al., 1979; Böing, 1982), and its advantages as well as limitations are described elsewhere (Kumar and Lonsane, 1989).

The production of bacterial thermostable α-amylase by the SmF process has been thoroughly investigated and is known to be affected by a variety of physicochemical factors. Most notable among these are the genetic improvement of the strain, composition of growth medium, pH, phosphate concentration, inoculum age, aeration, temperature, availability of an efficiently metabolizable carbon source, and the use of a complex nitrogen source (Aiba et al., 1983; Saito and Yamamoto, 1975; Chiang et al., 1979; Chandra et al., 1980; Zhang et al., 1983; Tobey and Yousten, 1976; Mahmoud et al., 1983; Kelly and Fogarty, 1976; Sekiguchi et al., 1975; Maruo and Tojo, 1985). Enzyme formation is known to be induced by oligosaccharides or repressed by exogenous glucose or low molecular weight metabolites (Saito and Yamamoto, 1975; Vortruba et al., 1984; Srivastava and Mathur, 1984; Welker and Campbell, 1963a,b), although it was also reported that the inducer is not required (Meers, 1972). The inclusion of carotenoids (Clark et al., 1977), glycine (Zhang et al., 1983), or cyclic AMP (Saito and Yamamoto, 1975) in the medium is reported to enhance the enzyme production many fold. Potentially productive cells are produced during trophophase, whereas the enzyme is produced in idiophase; these two phases are separated by a transition phase (Ingle and Erickson, 1978; Yuki, 1967; Nomura et al., 1956). In some cases, enzyme production in the exponential phase is also reported (Meers, 1972; Vortruba et al., 1984; Coleman and Grant, 1966; Davies et al., 1980). Studies were also carried out for production of bacterial thermostable α-amylase by continuous fermentation (Vortruba et al., 1984; Meers, 1972; Priest and Thirunavukkarasu, 1985; Heineken and O'Connor, 1972; Davies et al., 1980). Fed-batch culture (Baig et al., 1984; Pazlarova et al., 1984) and the use of immobilized bacterial cells (Chevalier and de la Noüe, 1987; Kokobu et al., 1978; Shinmyo et al., 1982) were also investigated for achieving a higher yield of the enzyme. A number of critical and exhaustive reviews on the production of bacterial thermostable α-amylases by SmF are also available (Windish and Mhatre, 1965; Fogarty and Kelly, 1979, 1980; Ingle and Erickson, 1978; Priest, 1977; Fogarty, 1988).

A typical fermentation medium used for production of the enzyme by B. subtilis includes (in g/dl): ground soybean meal, 1.85; autolysed brewer's yeast, 1.50; distillers dried solubles, 0.76; casein hydrolysate, 0.65; lactose, 4.75; $MgSO_4 \cdot 7H_2O$, 0.04; Hodag KG-1 antifoam agent, 0.05. The fermentation medium is sterilized at 121°C for 35 minutes and is inoculated, after cooling to 35°C, with 0.5% (vol/vol) inoculum. The fermentation is run at 35°C and the pressure maintained in the fermentor vessel of 200 liters working capacity is 3–5 psi. The medium is agitated at 170 rpm while air is supplied at a rate of 35 ft^3/min; the run is terminated after 48 hours (Underkofler, 1976). The scheme employed for downstream processing of the fermented medium to obtain crude enzyme powder is given in Fig. 1.

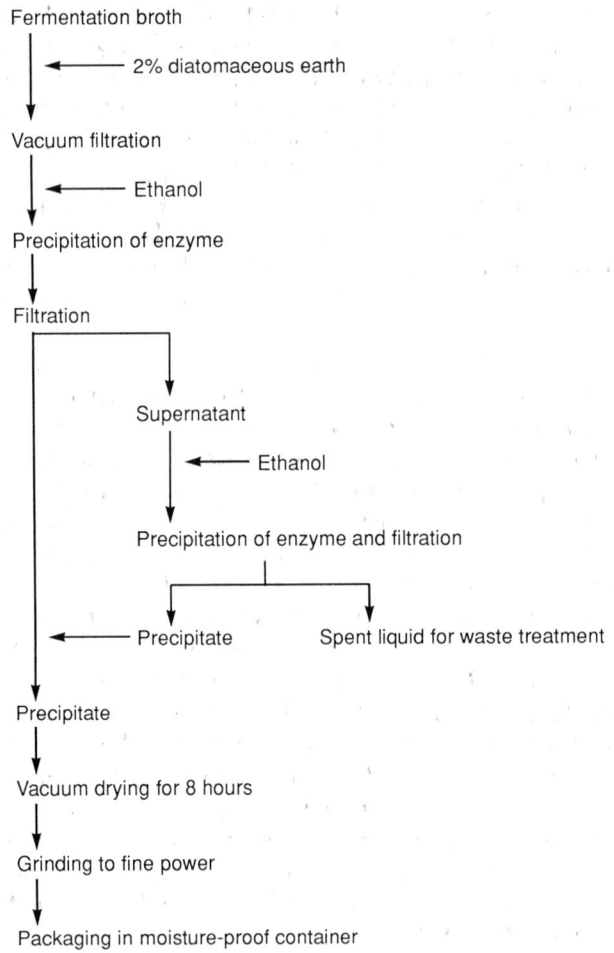

FIG. 1. Downstream processing of bacterial α-amylase produced by submerged fermentation for obtaining crude enzyme powder.

Bacterial thermostable α-amylase is marketed either in powder form or as a liquid concentrate. The latter is preferred by the industry because of its convenience in handling. The procedure followed for making liquid concentrate consists of dissolving the precipitate in buffered water, probably because it eliminates the cost-intensive step of vacuum concentration of the dilute enzyme titer present in cell-free fermentation broth.

The LSF technique was employed in earlier years for the production of bacterial themostable α-amylase. Because it has been completely replaced by the SmF technique, the information on production of the enzyme employing LSF is of historical importance only. However, the LSF can be classified as an intermediate technique between the SmF and SSF techniques. Therefore, the production of the enzyme in the LSF technique is outlined in the following paragraphs.

Liquid surface fermentation, referred to as surface fermentation by many workers (Kumar and Lonsane, 1989), consists of growing the culture on a shallow depth of the liquid medium held in a suitable container such as stainless-steel trays. The various steps or unit operations involved in LSF, the advantages as well as limitations of the technique, the difficulties involved in scaling up the technique to an industrial level, and the economic considerations are described elsewhere (Kumar and Lonsane, 1989).

The production of bacterial α-amylase on a commercial scale was pioneered by Boidin and Effront (1917a,b) based on the use of LSF. The medium employed contained a rich nitrogen source, soybean cake, peanut cake, salts, and a low concentration of starch. A special culture apparatus to accommodate 500 and 1000 gallons of fermentation medium was constructed to hold the medium in thin layers in a large number of trays. The apparatus was also provided with controls for maintaining temperature and aeration while operating it under pressure to eliminate contamination problems. Strong aeration was provided in the initial stages of fermentation and the culture was allowed to grow in the form of a film over the shallow layers of medium to achieve a better yield. Larger scale commercial production was initiated in improved culture vessels in subsequent years (Wallerstein, 1939; Schultz et al., 1931). The bacterial cultures commonly used include B. subtilis and B. diastaticus (Hoogerheide, 1954).

Considerable work on various aspects such as development of efficient medium, factors affecting the yield of the enzyme, improvements in the design of the culture vessel, optimum inoculum ratio, treatment of some of the substrates used in the fermentation medium, growth characteristics, importance of pellicle formation over the surface of the medium, application of the drip method for trickling the medium over wood chips covered with pellicle growth, and downstream processing was carried out by many workers (Boidin and Effront, 1917a,b; Wallerstein, 1939; Schultz et al., 1931; Imshenetskii and Solntseva, 1944; Kline et al., 1944; Peltier and Beckord, 1945; Kneen and Beckord, 1946; Surovaya, 1944). The availability of stirred and aerated fermentor vessels employing deeper layers of medium for SmF in the 1940s, as well as the superiority of SmF over LSF (Dunn et al., 1959; Kumar and Lonsane, 1989) have culminated in complete cessation of work on the latter technique.

VI. Bacterial α-Amylase Production Using Solid-State Fermentation: Present Status

A. Historical Highlights

It is always interesting, both from academic and industrial points of view, to study the historical developments of a fermentation process and techniques. It gives insight into the events and thinking that led to adopting or neglecting a particular fermentation technique, based on the knowledge available at that time. The reexamination of the events and decisions made may indicate, especially in the light of more knowledge gathered through the passage of time, that the path neglected or abandoned earlier actually deserved more attention as compared to the path actually followed at that time. It also opens up new vistas for achieving improvements and economy in the accepted technique or suggests its replacement with a neglected approach.

Boidin and Effront (1917a,b) were the first to use *B. subtilis* and *B. mesentericus* for the production of α-amylase on a commercial scale using large fermentors and LSF. In fact, the employment of bacterial cultures for the production of commercial enzymes was pioneered by them (Hoogerheide, 1954). It was also reported that the applicability of solid media such as soybean cake with one to two parts of water was not favorable (Boidin and Effront, 1917a,b). The culture vessel and process were further improved (Schultz et al., 1931; Wallerstein, 1939), and LSF became accepted industrial practice throughout the world for the production of bacterial α-amylase (Imshenetskii and Solntseva, 1944; Surovaya, 1944).

Prior to these developments, fungal enzymes were extensively produced in the United States by the SSF technique as pioneered by Takamine (1894, 1911, 1914). Workers with expertise in the SSF process must have realized the advantages of the SSF technique over the LSF process and must have felt an urge to explore the SSF technique for the production of bacterial α-amylase. Underkofler and co-workers with an interest in amylolytic enzyme production in the SSF technique studied the potential of the SSF technique for production of bacterial α-amylase (Underkofler et al., 1939). However, the degree of enzyme production was reported to be extremely low with the eight cultures studied.

The interest in the SSF process for bacterial α-amylase production probably got a boost from the success achieved by Christensen (1944) in increasing the diastatic power of barley malt or soybeans by cultivating *B. subtilis* on them in a culture system similar to the SSF technique. Similarly, the successful production of bacterial α-amylase by the drip method, a technique similar in many respects to the SSF process, was also reported (Kline et al., 1944). It involved the trickling of thin stillage over wood chips covered with pellicle growth of *B. subtilis*. These reports established the ability of *Bacillus* strains to grow under a technique very similar to the SSF process and to produce α-amylase.

The first report on the successful production of bacterial α-amylase by the SSF technique was by Beckord et al. (1945). At about the same time, unfortunately, interest in the Western world was nonexistent for the LSF or SSF processes because of the emergence of the SmF process (Smythe et al., 1950). Except for a report by Tobey and Yousten (1976), further work on the topic was mainly confined to Eastern and Oriental countries until 1980. Subsequently, the bacterial α-amylase produced under the SSF technique was evaluated by Park and Rivera (1982), in comparison to the enzymes obtained by SmF processes. The expertise in SSF processes at Central Food Technological Research Institute (CFTRI), Mysore, India, and the economic advantages of the SSF technique over SmF led us (Ramesh and Lonsane, 1986, 1987a,b, 1988) to revive interest in exploitation of the SSF process for production of bacterial α-amylase in late 1985. The chronological sequence of studies on bacterial α-amylase production under the SSF process is presented in Table I.

B. Bacterial Cultures

Bacterial cultures belonging to the genus *Bacillus* are used throughout the world in the commercial production of bacterial thermostable α-amylase in the SmF and LSF processes (Wallerstein, 1939; Hoogerheide, 1954; Aunstrup et al., 1979; Underkofler, 1976). The bacterial cultures reported for the production of α-amylases in the SSF process are also limited to the genus *Bacillus* (Table II).

Beckord et al. (1945) studied 16 starch-hydrolyzing bacterial cultures for their ability to produce the enzyme in the SSF technique and reported that 11 cultures were able to produce α-amylases in varying amounts. Among these, the capability to produce high titers of the enzyme was confined to three species (*B. subtilis*, *B. mesentericus*, and *B. vulgatus*). The rest of the cultures definitely produced the enzyme but in quantities below the practical limits of measurement. Among these were *B. graveoleus*, *B. mycoides*, *B. aterrimus*, and *B. polymyxa*. Qadeer

TABLE I

Chronological Sequence of the Exploration of SSF Processes for Production of Bacterial α-Amylases

Year	Result	Reference
1939	Exploratory work; extremely low yield of enzyme	Underkofler et al. (1939)
1945	Good yields, alternative technology	Beckord et al. (1945)
1954	Process optimization	Lulla and Subrahmanyan (1954)
1960	Comparison with LSF, better yield in SSF process	Feniksova et al. (1960)
1961	Process know-how development, downstream processing	Oji Paper Co., Ltd. (1961)
1962	Genetic improvement of the culture	Minagawa and Hamaishi (1962)
1964	Advantages and economy aspects	Arima (1964)
1976	Production by *Bacillus thuringiensis* var. *Kurstaki* (HD-1)	Tobey and Yousten (1976)
1980	Factors affecting enzyme production	Qadeer et al. (1980)
1982	Evaluation against enzyme produced under SmF conditions in starch hydrolysis	Park and Rivera (1982)
1987	Achieving yields that are unattainable in SmF	Ramesh and Lonsane (1987a)
1987	Enzyme system with three peaks for pH optima	Ramesh and Lonsane (1987b)
1987	Continuation of the search for enzyme with better thermostability	Ramesh and Lonsane (unpublished observations)
1988	Recovery of enzyme from bacterial bran	Ramesh and Lonsane (1988)

et al. (1980) also tested 50 strains of *B. subtilis* for α-amylase production under the SSF technique and reported that all of them were positive but that the degree of enzyme production varied between 4 and 370 units/g WB. It is interesting to note that *B. thuringiensis* var. *Kurstaki* HD-1, the culture used for commercial production of bacterial insecticide in both SmF and SSF processes, is also able to produce amylase in these fermentation techniques (Tobey and Yousten, 1976).

The ability of the strains of *B. megaterium* and *Bacillus* HOP-40 to grow in the SSF technique and to produce α-amylase was reported by Ramesh and Lonsane (1986, 1987a,b, 1988) after screening 51 bacterial cultures belonging to the genus *Bacillus*. It was also concluded that some strains belonging to the genus *Bacillus* have the specific ability to grow and to produce higher titers of α-amylase in the SSF technique as compared to the SmF process (Ramesh and Lonsane, 1987a,b). It was

TABLE II
Bacterial Cultures Capable of Growth and Production of α-Amylase in the SSF Technique

Culture	References
Bacillus subtilis	Beckord et al. (1945); Qadeer et al. (1980); Arima (1964); Oji Paper Co., Ltd. (1961); Minagawa and Hamaishi (1962); Feniksova et al. (1960); Park and Rivera (1982); Lulla and Subrahmanyan (1954); Tobey and Yousten (1976)
Bacillus mesentericus, Bacillus vulgatus, Bacillus gaveoleus, Bacillus mycoides, Bacillus aterrimus, Bacillus polymyxa	Beckord et al. (1945)
Bacillus thuringiensis var. Kurstaki	Tobey and Yousten (1976)
Bacillus megaterium	Ramesh and Lonsane (1987a)
Bacillus HOP-40	Ramesh and Lonsane (1987b)

also speculated that this ability may be related to the requirement of water activity by these strains (Ramesh and Lonsane, 1986).

The bacterial cultures used in the studies just mentioned were maintained chiefly on nutrient agar slants (Beckord et al., 1945; Qadeer et al., 1980; Park and Rivera, 1982), although nutrient agar containing 0.3% starch was also employed (Ramesh and Lonsane, 1987a,b).

C. Genetic Improvements of Cultures

The genetic improvement of a candidate culture, isolated after an extensive screening program and studied for factors influencing its ability, is one of the important aspects in the development of an efficient fermentation process. It is invariably performed to achieve various benefits for a commercial fermentation. The improvements possible are diverse in nature and may involve higher biomass formation, greater product yield, faster kinetic reactions, elimination of co-metabolite productions, formation of product with desired characteristics, and a variety of other beneficial effects (Curtis, 1976; Egorov, 1985). In fact, these improvements are considered a necessity before using the culture on an industrial scale. Improvement in the productivity of the culture may be many fold as compared to that of the naturally occurring culture (Egorov, 1985). The genetic improvement of a given culture is carried out by physicochemical means, plasmid conjugation, protoplast fusion, recombinant DNA technology, or transformation.

In spite of the importance of genetic improvement of the culture in fermentation processes, only two reports are available in the case of production of bacterial α-amylase in the SSF technique. Such minimal effort on this important aspect of fermentation process development is, however, not surprising in that it is linked to a neglect of the SSF process in the Western world for production of bacterial α-amylase. This is in contrast to the intense efforts that have been directed by various workers toward genetic improvement of cultures used in SmF processes for the production of bacterial α-amylase (Aiba et al., 1983).

Minagawa and Hamaishi (1962) claimed the isolation of an improved mutant by subjecting B. subtilis to prolonged X-ray treatment. The culture, grown at 33°C for 24 hours in a peptone–beef broth medium containing 0.5% NaCl and 3% agar, was then regrown on solid medium for 12 hours at 33°C. It was then subjected to X-ray treatment using an X-ray intensity of 24–50 roentgens, obtained by placing the generator ~30 cm away from the colony. The duration of exposure to the X radiation was at least 30 minutes in a range of 1–4 hours. The colony was then cultured in plane culture in a shallow tray at 33°C for 24 hours using solid medium of the same composition. One mutant with the capacity to produce α-amylase and another for production of protease were recovered. The X-ray treatment thus resulted in a mutant that produces α-amylase, without coproduction of protease, if a suitable medium such as one based on wheat was employed. The other mutant, when grown on cornmeal-based medium, produced protease. The mutant colony for α-amylase production was hairless, rough, jagged, spotted, and dull white.

The genetic manipulations carried out by Qadeer et al. (1980) on B. subtilis involved the use of ultraviolet (UV) light and chemical mutagens, either singly, in combination, or in sequence. The general methodology followed included the use of 24-hour culture, centrifugal separation of cells, washing, resuspension in 0.9% saline, placing in a petri dish, and irradiation under a UV lamp for varying times. The chemical mutagen used (6-mercaptopurine at 0.1–1.0 mg/liter) was added to the saline used for resuspension of cells. Many mutants were obtained in these studies. For example, mutant PCSIR-20 was isolated after treatment with the chemical mutagen for 20 hours; it produced 5850 units/g WB against 360 units/g WB produced by the parent culture. The mutant PCSIR-20 was further exposed alternately to chemical mutagen and UV radiation to obtain mutant 6. It produced 6300 units/g WB but was found to be unstable and produced only ~3500–4000 units/g WB after two to three generations. Another mutant, PCSIR-29, was obtained by exposure to UV radiation for 40 minutes. It produced 4000 units/g WB and showed no further improvement by further treatment with UV radiation or chemical mutagen. The mutant, PCSIR-29, was reported to be stable and was selected for further studies on optimization of parameters.

Recent work on the establishment of the existence of a regulatory gene for α-amylase production in SmF processes is of commercial significance (Sekiguchi et al., 1975; Green and Colarusso, 1964; Yamaguchi et al., 1969; Yuki, 1968). Coupled with the fact that the α-amylase production in the SSF technique is greater than that by SmF, the use of genetically improved cultures in the SSF process may lead to achieving still higher yields of the enzyme.

D. Solid Substrates

In the SSF process, the solid substrate not only supplies the nutrients to the culture but also serves as an anchorage for the microbial cells. Therefore, the particle size and the chemical composition of the substrate are of critical importance (Lonsane et al., 1985). The ideal solid substrate is one that provides all the necessary nutrients to the microorganism for optimum function. The cost and availability are other important considerations. In some cases, the extent of mechanical or chemical treatment necessary to improve the amenability of the solid substrate to microbial attack and the physicochemical changes occurring after sterilization of the solid substrate determine its suitability for use in the SSF process. Thus, the selection of an appropriate solid substrate plays an important role in the development of efficient SSF processes (Lonsane et al., 1985).

Various solid substrates used by different workers in the production of bacterial α-amylase in the SSF technique are listed in Table III. These substrates were employed individually without any supplementation with other carbon and nitrogen sources. Among these, WB was reported to be the best (Qadeer et al., 1980) and most satisfactory for manipulations (Beckord et al., 1945). Qadeer et al. (1980) were unable to detect enzyme production with rice bran or peanut meal. In another study, rice bran was found to produce reasonably good titers of the enzyme, though much less than those using wheat bran (Lulla and Subrahmanyan, 1954). The inability of the culture to grow on rice husk was also reported (Qadeer et al., 1980).

It appears that the suitability of a particular solid substrate for bacterial α-amylase production in the SSF process is determined by the culture used and its physicochemical requirements. Thus, it is essential to evaluate a large number of potential solid substrates with any given

TABLE III

SOLID SUBSTRATES EMPLOYED IN BACTERIAL α-AMYLASE PRODUCTION IN THE SSF TECHNIQUE

Solid substrate	Reference
Wheat bran	Beckord et al. (1945); Qadeer et al. (1980); Arima (1964); Oji Paper Co. Ltd. (1961); Minagawa and Hamaishi (1962); Feniksova et al. (1960); Park and Rivera (1982); Tobey and Yousten (1976); Lulla and Subrahmanyan (1954)
Soybean meal, potato meal, alfalfa meal	Beckord et al. (1945)
Maize bran	Qadeer et al. (1980)
Cornmeal base	Oji Paper Co., Ltd. (1961)
Defatted groundnut cake, defatted castor seed cake, rice bran, lucerne leaf powder, lucerne stalks, rice husk, sweet potato, maize gluten (free from starch), soya residue obtained after extraction of milk	Lulla and Subrahmanyan (1954)

culture to achieve higher enzyme production. The universal suitability of WB is also apparent from literature data. It was reported to contain a sufficient amount of nutrients and was found to remain loose even in the moist state, thereby providing a large surface area (Feniksova et al., 1960).

The presence of amylase inhibitor in cereals and WB is well documented (Kneen and Standstedt, 1946; Novotel'nov et al., 1964). Studies on partially purified inhibitor, obtained by extracting WB with water, showed that it acts by oxidizing SH groups of amylases (Novotel'nov et al., 1964). It is able to inhibit the growth of B. subtilis when used at the 6.25 mg/ml level. However, much higher concentrations were necessary to inhibit the growth of A. oryzae and A. niger. Its ability to inhibit salivary, pancreatic, and bacterial amylases was also reported (Kneen and Standstedt, 1946). Although the inhibitor showed resistance to temperatures below that of the atmospheric boiling point of water, it was destroyed by autoclaving (Kneen and Standstedt, 1946). Therefore, the inhibitor present in WB is expected to be destroyed when moist WB medium is autoclaved before inoculation in the SSF technique, although some residual activity may still persist. The ability of the inhibitor to activate amylases at lower concentrations is also known (Novotel'nov et al., 1964). No data on these aspects are available in the case of bacterial α-amylase production in the SSF process.

E. SUPPLEMENTARY NUTRIENTS

Commercial wheat bran contains 8.5 and 9.5% starch and proteins, respectively (Fisher, 1973), in addition to various minerals. These nutrients are present in the numerous plant cells that are surrounded by a rigid cell wall in WB particles. The uptake of nutrients from the WB cells by fungi, due to their ability to penetrate deeply into the WB particles, is well established in SSF processes (Lonsane et al., 1985; Chahal, 1983). The nutrients present in those plant cells of WB that are not penetrated by the mycelial cells are thus not available to the culture during fermentation. Some of the vital nutrients necessary for optimum growth and product formation processes may also be present in WB at suboptimal levels. Hence, the supplementation of WB with other solid and/or water-soluble nutrients was found to lead to enhanced product formation in SSF processes (Kumar and Lonsane, 1987a,b).

The various solid and water-soluble supplementary nutrients used along with WB for improving enzyme yield are listed in Tables IV and V, respectively. The supplementation of WB with other solid substrates was found to increase the enzyme yield in general. About a 41% increase in enzyme titer was achieved by Lulla and Subrahmanyan (1954) by supplementing WB with castor seed cake. Supplementation with peanut meal also improved yield of the enzyme (Qadeer et al., 1980). Maize bran, rice husk, and penicillium waste mycelia did not improve the yield (Qadeer et al., 1980), while rice bran and gluten were found to bring about a definite drop in enzyme production (Lulla and Subrahmanyan, 1954) when used to supplement WB.

Combinations of solid substrates employed to obtain higher enzyme production include corn, WB, and rice bran in a 40 : 40 : 20 ratio (Oji Paper Co., Ltd., 1961), or WB with soybean cake, casein, and rice at 20, 5, and 20% levels, respectively (Arima, 1964). A combination of two different solid substrates was also investigated (Lulla and Subrahmanyan, 1954). Among these, groundnut cake and lucerne were reported to give better yields than were obtained with WB alone. It was stated that lucerne exerted a beneficial influence in enhancing α-amylase production when admixed with WB or groundnut cake.

The supplementation of WB with various water-soluble nutrients was also found to lead to enhanced production of α-amylase in the SSF technique (Table V). The use of nitrogen sources such as urea (Qadeer et al., 1980) or diammonium phosphate (Ramesh and Lonsane, 1987b) resulted in increased enzyme production. Other water-soluble nutrients used were diverse in nature and included casein, liquefied starch, ethanol, soybean extract, peptone, beef extract, protease, and various minerals (Table V). Calcium carbonate was also used by Tobey and Yousten

TABLE IV

Supplementary Solid Substrates Used along with Wheat Bran and Other Combinations of Substrates

Supplementary nutrient	Concentration (% or ratio)	Reference
Solid substrates		
Soybean cake/residue	10–20	Oji Paper Co., Ltd. (1961); Arima (1964); Tobey and Yousten (1976); Lulla and Subrahmanyan (1954)
Peanut meal	1 : 0.11	Qadeer et al. (1980)
Rice	20	Arima (1964)
Corn	50	Oji Paper Co., Ltd. (1961)
Rice bran	10–50	Oji Paper Co., Ltd. (1961); Lulla and Subrahmanyan (1954)
Cornmeal	—	Minagawa and Hamaishi (1962)
Groundnut cake, castor seed cake, maize gluten, lucerne stalk or leaf powder	10	Lulla and Subrahmanyan (1954)
Combination of substrates other than wheat bran		
Groundnut cake + soya residue	1 : 1	Lulla and Subrahmanyan (1954)
Groundnut cake + castor seed cake	1 : 1	Lulla and Subrahmanyan (1954)
Groundnut cake + lucerne stalk or leaf powder	1 : 1	Lulla and Subrahmanyan (1954)
Castor seed cake + soya residue	1 : 1	Lulla and Subrahmanyan (1954)

(1976), probably with the dual objective of providing Ca^{2+} and pH maintenance.

F. Moisture Level

Considerable work has been done on the effect of moisture levels on the extent of bacterial α-amylase production in the SSF technique. In fact, it is the parameter on which more workers have expended their efforts than any other parameter. However, the aspects studied involve only two factors: the ratio of solid substrate to moistening agent and the nature of the latter. The amount of moistening agent used was shown to influence the physical properties of the moist solids (Feniksova et al., 1960). For example, too much of the moistening agent was reported to make the WB less suitable for bacterial growth and enzyme production

TABLE V

WATER-SOLUBLE NUTRIENTS USED ALONG WITH WHEAT BRAN

Supplementary nutrient	Concentration (% or ratio)	Reference
Urea	1 g N per liter diluent	Qadeer et al. (1980)
Casein	5	Arima (1964)
Soybean extract	5	Park and Rivera (1982)
Liquefied starch (by bacterial α-amylase)	3	Park and Rivera (1982)
Ethanol	2	Park and Rivera (1982)
$(NH_4)_2HPO_4$	1.3	Park and Rivera (1982); Ramesh and Lonsane (1987b).
$MgSO_4 \cdot 7H_2O$	0.02	Park and Rivera (1982)
KCl	0.02	Park and Rivera (1982)
0.015 M Phosphate buffer (pH 7.0)	1 : 2.5	Beckord et al. (1945); Qadeer et al. (1980)
1/15 M Phosphate buffer (pH 7.17)	1 : 4.5	Lulla and Subrahmanyan (1954)
$CaCl_2 \cdot H_2O$	0.41	Tobey and Yousten (1976)
$MnCl_2$	0.04	Tobey and Yousten (1976)
$MgCl_2$	0.81	Tobey and Yousten (1976)
Nutrient broth containing peptone and beef extract	2.5	Tobey and Yousten (1976)
Protease	3 units	Oji Paper Co., Ltd. (1961)
$Ca(OAc)_2$	0.01	Oji Paper Co., Ltd. (1961)

because of a reduction in substrate porosity, loss of the structure of WB, development of stickiness in the moist medium, and difficulty in oxygen transfer thereby making such a medium difficult to use, especially on a large scale. On the other hand, the solubility of the nutrients present in WB were reduced by insufficient moisture (Feniksova et al., 1960).

Various agents used for moistening the solid substrates include water, phosphate buffer, or diluent solution (Beckord et al., 1945; Oji Paper Co., Ltd., 1961; Park and Rivera, 1982; Tobey and Yousten, 1976). The phosphate buffer solution of 0.015 M strength (pH 6.0–7.0) contained 1.5 g KH_2PO_4 and 3.5 g $K_2HPO_4 \cdot 3H_2O$ per liter of the water (Beckord et al., 1945; Qadeer et al., 1980; Ramesh and Lonsane, 1987a), while a 1/15 M phosphate buffer solution (pH 7.17) was employed by Lulla and Subrahmanyan (1954). The diluent solutions used included water containing 0.01% $Ca(OAc)_2$ plus 3 units of protease/ml (Oji Paper Co.,

Ltd., 1961), or nutrient sporulation broth containing 8 g/liter DIFCO nutrient broth with 7×10^{-4} M $CaCl_2$, 5×10^{-5} M $MnCl_2$, and 1×10^{-3} M $MgCl_2$ (Tobey and Yousten, 1976). No noticeable effect on the production of the enzyme was reported by Feniksova et al. (1960) with the use of tap water or buffer as the moistening agent. In addition, the use of phosphate buffer was reported to complicate the subsequent purification of the enzyme because of the introduction of the salts in the system in the form of buffers. On the contrary, the use of dilute buffer for moistening solids was found to be better than distilled water both for growth of the organism and enzyme production by Beckord et al. (1945). When distilled water was used, enzyme production was 50% of that obtained when dilute buffer was used for moistening the solid substrate. It is interesting to note that most of the earlier workers ignored the role played by phosphate in enhancing enzyme production through the use of higher ratios of solids and phosphate buffer. The inferences drawn were based on the water content of the medium alone (Beckord et al., 1945; Qadeer et al., 1980). The resulting increased concentration of buffer constituents at lower solid–buffer ratios may have also played a role in enhancing enzyme production.

The ratios of the solids to the moistening agents employed by various workers are presented in Table VI. It is evident that the ratio used varies with the bacterial culture used in the process. Wheat bran–moistening agent ratios ranging from 1 : 2 to 1 : 6 were studied by Lulla and

TABLE VI

RATIOS OF SOLIDS TO MOISTENING AGENT USED IN BACTERIAL α-AMYLASE PRODUCTION IN THE SSF TECHNIQUE

Solid substrate	Moistening agent	Ratio of solid to moistening agent	Reference
Wheat bran	Buffer	1 : 1	Ramesh and Lonsane (1987a)
		1 : 2.5	Qadeer et al. (1980); Lulla and Subrahmanyan (1954); Ramesh and Lonsane (1987b)
		1 : 1.75	Beckord et al. (1945)
		1 : 4.5	Lulla and Subrahmanyan (1954)
Wheat bran	Water	1 : 0.5	Oji Paper Co., Ltd. (1961)
		1 : 1.2	Arima (1964)
		1 : 1.6–3.0	Feniksova et al. (1960)
Wheat bran	Diluent	1 : 1	Tobey and Yousten (1976)
		1 : 2	Park and Rivera (1982)
Cornmeal base	Water	1 : 0.5	Minagawa and Hamaishi (1962)

Subrahmanyan (1954), and it was reported that the ratio of 1.0 : 4.5 was optimum and nearly doubles enzyme titers as compared to a ratio of 1 : 2. Buffer–WB ratios ranging from 1 to 3 were also studied by Ramesh and Lonsane (1987a), and the ratio of 2 : 1 was found to be optimum. Further increase in the ratio resulted in higher enzyme production, but the increase was not significant. The presence of moisture in the fermenting solid medium within the range of 62–75% was reported to produce the enzyme at maximum levels (Feniksova et al., 1960). The effect of different ratios of WB and moistening agent on growth of the bacterial cells and degree of enzyme production was also studied by Ramesh and Lonsane (1986) using different species of the genus *Bacillus*. The requirement of water activity for growth and enzyme production was shown to differ among these cultures. It was even speculated that the ability of the culture to grow in the SSF technique was related to the water activity requirement of the culture (Ramesh and Lonsane, 1986).

The pioneering work carried out by Beckord et al. (1945) on the effect of WB–moistening agent ratios on the extent of bacterial α-amylase production in the SSF process showed that the ratio at 1.0–1.5 is the lowest solid–liquid ratio that can be employed with satisfactory results. Although a ratio of 1.0 : 1.75 was found to be optimum, the use of a 1.0 : 2.5 ratio was shown to facilitate the spreading of the moist solid medium on the inner surfaces of the culture vessel.

The moisture content of the moist fermenting solid medium is one of the key factors in the SSF process and governs the success of the technique (Lonsane et al., 1985). It is thus not surprising that a large body of such data related to the production of bacterial α-amylase in the SSF technique is available in the literature. The studies just reported are based on the ratio of solid substrate and moistening agent used initially in the preparation of the moist solid medium. The moisture content of the medium is known to change during fermentation as a result of evaporation and metabolic activities (Nishio et al., 1979). It is essential to control the moisture content of the fermenting medium at the desired level, and this is achieved on a large scale by humidification of the fermentation chamber (Ghildyal et al., 1981). However, it is difficult to control moisture level in flasks used in laboratory-scale experiments, and this explains the paucity of information on the effect of moisture control during fermentation.

G. pH

The monitoring and control of pH in SSF processes is not usually attempted, probably because of the difficulties in measuring the pH of the moist solids and the problem of mixing a small quantity of acidic or alkaline solutions with the large body of the moist fermenting solids to achieve pH control throughout the duration of the fermentation experiment. In fungal SSF processes, the buffering capacity of some medium constituents is employed to eliminate the need for pH control (Chahal, 1983). Another approach to minimize the acidification of the fermenting medium is to use urea instead of ammonium salts for nitrogen supplementation of the medium (Raimbault and Alazard, 1980). Both these approaches were relied on by workers in the production of bacterial α-amylase in the SSF process. The incorporation of 2% $CaCO_3$ in the fermenting solids was also employed (Tobey and Yousten, 1976) for control of pH at the desired level during fermentation. The use of buffer to moisten the medium and adjust the initial pH at 6.0 as well as to extract the enzyme from fermented bacterial bran was found to give an enzyme extract with an approximate pH of 6.0 (Beckord et al., 1945). This indicates that the pH probably is not altered during the course of fermentation or that the change in pH of the fermenting solid medium is effectively counteracted by the buffer.

Most of the *Bacillus* strains used commercially for production of bacterial α-amylase by SmF have an optimum pH between 6.0 and 7.0 for the growth and enzyme production. This may also be true in the case of strains used in production of the enzyme in the SSF process, as the pH range used by many workers was between 6.0 and 7.0. In most cases, the pH used is not specified except in case of the use of pH 6.0 (Beckord et al., 1945), pH 6.3–6.4, or pH 6.0–6.3 (Arima, 1964) and pH 7.0 (Park and Rivera, 1982).

In the SSF process, greater attention is given to optimizing the initial pH value of the moist solid medium (Lonsane et al., 1985). No information on this subject is available in the literature except for a report by Ramesh and Lonsane (1987a). The effect of the initial pH of the medium on the yield of the enzyme was studied by suitably altering the pH of the buffer solution with 1 N NaOH or HCl before using it for moistening WB. The results obtained were interesting and of commercial importance. The degree of enzyme production was found to be highest at an initial pH of 7.0 at 24 hours and it was reduced with any increase or decrease in this pH value. The destruction of the enzyme after >24 hours of fermentation was reported to be comparatively faster at lower pH values (Ramesh and Lonsane, 1987a).

H. Culture Vessels

Most of the work carried out on the production of bacterial α-amylase using the SSF technique was confined to laboratory scale. Hence, the culture vessels employed in these cases were either 250- or 500-ml Erlenmeyer flasks. The quantity of moist solid medium employed was 35 or 50 g in 500-ml flasks (Qadeer et al., 1980; Park and Rivera, 1982; Lulla and Subrahmanyan, 1954) or 20 g in 250-ml flasks (Feniksova et al., 1960; Tobey and Yousten, 1976; Ramesh and Lonsane, 1987a,b). The screw-capped French squares of 6 × 6 × 14 cm size or 64-ounce capacity round bottles, 11.5 cm diameter and 20 cm height, each charged with 100 g moist solid medium, were also used (Beckord et al., 1945). Only one report mentioned the use of trays for large-scale cultivation, but details are not available because it is in the form of a patent (Minagawa and Hamaishi, 1962). The use of vessels to evaluate industrial applicability was also reported but without any details (Feniksova et al., 1960).

I. Autoclaving

The moist solid medium requires cooking for physical modification of solid particles, softening and partial hydrolysis of starches and proteins present in the solids, as well as sterilization for killing the contaminant microbial cells and spores. The autoclaving time and temperature are known to have a significant effect on the overall fermentation in the case of SSF processes involving fungal cultures (Lonsane et al., 1985). Negligible work on this subject was carried out in the case of α-amylase production by bacterial cultures in the SSF technique, and consequently parameters were either selected arbitrarily or based on practice followed in fungal SSF processes. The cooking–sterilization temperature used by most workers was 121°C, but the time varied between 15 minutes (Qadeer et al., 1980), 30 minutes (Park and Rivera, 1982), and 60 minutes (Beckord et al., 1945; Ramesh and Lonsane, 1987a,b). Autoclaving at 20 psi for 30 minutes (Lulla and Subrahmanyan, 1954) and 60 minutes at 1 atm (Feniksova et al., 1960) was also performed, but no information on autoclaving time and temperature was given by Tobey and Yousten (1976).

J. Nature and Ratio of Inoculum

Diverse methodology and media were used by different workers for preparation of inoculum. The stock culture was subcultured on nutrient agar slants (Beckord et al., 1945; Park and Rivera, 1982), beef broth–peptone–NaCl–agar slants (Minagawa and Hamaishi, 1962), or agar slants consisting of meat–peptone broth and brewing wort at a 1 : 1 ratio (Feniksova et al., 1960). The culture was allowed to grow on the agar slants at optimum temperature for 24–48 hours and the growth was collected in distilled water to obtain inoculum (Beckord et al., 1945; Tobey and Yousten, 1976). Alternatively, the growth from freshly grown nutrient agar slants was subcultured in nutrient broth (Qadeer et al., 1980; Tobey and Yousten, 1976; Lulla and Subrahmanyan, 1954; Ramesh and Lonsane, 1987a,b), nutrient sporulation broth (Tobey and Yousten, 1976), fluid WB extract medium (Beckord et al., 1945), beef broth containing peptone and NaCl (Minagawa and Hamaishi, 1962), peptone–corn steep liquor–starch medium (Ramesh and Lonsane, 1987b), and was used as inoculum after growth at the desired temperature and time under static or shaking conditions. The ratio of inoculum used was 2.5 ml/10 g commercial bran (Beckord et al., 1945), 2 ml/50 g moist WB medium (Park and Rivera, 1982), 0.2 ml of midlogarithmic culture/10 g commercial WB (Tobey and Yousten, 1976), or 20% based on the weight of commercial WB (Ramesh and Lonsane, 1987a,b).

Preparation of the inoculum in a medium of essentially the same composition as that employed for final culture has been demonstrated to have a slight beneficial effect (Beckord et al., 1945). However, a single serial transfer in such a medium from a nutrient agar slant was found to be adequate, in that no further improvement in enzyme production could be achieved by further serial transfers. Similarly, the ratio and age of inoculum were shown to have little effect on the enzyme production. For example, the use of inoculum grown on nutrient agar slants at 37°C for 12, 24, 42, 48, and 180 hours, as well as the use of 2.5 ml inoculum containing 5–80 million cells, were found to have negligible influence on enzyme production (Beckord et al., 1945). These workers have stated that inoculum grown for 12 hours was best and that the small differences between inoculum grown for 12–42 hours were within the limits of experimental error. In addition, inoculum grown for 180 hours was not quite satisfactory. In recent studies, the use of peptone–corn steep liquor–starch medium instead of nutrient broth for inoculum development was shown to give a 22-fold increase in enzyme production when used in combination with other factors (i.e., enrichment of WB with 0.3% diammonium phosphate and a 1.0 : 2.5 ratio of WB and phosphate buffer (Ramesh and Lonsane, 1987b).

K. Incubation Temperature and Time

The selection of incubation temperature is guided by the optimum growth temperature of the culture. Based on this criterion, the temperature employed by various workers ranged from 30° to 45°C. The effect of incubation temperature on the level of enzyme production was shown to be similar to that observed in SmF (Beckord et al., 1945). The incubation was carried out by keeping the culture flask upright (Tobey and Yousten, 1976), in a slanted position (Beckord et al., 1945; Ramesh and Lonsane, 1987a,b), or by placing the culture bottles on their sides (Beckord et al., 1945). In the latter two cases, the fermenting solids spread over the available surface of the culture vessel (Beckord et al., 1945). The incubation temperature used in individual cases was 33°C (Tobey and Yousten, 1976), 35°C (Park and Rivera, 1980), 35°–40°C (Beckord et al., 1945; Arima, 1964), 37°C (Lulla and Subrahmanyan, 1954; Feniksova et al., 1960), 40°C (Ramesh and Lonsane, 1987a), and 45°C (Ramesh and Lonsane, 1987b).

Feniksova et al. (1960) studied enzyme production by B. subtilis at 30°C and 37°C and reported that the higher temperature (37°C) promoted better growth and a greater accumulation of enzyme. Recently, incubation temperature was found to have a profound effect on enzyme yield and the duration of the enzyme synthesis phase by B. megaterium 16M. The enzyme synthesis phase was found to last as long as 48 hours at 35°C, thereby leading to the highest yield of the enzyme (Ramesh and Lonsane, 1987a). The rate of enzyme production and the yield were much lower at 30°C and 40°C. It was also shown that enzyme production was optimum at 35°C, although the optimum temperature for growth of the strain was 40°C.

The incubation time is governed by characteristics of the culture and is mainly based on the growth rate and the peak in enzyme production. In most cases, the time employed varied between 30 hours (Beckord et al., 1945) and 72 hours (Park and Rivera, 1982; Lulla and Subrahmanyan, 1954), although an average of 48–50 hours is most common (Beckord et al., 1945; Qadeer et al., 1980; Minagawa and Hamaishi, 1962; Feniksova et al., 1960; Ramesh and Lonsane, 1987a,b). Enzyme yield was reported to be highest by 48 hours with a marked increase in enzyme production between 24 and 48 hours (Beckord et al., 1945). The enzyme titer was increased almost 10-fold at 48 hours as compared to 24 hours. There was no further increase in enzyme yield even when the incubation time was extended to 8 days at 37°C (Beckord et al., 1945).

L. Aeration

Limited aeration is known to enhance the yield of bacterial α-amylase in the SmF technique (Markkanen and Bailey, 1975). Aeration of the fermenting moist solid medium was also reported to result in enhancement of enzyme production (Minagawa and Hamaishi, 1962). The protocol followed was to initiate aeration of the fermenting mass as soon as growth of the culture leveled off. Enzyme production of the culture employed was reported to begin at the end of the growth phase, and thus aeration was provided during the enzyme production phase only. The effect of lower oxygen availability on enzyme yield was recently evaluated under optimal parameters by closing culture flasks with rubber stoppers immediately after inoculation (Ramesh and Lonsane, 1987a). The rate and degree of enzyme production was found to be reduced by ~5-fold as compared to the control experiment. Thus, it is evident that the requirement of the extent of the aeration of culture medium is determined by the culture used in the SSF technique.

It is interesting to note that culture flasks were incubated in a stationary state in almost all cases for production of bacterial α-amylase in the SSF technique (Lulla and Subrahmanyan, 1954; Ramesh and Lonsane, 1987a,b). However, a more extensive exposure of the fermenting porous medium to air contained in the culture flask was achieved by spreading the inoculated bran on the available inner surfaces to obtain better yields (Beckord et al., 1945). Placing the culture flasks in a slanted position during incubation (Ramesh and Lonsane, 1987a,b) also provided higher exposure of the fermenting mass to air.

M. Growth Characteristics

The ability of filamentous fungi to penetrate deep into the intercellular and intracellular space in solid substrates and the lack of this ability in bacteria, yeasts, and even single-celled fungi are well known (Chahal, 1983; Lonsane et al., 1985). Consequently, bacteria were considered unsuitable for use in SSF processes (Lonsane et al., 1985). The successful exploitation of *Bacillus* cultures in the production of α-amylase in the SSF technique (Beckord et al., 1954; Qadeer et al., 1980; Ramesh and Lonsane, 1987a,b), therefore, indicates the involvement of some other mechanism for attachment of the cells to the solid particles. However, no information on these aspects is available in the literature except for the recent report on the growth characteristics of *B. megaterium* 16M and *Bacillus* HOP-40 on WB solid media (Ramesh and Lonsane, 1987a,b). The microscopic examination of the fermenting moist medium showed growth of the culture throughout the medium either in the form of thin flocculant growth or as small pellicles on the upper surfaces of the solid substrate particles.

The attachment and adsorption of bacterial cells to the solid surfaces and the consequent development of biofilms are well known (ZoBell and Anderson, 1936; Costerton et al., 1985; Rutter, 1984; Shales, 1985; Wardell et al., 1983). Various environmental and bacterial cell-associated factors as well as physicochemical interactions between the cells and the solids are known to influence the development of biofilms (Rutter and Vincent, 1984; Shales and Kumarasingham, 1987). An initial attachment of the cells to solid particles and growth on solid surfaces are involved in biofilm development (Shales and Kumarasingham, 1987). The adhesion is also known to be made irreversible by the exopolysaccharide produced by the cells during growth phase, while components of the cell surface further aids in the attachment (Abbott et al., 1983; Costerton and Irvin, 1981; Rutter and Vincent, 1984). The adsorption of B. subtilis cells on anion exchange resin and kaolin was also studied (Daniels, 1967; Krishnamurti and Soman, 1951). About 0.61×10^{10} and 50×50^{10} cells are found to be adsorbed onto resin and kaolin, respectively, thereby covering 31 and 15% of the total available adsorbent surface. It is felt that these mechanisms of attachment and adsorption of the bacterial cells to the solid surfaces may be operative in the growth of Bacillus cultures on solid surfaces.

N. Enzyme Yields

Diverse types of enzyme assays, including hydrolysis of starch until dextrins unstained by iodine are obtained (Feniksova et al., 1960), measurement of a specific reduction in the blue color of the starch–iodine complex (Ramesh and Lonsane, 1987a), determination of the liberation of reducing sugars from starch under specific conditions (Beckord et al., 1945; Qadeer et al., 1980; Tobey and Yousten, 1976; Ramesh and Lonsane, 1987b), and measurement of time to reach the red-brown colour in a reaction mixture with iodine (Beckord et al., 1945), were employed to estimate the titer of the bacterial α-amylase in the SSF technique. Hence, it is impractical to compare the enzyme yields reported in the literature. However, the enzyme yields can be compared to each other based on the measurement of the micromoles of reducing sugar released per minute under the assay conditions. Such yields reported by various workers for the Bacillus strains include 1200–1400 units along with 300 proteinase units/g bran (Feniksova et al., 1960), 9.2 and 12,690 units with B. thuringiensis HD-1 and B. subtilis SJT, respectively, per gram of dry bacterial bran (Tobey and Yousten, 1976), and 22 units of the enzyme with novel characteristics by Bacillus HOP-40 (Ramesh and Lonsane, 1987b). The enzyme yields reported by other workers are found to be much lower than those with B. subtilis SJT (Tobey and Yousten, 1976) when approximately equated on an activity basis. The highest titer of enzyme, amounting to ~33,000 units/g dry bacterial bran, was reported by Ramesh and Lonsane (1987a). The culture used was B. megaterium 16M, while the enzyme was estimated by measuring the amount that produces a 10% reduction in the intensity of the blue color of the starch–iodine complex under the assay conditions (i.e., 10-minute reaction time at 70°C and pH 6.0). The production of such a high titer of enzyme in the SSF process is of significant commercial importance in that the authors stressed that it was unattainable in SmF.

The coproduction of proteases along with α-amylase is a general characteristic of *Bacillus* strains in SmF. A similar characteristic was also reported in the SSF process, and it was even stressed that the preferential production of either of these enzymes can be achieved by using appropriate solid substrate or cultural parameters (Minagawa and Hamaishi, 1962). The absence of the coproduction of protease in *B. megaterium* 16M when grown in the SSF technique (Ramesh and Lonsane, 1987a) may provide commercial advantages because it will eliminate the processing steps for separation of protease and may also confer greater stability to α-amylase preparations.

It is interesting to note that the comparative production of bacterial α-amylase under SmF and SSF techniques was evaluated by some workers. *Bacillus thuringiensis* HD-1 was reported to produce merely 0.8 units/ml in nutrient broth–salt medium or <2 units/ml after optimization of the medium by an SmF process, in contrast to the production of 9.2 units/g dry bacterial bran in the SSF technique (Tobey and Yousten, 1976). Production of ~3.5-fold more enzyme by *Bacillus* HOP-40 in the SSF process as compared to that by SmF based on units of enzyme produced per milliliter of liquid medium or per gram of bran was also reported by Ramesh and Lonsane (1987b). The yield of bacterial α-amylase in WB extract medium or in peptone–corn steep liquor–starch medium when grown by SmF was <3 units/ml with *B. megaterium* 16M compared to ~33,000 units/g dry bacterial bran in the SSF process (Ramesh and Lonsane, 1987a). Data on the comparative yield of the enzyme by LSF and SSF techniques are also available. Based on the yield per gram of dry bran, the enzyme production by *B. subtilis* in the SSF technique was three times higher than that by LSF (Feniksova *et al.*, 1960). These results establish the formation of higher titers of enzyme in the SSF technique as compared to LSF and SmF processes.

O. Recovery of Enzyme

Methodology, developed over a long period of time and employed presently for extraction of fungal α-amylase from moldy substrates produced in the SSF technique, will also be most appropriate and effective for the extraction of bacterial α-amylase from bacterial bran. This is also true, in fact, in the case of any water-soluble enzyme extraction from fermented solid substrates. Thus, since knowledge on the extraction of enzymes from moldy substrate is of scientific and industrial importance in the present case, the salient features are presented here.

1. Meager information on the factors affecting enzyme extraction from dry moldy substrate and methodology standardization is available in the literature (Ramakrishna et al., 1982; Dwarkanath and Rao, 1971).

2. The enzymes are extracted by employing a percolation technique, serial extraction, and countercurrent leaching (Ramakrishna et al., 1982; Dwarkanath and Rao, 1971; Underkofler et al., 1958; Meyrath and Volavsek, 1975).

3. The percolation technique dictates the use of several volumes of solvent based on the weight of moldy substrate (Waksman and Allen, 1933; Ramakrishna et al., 1982). The commonly used ratio of moldy bran to solvent is 1 : 10. Consequently, the extract is too diluate and requires concentration by costly and energy-intensive evaporation under vacuum (Ramakrishna et al., 1982).

4. Serial extraction (Dwarkanath and Rao, 1971) gives a slightly more concentrated extract by reducing the volume of the extract but leads to lower extraction efficiency.

5. The multiple-contact countercurrent leaching technique leads to a highly concentrated enzyme extract by combined use of the lowest possible volume of solvent and a reasonably practical number of contact stages (Beckhorn, 1960; Fogarty and Ward, 1974; Ramakrishna et al., 1982). It thus eliminates the need for vacuum concentration of the dilute extract usually obtained by a percolation technique. The use of a five-contact stages countercurrent leaching technique was found to result in ~85% extraction efficiency (Ramakrishna et al., 1982).

6. Various solvents such as water, dilute salt solution, buffer, dilute solution of glycerol, tap water, aqueous solution of dilute ethanol, and toluene-saturated water were used for extraction of the enzyme (Wallerstein, 1939; Waksman and Allen, 1933; Beckhorn, 1960; Sreekantiah, 1976; Rao et al., 1947). Among these, distilled or tap water alone or with 0.5–1.0% glycerine or NaCl was found to be the best (Dwarkanath and Rao, 1971; Rao et al., 1947; Ramakrishna et al., 1982). Enzyme extraction by an alcoholic solvent was found to be less efficient than with aqueous solvents (Rao et al., 1947). However, the enzyme purity was ~150% with the former system. Nonenzymatic solubles were extracted more efficiently with aqueous solvents as compared to alcoholic solvents (Rao et al., 1947).

7. A mechanically agitated system recovered ~8% more enzyme than

a stationary system as a result of enhanced contact of the solids with solvent (Ramakrishna et al., 1982).

8. The moldy fermented solids are subjected to extraction immediately at the end of the fermentation or are air or vacuum dried. The dried bran can be stored for several months without any appreciable decrease in enzyme titer. It may be used as such, or the enzyme may be extracted whenever required.

A survey of literature on the methodology adopted by various workers for extraction of bacterial α-amylase from bacterial bran produced in the SSF technique indicates that it is similar to the trends just described. The air or vacuum drying of the fermented bran (Beckord et al., 1945; Minagawa and Hamaishi, 1962), use of dry bran as such without extraction (Park and Rivera, 1982), storage of dry bran for several months without appreciable loss in potency (Beckord et al., 1945), use of water or buffer for extraction of the enzyme (Beckord et al., 1945; Qadeer et al., 1980; Oji Paper Co., Ltd., 1961; Minagawa and Hamaishi, 1962), use of percolation techniques with large volume of solvents (Beckord et al., 1945; Qadeer et al., 1980; Minagawa and Hamaishi, 1962), and the use of a multiple-contact countercurrent leaching technique (Ramesh and Lonsane, 1988) are some of the techniques or parameters followed in the extraction. The bacterial bran, when air dried at room temperature, became very much darker in color and retained an offensive odor (Beckord et al., 1945). The methodology followed by Lulla and Subrahmanyan (1954) consisted of autolysis of bacterial bran grown for 72 hours by adding 2 ml toluene, holding it for 24 hours, at room temperature, and extraction with 50 ml water. The fermented bran cake of ~60–150 g moist weight was shaken at 30°C for 1 hour on a rotary shaker after addition of 200 ml phosphate buffer at pH 7.0 (Qadeer et al., 1980).

The ratio of bacterial bran to solvent used by some workers is too high. For example, the dried bacterial bran was contacted with 16 parts water and the extract was separated by filtration (Minagawa and Hamaishi, 1962). The residue was subsequently washed with 14 parts water and the resulting extract was again filtered. Thus a total of a 30-fold volume of solvent was employed to extract the enzyme from 1 part of dry bacterial bran. For achieving optimum efficiency, the different extraction parameters such as selection of solvent, ratio of solids to solvent, contact time and temperature of the solids with solvent, and the drying temperature of moist fermented bran were investigated by Ramesh and Lonsane (1988). The data indicate that the degree of extraction of the enzyme increased with an increase in the ratio of dry bacterial bran to buffer up to 1 : 9. The recovery of enzyme was ~12% higher at an extraction temperature of 28°C as compared to that at 4°C. The use of a 1% solution of glycerol as solvent was found to extract more enzyme as compared to other solvents evaluated. However, the degree of extraction of the enzyme was more or less equal at 5–90 minutes contact time of the solvent with bacterial bran. The retention of the enzyme activity was highest when the moist fermented bran was dried at 25°C, while the enzyme activity decreased gradually with an increase in drying temperature (Ramesh and Lonsane, 1988).

The effect of the use of phosphate buffer at pH 5.9 or 7.17, acetate buffer at pH 5.9, or water for extraction of the enzyme using 50 ml solvent with 5 g dry bacterial bran, previously ground thoroughly with quartz sand, were also evaluated by Feniksova et al. (1960). It was reported that the enzyme was completely extracted by using a 1-hour contact time with phosphate buffer of pH 5.9. The claim of complete extraction by these workers is probably based on the fact that the highest titers were given by this system as compared to others studied. Retention of the enzyme in the solvent that adheres to the WB was ignored by them. It is interesting to note that solvent amounting to about twice the weight of dry fermented bran adheres to WB during the extraction process and it cannot be recovered easily by normal processing techniques (Kumar and Lonsane, 1987c).

Studies carried out by Ramesh and Lonsane (1988) also showed that the use of a multiple-contact countercurrent leaching technique for extraction of α-amylase produced by B. megaterium 16M in the SSF technique resulted in an 86% extraction efficiency and higher concentration of the enzyme in the extract. The extraction efficiency with a 1 : 3 ratio of bacterial bran to solvent was only 50% but was improved to 86% by the application of three contact stages because of simultaneous extraction and enrichment during countercurrent leaching. The method yields 1 liter of extract from 1 kg dry bacterial bran, and the enzyme concentration in it is ~6-fold higher than the extract obtained using a 1 : 9 ratio in a single-step percolation technique (Ramesh and Lonsane, 1988).

P. Clarification and Purification

The enzyme extract obtained from bacterial bran, produced in the SSF technique, was pale to dark brown in color (Ramesh and Lonsane, 1988) and contained suspended matter, including smaller particles of WB (Ramesh and Lonsane, 1987a). In addition, it also contained polysaccharides, which were also coextracted from WB during extraction of the enzyme. It was stressed by Beckord et al. (1945) that clarification of the extract is necessary because the suspended matter resists removal by centrifugation or filtration and also obscures the end point in the iodine test for estimation of the dextrinizing activity of the enzyme. Clarification of the extract was achieved by adding 1 ml of 20% $CaCl_2 \cdot 2H_2O$ solution per 10 ml buffer, and after a 1-hour contact time of buffer with bacterial bran, the resulting precipitate was centrifuged or filtered off along with WB particles (Beckord et al., 1945; Qadeer et al., 1980). The clarification leads to a clear extract with undiminished enzyme activity (Beckord et al., 1945). The extract was strained through cheesecloth for clarification by Lulla and Subrahmanyan (1954).

The enzyme extract was centrifuged at 0°–4°C for 20 minutes for clarification by Ramesh and Lonsane (1987a,b, 1988). The clarity of the extract achieved by following any one of the foregoing methods was similar and the clarified extract thus obtained was found to pose no interference either in enzyme assay or in liquefaction of starch (Ramesh and Lonsane, 1988). Some workers have also used Ca acetate and Pb acetate to precipitate unwanted proteins from the enzyme extract (Oji Paper Co., Ltd., 1961; Minagawa and Hamaishi, 1962).

Further downstream processing of the extract to obtain pure enzyme or to achieve partial purification was not attempted by most workers except in three cases. The methodology followed is shown schematically in Fig. 2. It is also necessary to eliminate proteases, which are often coproduced with bacterial α-amylase. The dried product so obtained was reported to be nonhygroscopic and soluble in water, and it was also claimed to have the highest activity ever achieved (Oji Paper Co., Ltd., 1961).

The bacterial α-amylase is used in industry in crude form and, therefore, the extensive downstream processing need not be applied if the enzyme does not pose any difficulties in its industrial use (Ramesh and Lonsane, 1988). Enzyme of such quality can be obtained by centrifugation and filtration of the enzyme extract. The latter involves the lowest possible expenditure in downstream processing (Ramesh and Lonsane, 1988). It may be of commercial interest to have a higher concentration of enzyme in the product, and this could be most economically achieved by precipitating the enzyme from the clear extract and redissolving the precipitate in an appropriate volume of solvent to obtain the desired potency of the enzyme in the product.

Q. Enzyme Characteristics

The characterization of bacterial α-amylase produced in the SSF technique was not attempted by most of the workers in spite of the commercial importance of these characteristics. The enzyme produced by *B.*

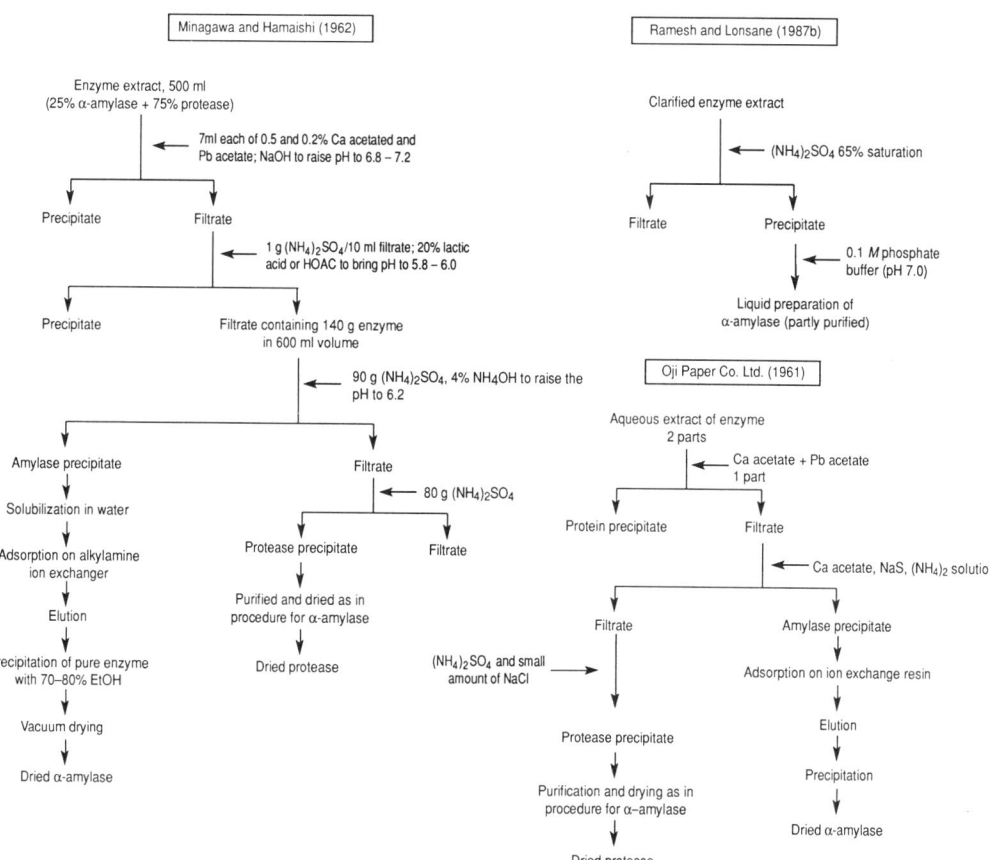

FIG. 2. Downstream processing of the enzyme extract obtained from dry bacterial bran.

subtilis was reported to have optimal pH values in the range of 7.0–7.6 and 6.6–7.0 for starch dextrinization and starch saccharification, respectively (Beckord et al., 1945). The enzyme was found to be most resistant to thermal inactivation at pH 7.0 as compared to pH 6.0 and 8.0. The enzyme was stated to have a high degree of resistance to thermal inactivation when its dextrinization rate at 75°C was compared with that of barley malt α-amylase. The thermostability was enhanced by Ca^{2+} when used at a concentration of 4 mg of $CaCl_2 \cdot 2H_2O$ per milliliter of reaction mixture. However, degradation of starch by the enzyme was found to be different from that of typical commercial bacterial α-amylase preparations. At equal starch-dextrinizing activity, it showed a lower saccharification rate than that of commercial bacterial α-amylase in the early phase of the process. However, postdextrinization saccharification was more pronounced (Beckord et al., 1945). These observations indicate that the enzyme had both dextrinizing and saccharifying activities. On the contrary, the enzyme produced in the SSF technique was reported to show results similar to those usually obtained with commercial enzymes in the saccharification of starch (Park and Rivera, 1982).

The enzyme produced by B. megaterium 16M in the SSF technique was reported to have pH and temperature optima at pH 6.0 and 70°C, respectively (Ramesh and Lonsane, 1987a). A sharp drop in enzyme activity at lower and higher temperatures as well as at acidic pH values was reported. The products of starch hydrolysis were found to be higher oligosaccharides > maltose > glucose. Enzyme activity was increased by 72% at 85°C by adding zinc sulfate at 10 μg level in the reaction mixture. Salts of Ca^{2+} and Mg^{2+} were less effective, even at higher concentrations.

The characteristics of a novel α-amylase system produced by Bacillus HOP-40 in the SSF technique were reported recently by Ramesh and Lonsane (1987b). The enzyme showed three peaks of activity at pH 5.0, 7.0, and 8.5–9.0 when used in crude and partially purified forms as well as with different buffers. Peak activity was highest at pH 5.0 and lowest at pH 7.0. This finding and 76% relative activity at pH 4.5 indicates the potential of the enzyme system for achieving improved yields of dextrose from starch due to decreased formation of maltulose precursors (Luenser, 1983). More than 90% relative activity at pH 8.5–9.0 was also exhibited by the enzyme system. Such a novel enzyme system was not known earlier, as the characteristics are distinct from all the known alkaline amylases reported by various workers (Horikoshi and Akiba, 1982; Boyer et al., 1979; Saito, 1973; Medda and Chandra, 1980; Kelly and Fogarty, 1976).

Among all the bacterial α-amylases produced in the SSF technique, the enzyme produced by Bacillus HOP-40 (Ramesh and Lonsane, 1987b) showed the highest temperature optimum, as well as other industrially important novel characteristics. It would have had potential for industrial exploitation but for the very low yields.

R. Economic Considerations

Most workers have not discussed the economic considerations of the production of bacterial α-amylase in the SSF technique. The reasons usually given for investigating the applicability of the SSF technique to bacterial α-amylase production include the numerous advantages (Beckord et al., 1945; Ramesh and Lonsane, 1987a, 1988) and an easier operation (Qadeer et al., 1980) of the technique as compared to LSF and

SmF processes. The advantage of the inclusion of the plant extracts or meals in solid-state media, without resorting to extraction of active principles from the plant material, was also pointed out (Beckord et al., 1945). However, the economic considerations have been given due attention by Ramesh and Lonsane (1987a, 1988) in light of the recent awareness of the inherent limitations of the SmF technique and the capability of the SSF process to overcome these shortcomings. It was stressed that the higher enzyme titers achieved in the SSF technique and the consequent lower expenses for downstream processing and effluent disposal, as well as other well-known economic advantages of the SSF process (Lonsane et al., 1985), provide sufficient ground for exploitation of SSF processes on an industrial scale. The industrial value of SSF processes in replacing the currently used SmF processes throughout the world was also emphasized (Ramesh and Lonsane, 1987a, 1988). However, no calculations to support these views were given.

S. Research and Development Needs

In order to exploit efficiently the SSF technique on an industrial scale to produce bacterial α-amylase and also to generate more scientific information, many different research and development needs should be met.

1. The search for a *Bacillus* culture with the ability to produce a highly thermostable enzyme, preferably with a temperature optimum for enzyme activity around 90°–95°C (or at still higher temperatures) in 1% starch solution is essential to compete with the best of the enzymes available in the international market.
2. No report exists on the ability of *B. licheniformis* to grow and produce α-amylase in the SSF technique. A highly thermostable enzyme is produced on an industrial scale by SmF using this culture. It would be of industrial value to isolate strains of this species with the capability to produce highly thermostable α-amylase using the SSF technique.
3. The bacterial α-amylases available in the international market have a narrow pH range for enzyme activity and are largely inactivated when used in more acidic or alkaline pH ranges. The production under SSF conditions of an enzyme with a broad optimum for pH in acidic, neutral, and alkaline ranges was reported by Ramesh and Lonsane (1987b), but the yield was too low. Attempts to produce such an enzyme with higher yield in the SSF process are indicated.
4. Proteases are often produced along with bacterial α-amylase in SmF processes. However, *B. megaterium* 16M produced α-amylase in the absence of any protease or cellulases in the SSF technique (Ramesh

and Lonsane, 1987a). This eliminated the need to separate the proteases in downstream processing and also reduced the possibility of the destruction of α-amylase preparations during the storage by the action of proteases. More work to discover a culture of similar characteristics, combined with higher thermostability and broader pH optima, is of economic importance.

5. Investigations on the mechanism of attachment of *Bacillus* cultures to solid substrates, the physiological and morphological changes in the culture due to the growth in the environment existing in the SSF technique, the effect of aeration and agitation of the fermenting moist medium in a rotary-drum type of fermentor on yield of the enzyme may provide means to achieve better growth and higher enzyme yields in the SSF process.

6. Most of the *Bacillus* cultures used for production of α-amylase by the SmF process exhibit catabolic repression or substrate inhibition. This necessitates the use of lower substrate concentration or the utilization of specific fermentation strategies to overcome the problem that lead to a higher cost of production. In the SSF technique, the substrate inhibition problems are known to be minimized or eliminated. Studies of comparative substrate inhibition in SmF and SSF processes will be highly useful from an industrial point of view.

7. Comparison of enzyme yields in SmF and SSF processes and the cost of production as well as capital and recurring expenses will be helpful to the fermentation industry. It will aid in selecting the most appropriate technique whenever a new fermentation plant for production of bacterial α-amylase is contemplated or in replacing the presently used SmF process by the SSF technique.

8. Scientific data to link the requirement of water activity by various species of the genus *Bacillus* to the ability or inability to grow and produce enzyme in the SSF technique will have both scientific and industrial importance.

9. Detailed studies on factors influencing extraction of the enzyme from bacterial bran, development of an extraction process to obtain the highest possible extraction efficiency with a higher titer of the enzyme in the extract and better shelf-life of the extract as well as dried bacterial bran are indicated.

10. The level of polysaccharides that are coextracted with the enzyme from WB should be measured, and an efficient methodology to eliminate them from the extract needs to be developed from an industrial point of view.

11. The possibility of precipitating the enzyme from the extract and dissolving the wet precipitate in buffer to achieve the desired potency

and any shortcoming of such an enzyme preparation in the hydrolysis of starch on an industrial scale are worth investigating.

12. Engineering studies on transport phenomena are also indicated, since these will provide useful information for optimization of the process technology, formulation of scale-up criteria, design of an industrial plant, and the development of kinetic and other mathematical models for the growth and enzyme production of *Bacillus* cultures in the SSF technique.

13. The application of modern fermentor operation strategies such as fed-batch culture, continuous fermentation, and repeated batch fermentation using one or two fermentors may further improve the yield of bacterial α-amylase in the SSF technique.

14. A mixed-culture fermentation or successive fermentation, using bacterial and fungal cultures or two different species of *Bacillus* in succession, may also offer a better enzyme preparation. It was shown by Christensen (1944) that the combination of amylases from different sources or origin was better for industrial application.

15. The automation of the SSF technique for industrial-scale operation, the development of scaled-up know-how for efficient and economic production, and the establishment of a demonstration pilot plant are essential for generating the interest of the fermentation industry in the SSF technique for production of bacterial α-amylase.

16. α-Amylases are reported to have resistance to proteolytic enzymes (Stein and Fischer, 1958). It would be interesting to study the comparative resistance of bacterial thermophilic α-amylases produced using SmF and SSF techniques.

17. The development of accurate, indirect methods for estimation of bacterial mass in fermenting solids is necessary.

18. A comparative study of the purification of α-amylase produced by *Bacillus* cultures in both SmF and SSF processes and the characterization of the purified enzymes may provide interesting results from both scientific and industrial points of view.

19. It would be interesting to study the effect of the inhibitors or activators of bacterial α-amylases present in WB and other substrates used in SSF processes.

20. The fermented moist solids could be dried slowly or rapidly in a variety of different ways such as dehydration at room temperature in an evacuated room or under forced-air current, heating in a suitable dryer up to 41°C (Weller *et al.*, 1984), or in various other modes and combination of modes. Studies on the comparative retention of enzyme activity in the fermented bran dried under different modes are of industrial importance.

21. It is of commercial interest to study factors capable of enhancing the saccharifying activity of bacterial thermostable α-amylase, without any change in the dextrinizing activity, when produced in the SSF technique.

VII. Epilogue

The demand for bacterial thermostable α-amylase has increased many fold in recent years mainly because of the launching of biotechnological plants and policies by many countries for the utilization of renewable starch and starchy substrates to obtain a variety of products of industrial and economic importance. Simultaneously, it has been agreed by internationally recognized scientific bodies that reduction in the capital and recurring expenses of the saccharification process for starch is necessary for achieving acceptable economics and profit.

The critical and evaluative analysis of various possible modes for reduction in the cost of starch hydrolysis, considered in the present paper, identifies the liquefaction of starch by bacterial α-amylase as one of the costliest unit operations in starch hydrolysis mainly because of the cost of the enzyme. It also reveals the potential for cost reduction in bacterial α-amylase production by substituting the SSF technique for the SmF process presently used. Other possible modes for economy are found to offer a negligible impact for further economy in starch hydrolysis. Glucoamylase, another enzyme used in saccharification of starch, already enjoys the benefit of the advantages of SSF and is known to be one of the cheapest enzymes available in the world. Substantial reduction in its cost since 1984 was accomplished primarily through lowered production costs resulting from yield improvements and other changes in process technology.

Bacterial strains are generally considered to be unsuitable for growth in the SSF technique. Hence, cases are cited to show their successful growth and metabolic activities in SSF processes. As many as 15 specific advantages are enumerated for production of bacterial thermostable α-amylase in the SSF technique.

Presently available information on the production of bacterial thermostable α-amylase in the SSF process is critically and exhaustively analyzed to pinpoint the state of the art. The economic and industrial feasibility of the production of bacterial thermostable α-amylase in the SSF technique as well as its superiority over the presently used SmF process are also established in the present chapter beyond any doubt.

It is apparent that the SSF technique not only provides a potential tool for economy in the production of bacterial thermostable α-amylase as well as in starch hydrolysis but also opens a whole new exciting area for production of bacterial extracellular metabolites in SSF processes with practical and economic advantages over presently employed SmF processes.

There is, however, a considerable need for further R and D efforts for effective exploitation of the SSF technique in the production of bacterial α-amylase on an industrial scale. A total of 21 such needs, both from industrial and academic points of view, are identified in this chapter.

ACKNOWLEDGMENTS

The authors are thankful to Mr. S. Y. Ahmed, Dr. N. G. Karanth, Mr. S. K. Majumder, and Dr. B. L. Amla, for valuable suggestions and interest in the work. M. V. Ramesh is grateful to Council of Scientific and Industrial Research, New Delhi, India, for the award of a fellowship.

REFERENCES

Abbott, A., Rutter, P. R., and Berkley, R. C. W. (1983). The influence of ionic strength, pH and a protein layer on the interaction between *Streptococcus mutans* and glass surfaces. *J. Gen. Microbiol.* **129,** 439–445.

Aiba, S., Kitai, K., and Imanaka, T. (1983). Cloning and expression of thermostable α-amylase gene from *Bacillus stearothermophilus* in *Bacillus stearothermophilus* and *Bacillus subtilis*. *Appl. Env. Microbiol.* **46,** 1059–1065.

Aidoo, K. E., Hendry, R., and Wood, B. J. B. (1982). Solid substrate fermentations. *Adv. Appl. Microbiol.* **28,** 201–237.

Akamatsu, T., and Sekiguchi, J. (1984). An improved method of protoplast regeneration for *Bacillus* species and its application to protoplast fusion and transformation. *Agric. Biol. Chem.* **48,** 651–655.

Arima, K. (1964). Microbial enzyme production. *In* "Global Impacts of Applied Microbiology" (M. P. Starr, ed.), pp. 277–294. Wiley, New York.

Aunstrup, K. (1974). Industrial production of proteolytic enzymes. *Proc. Fed. Eur. Biochem. Soc. Meet.* **30,** 23–46.

Aunstrup, K. (1979). Production, isolation and economics of extracellular enzymes. *Appl. Biochem. Bioeng.* **2,** 27–69.

Aunstrup, K., Anderson, O., Falch, E. A., and Nielsen, T. K. (1979). Production of microbial enzymes. *In* "Microbial Technology" (H. J. Peppler and D. Perlman, eds.), 2nd Ed., Vol. 1, pp. 281–309. Academic Press, New York.

Babbar, I., Behki, R. M., and Srinivas, M. C. (1960). Bacterial diastase. Indian Pat. 66 096.

Baig, M. A., Pazlarova, J., and Vortruba, J. (1984). Kinetics of α-amylase production in a batch and a fed-batch culture of *Bacillus subtilis*. *Folia Microbiol (Prague)* **29,** 359–364.

Balls, A. K., and Schwimmer, S. (1945). Digestion of raw starch. *J. Biol. Chem.* **156,** 203–210.

Barfoed, H. (1967). Die Verwendung von Enzymen bei der Herstellung von Dextrose und Stärkesirup. *Staerke* **19,** 2–8.

Barton, L. L., Georgi, C. E., and Lineback, D. R. (1972). Effect of maltose on glucoamylase formation by *Aspergillus niger*. *J. Bacteriol.* **111,** 771–777.

Beckhorn, E. J. (1960). Production of industrial enzymes. *Wallerstein Lab. Commun.* **23**, 201–212.
Beckord, L. D., Kneen, E., and Lewis, K. H. (1945). Bacterial amylases production on wheat bran. *Ind. Eng. Chem.* **37**, 692–696.
Beuchat, L. R. (1983). Indigenous fermented foods. In "Biotechnology. Vol. 5: Food and Feed Production With Microorganisms" (G. Reed, ed.), pp. 477–528. Verlag Chemie, Weinheim.
Biddlestone, A. J., and Gray, K. R. (1985). Composting. In "Comprehensive Biotechnology. Vol. 4: The Practice of Biotechnology, Speciality Products and Service Activities" (C. W. Robinson and J. A. Howell, eds.), pp. 1059–1070. Pergamon, Oxford.
Board of Science and Technology for International Development, Office of International Affairs and the National Research Council, Washington, D.C. (1982). "Priorities in Biotechnology Research for International Development: Proceedings of a Workshop," pp. 13–14. Natl. Acad. Press, Washington, D.C.
Böing, J. P. T. (1982). Enzyme production. In "Prescott and Dunn's Industrial Microbiology" (G. Reed, ed.), 4th Ed., pp. 634–708. AVI, Westport, Connecticut.
Boidin, A., and Effront, J. (1917a). Process for treating amylaceous substances. U.S. Patent 1,227,374.
Boidin, A., and Effront, J. (1917b). Process for manufacturing diastase and toxins by oxidizing ferments. U.S. Patent 1,227,525.
Boidin, A., and Effront, J. (1924). Nitrogenous extracts and modified demineralized starchy products from cereals. U.S. Patent 1,509,467.
Boyer, E. W., Ingle, M. B., and Mercer, G. D. (1979). Isolation and characterization of unusual bacterial amylases. *Starch/Staerke* **31**, 166–171.
Bull, A. T., Holt, G., and Lilly, M. D. (1982). "Biotechnology—International Trends and Perspectives," pp. 1–84. OECD, Paris.
Casida, L. E., Jr. (1964). "Industrial Microbiology," pp. 390–401. Wiley, New York.
Castelein, J., and Verachtert, H. (1983). Coffee fermentation. In "Biotechnology. Vol. 5: Food and Feed Production with Microorganisms" (G. Reed, ed.), pp. 587–598. Verlag Chemie, Weinheim.
Chahal, D. S. (1983). Growth characteristics of microorganisms in solid state fermentation for upgrading of protein values of lignocelluloses and cellulase production. In "Foundations of Biochemical Engineering Kinetics and Thermodynamics in Biological Systems" (H. W. Blanch, E. T. Papoutsakis, and G. Stephanopoulos, eds.), ACS Symposium Series, No. 207, pp. 421–442. Am. Chem. Soc., Washington, D.C.
Chandra, A. K., Medda, S., and Bhadra, A. K. (1980). Production of extracellular thermostable α-amylase by *Bacillus licheniformis*. *J. Ferment. Technol.* **58**, 1–10.
Chang, L. T., and Terry, C. A. (1973). Intergenic complementation of glucoamylase and citric acid production in two species of *Aspergillus*. *Appl. Microbiol.* **25**, 890–895.
Chevalier, P., and de la Noüe, J. (1987). Enhancement of α-amylase production by immobilized *Bacillus subtilis* in air-lift fermentor. *Enzyme Microb. Technol.* **9**, 53–56.
Chiang, J. P., Alter, J. E., and Sternberg, M. (1979). Purification and characterization of a thermostable alpha-amylase from *Bacillus licheniformis*. *Starch/Staerke* **31**, 86–92.
Christensen, L. M. (1944). Increasing the diastase content of diastatic materials. U.S. Patent 2,359,356.
Clark, R. J., III, King, M. L., and Gainer, J. L. (1977). Increasing the production of enzyme via fermentation. In "Enzyme Engineering" (G. B. Broun, G. Manecke, and L. B. Wingard, Jr., eds.), Vol. 4, pp. 19–23. Plenum, New York.
Coker, L. E., and Venkatasubramanian, K. (1985). Starch conversion processes. In "Comprehensive Biotechnology. Vol. 3: The Practice of Biotechnology—Current Commod-

ity Products" (H. W. Blanch, S. Drew, and D.I.C. Wang, eds.), pp. 777–787. Pergamon, Oxford.
Coleman, G., and Grant, M. A. (1966). Characteristics of α-amylase formation by *Bacillus subtilis*. *Nature (London)* **211**, 306–307.
Costerton, J. W., and Irvin, R. T. (1981). The bacterial glycocalyx in nature and disease. *Annu. Rev. Microbiol.* **35**, 299–324.
Costerton, J. W., Marrie, T. J., and Cheng, K. J. (1985). Phenomena of bacterial adhesion. *In* "Bacterial Adhesion, Mechanism and Physiological Significance" (D. C. Savage and M. Fletcher, eds.), pp. 3–43. Plenum, New York.
Cragnoline, G., and Tuovinen, O. H. (1984). The role of sulfate reducing and sulfur oxidizing bacteria in the localized corrosion of iron-base alloys—a review. *Int. Biodeterior. Bull.* **20**, 9–26.
Cundell, A. M., and Mulcock, A. P. (1976). The biodeterioration of natural rubber-pipe joint rings in sewer mains. *Proc. Int. Biodegradation Symp., 3rd, Kingston, R.I., 1975* pp. 659–664.
Curtis, R., III (1976). Genetic manipulation of microorganisms: Potential benefits and biohazards. *Annu. Rev. Microbiol.* **30**, 507–533.
Daniels, S. L. (1967). Separation of Bacteria by Adsorption onto Ion Exchange Resins. Ph.D. Thesis, Univ. of Michigan, Ann Arbor [*Diss. Abstr. B* **29**, 1336 (1968)].
Datar, R. (1986). Economics of primary separation steps in relation to fermentation and genetic engineering. *Process Biochem.* **21**(1), 19–26.
Davies, P. E., Cohen, D. L., and Whitaker, A. (1980). The production of α-amylase in batch and chemostat culture by *Bacillus stearothermophilus*. *Antonie Van Leeuwenhoek J. Microbiol. Serol.* **46**, 391–398.
Davis, J. G. (1965). "Cheese," pp. 17–38. Churchill, London.
Dunn, C. G., Fuld, G. J., Yamada, K., Uriostle, J. M., and Casey, P. R. (1959). Production of amylolytic enzymes in natural and synthetic media. *Appl. Microbiol.* **7**, 212–218.
Dwarkanath, C. T., and Rao, T. N. R. (1971). Studies in the extraction of amylase from the mouldy wheat bran. *Ind. Food Packer* **25**, 5–9.
Egorov, N. S. (1985). "Antibiotics—A Scientific Approach," pp. 148–153. Mir, Moscow.
Feniksova, R. V., Tikhomirova, A. S., and Rakhleeva, B. E. (1960). Conditions for forming amylase and proteinase in surface culture of *Bacillus subtilis*. *Mikrobiologia* **29**, 745–748.
Fisher, N. (1973). Indigestible constituents of cereals and other foodstuffs. *In* "Molecular Structure and Function of Food Carbohydrates" (G. G. Birch and L. F. Green, eds.), pp. 275–295. Appl. Sci., London.
Fogarty, W. M. (1983). Microbial amylases. *In* "Microbial Enzymes and Biotechnology" (W. M. Fogarty, ed.), pp. 1–92. Appl. Sci., London.
Fogarty, W. M., and Kelly, C. T. (1979). Starch-degrading enzymes of microbial origin. *Prog. Ind. Microbiol.* **15**, 87–150.
Fogarty, W. M., and Kelly, C. T. (1980). Amylases, amyloglucosidases and related glucanases. *In* "Economic Microbiology. Vol. 5: Microbial Enzymes and Bioconversions" (A. H. Rose, ed.), pp. 115–170. Academic Press, London.
Fogarty, W. M., and Ward, O. P. (1974). Pectinases and pectic polysaccharides. *Prog. Ind. Microbiol.* **13**, 59–119.
Fullbrook, P. D. (1984). The enzymic production of glucose syrups. *In* "Glucose Syrups: Science and Technology" (S. Z. Dziedzic and M. W. Kearsley, eds.), pp. 65–115. Elsevier, London.
Gandhi, A. P., and Kjaergaard, L. (1975). Effect of carbon dioxide on the formation of α-amylase by *Bacillus subtilis* growing in continuous and batch cultures. *Biotechnol. Bioeng.* **17**, 1109–1118.

Ghildyal, N. P., Ramakrishna, S. V., Devi, P. N., Lonsane, B. K., and Asthana, H. N. (1981). Large scale production of pectolytic enzyme by solid state fermentation. *J. Food Sci. Technol.* **18,** 248–251.

Ghildyal, N. P., Lonsane, B. K., Sreekantiah, K. R., and Murthy, V. S. (1985). Economics of submerged and solid state fermentations for the production of amyloglucosidase. *J. Food Sci. Technol.* **22,** 171–176.

Gibson, T., and Gordon, R. E. (1974). Genus I. *Bacillus*. In "Bergey's Manual of Determinative Bacteriology" (R. E. Buchanan and N. E. Gibbons, eds.), pp. 529–550. Williams & Wilkins, Baltimore, Maryland.

Green, D. M., and Colarusso, L. J. (1964). The physical and genetic characterisation of a transformable enzyme: *Bacillus subtilis* α-amylase. *Biochim. Biophys. Acta* **89,** 277–290.

Hahn-Hägerdal, B. (1986). Water activity: A possible external regulator in biotechnical processes. *Enzyme Microb. Technol.* **8,** 322–327.

Hayashida, S. (1975). Selective submerged productions of three types of glucoamylases by a black-koji mold. *Agric. Biol. Chem.* **39,** 2093–2099.

Heineken, F. G., and O'Connor, R. J. (1972). Continuous culture studies on the biosynthesis of alkaline protease and α-amylase by *Bacillus subtilis* NRRL-B 3411. *J. Gen. Microbiol.* **73,** 35–44.

Hesseltine, C. W. (1977). Solid state fermentation—Part 1. *Process Biochem.* **12**(6), 24–27.

Hesseltine, C. W. (1981). Future of fermented foods. *Process Biochem.* **16,** 2–6, 13.

Hoogerheide, J. C. (1954). Microbial enzymes other than fungal amylases. In "Industrial Fermentations" (L. A. Underkofler and R. J. Hickey, eds.), Vol. 1, pp. 122–154. Chem. Publ. Co., New York.

Horikoshi, A., and Akiba, T. (1982). "Alkalophilic Microorganisms—A New Microbial World," pp. 93–142. Jpn. Sci. Soc. Press, Tokyo.

Howling, D. (1984). Introduction: Glucose syrups—past, present and future. In "Glucose Syrups: Science and Technology" (S. Z. Dziedzic and M. W. Kearsley, eds.), pp. 1–7. Elsevier, London.

Imshenetskii, A. A., and Solntseva, L. T. (1944). Production of amylase from cultures of thermophilic bacteria. *Mikrobiologiya* **13,** 54–64.

Ingle, M. B., and Erickson, R. J. (1978). Bacterial α-amylases. *Adv. Appl. Microbiol.* **24,** 257–278.

Keay, L. (1970). The action of *Bacillus subtilis* saccharifying amylase on starch and β-cyclodextrin. *Staerke* **22,** 153–157.

Kelly, C. T., and Fogarty, W. M. (1976). Microbial alkaline enzymes. *Process Biochem.* **11**(6), 3–9.

Kerr, R. W. (1950). "Chemistry and Industry of Starch," pp. 157–158. Academic Press, New York.

Kline, L., Mac Donnell, L. R., and Lineweaver, H. (1944). Bacterial proteinase from asparagus butts. *Ind. Eng. Chem.* **36,** 1152–1158.

Kneen, E., and Beckord, L. D. (1946). Quantity and quality of amylase produced by various bacterial isolates. *Arch. Biochem.* **10,** 41–54.

Kneen, E., and Standstedt, R. M. (1946). Distribution and general properties of an amylase inhibitor in cereals. *Arch. Biochem.* **9,** 235–249.

Kokobu, T., Karube, I., and Suzuki, S. (1978). α-Amylase production by immobilised whole cells of *Bacillus subtilis*. *Eur. J. Appl. Microbiol. Biotechnol.* **5,** 233–240.

Krishnamurti, K., and Soman, S. V. (1951). Studies in the adsorption of bacteria. Part I. Adsorption of *B. subtilis* and *E. coli* by kaolin and charcoal. *Proc. Indian Acad. Sci., Sect. B* **34,** 81–91.

Kumar, P. K. R., and Lonsane, B. K. (1987a). Gibberellic acid by solid state fermentation. Consistent and improved yields. *Biotechnol. Bioeng.* **30**, 267–271.
Kumar, P. K. R., and Lonsane, B. K. (1987b). Potential of fed-batch culture in solid state fermentation for production of gibberellic acid. *Biotechnol. Lett.* **9**, 179–182.
Kumar, P. K. R., and Lonsane, B. K. (1987c). Extraction of gibberellic acid from dry mouldy bran produced under solid state fermentation. *Process Biochem.* **22**, 139–143.
Kumar, P. K. R., and Lonsane, B. K. (1989). Microbial production of gibberellins: State of the art. *Adv. Appl. Microbiol.* **34**, 29–139.
Lee, D. D., Lee, Y. Y., and Tsao, G. T. (1975). Continuous production of glucose from dextrin by glucoamylase immobilized on porous silica. *Staerke* **27**, 384–387.
Lee, D. D., Lee, Y. Y., Reilly, P. J., Collins, E. N., Jr., and Tsao, G. T. (1976). Pilot plant production of glucose with glucoamylase immobilized to porous silica. *Biotechnol. Bioeng.* **18**, 253–267.
Lehrian, D. W., and Patterson, G. R. (1983). Cocoa fermentation. *In* "Biotechnology. Vol. 5: Food and Feed Production with Microorganisms" (G. Reed, ed.), pp. 529–575. Verlag Chemie, Weinheim.
Liepe, H.-U. (1983). Starter cultures in meat production. *In* "Biotechnology. Vol. 5: Food and Feed Production with Microorganisms" (G. Reed, ed.), pp. 399–424. Verlag Chemie, Weinheim.
Linko, Y. Y., Saarinen, P., and Linko, M. (1975). Starch conversion by soluble and immobilized α-amylase. *Biotechnol. Bioeng.* **17**, 153–165.
Lockwood, L. B. (1952). *Trans. N.Y. Acad. Sci.* **15**, 1–5. Cited in S. C. Prescott and C. G. Dunn, "Industrial Microbiology," 3rd Ed., pp. 497–514. McGraw-Hill, New York, 1959.
Lonsane, B. K., Ghildyal, N. P., Budiatman, S., and Ramakrishna, S. V. (1985). Engineering aspects of solid state fermentation. *Enzyme Microb. Technol.* **7**, 258–265.
Luenser, S. J. (1983). Microbial enzymes for industrial sweetener production. *Dev. Ind. Microbiol.* **24**, 79–96.
Lulla, B. S., and Subrahmanyan, V. (1954). Influence of culture media on the development of bacterial amylase. *J. Sci. Ind. Res., Sect. B.* **13**, 410–412.
Madsen, G. B., and Norman, B. E. (1973). New speciality glucose syrups. *In* "Molecular Structure and Function of Food Carbohydrates" (G. G. Birch and L. G. Green, eds.), pp. 50–64. Appl. Sci., London.
Madsen, G. B., Norman, B. E., and Slott, S. (1973). A new heat stable bacterial amylase and its use in high temperature liquefaction. *Staerke* **25**, 304–308.
Mahmoud, S. A. Z., Attia, R. M., and Abdel-Nasser, M. (1983). Some factors affecting the production of bacterial α-amylase. *Egypt. J. Microbiol.* Spec. issue, 109–119.
Markkanen, P. H., and Bailey, M. J. (1975). Effect of alteration of aeration and temperature on production of α-amylase by *Bacillus subtilis*. *J. Appl. Chem. Biotechnol.* **25**, 863–865.
Marshall, K. C., ed. (1984). "Microbial Adhesion and Aggregation—Dahlem Conference." Springer-Verlag, Berlin.
Maruo, B., and Tojo, D. (1985). Stepwise enhancement of productivity of thermostable amylase in *Bacillus licheniformis* by a series of mutations. *J. Gen. Appl. Microbiol.* **31**, 323–328.
Matsumoto, N., Fukushi, O., Miyanaga, M., Kakihara, K., Nakajima, E., and Yoshizumi, H. (1982). Industrialization of noncooking system for alcoholic fermentation from grains. *Agric. Biol. Chem.* **46**, 1549–1558.
Matsuoka, H., Koba, Y., and Ueda, S. (1982). Alcoholic fermentation of sweet potato without cooking. *J. Ferment. Technol.* **60**, 599–602.

Medda, S., and Chandra, A. K. (1980). New strains of *Bacillus licheniformis* and *Bacillus coagulans* producing thermostable α-amylase active at alkaline pH. *J. Appl. Bacteriol.* **48**, 47–58.
Meers, J. L. (1972). The regulation of α-amylase production in *Bacillus licheniformis*. *Antonie van Leeuwenhoek J. Microbiol. Serol.* **38**, 585–590.
Meyrath, J., and Volavsek, G. (1975). Production of microbial enzymes. In "Enzymes in Food Processing" (G. Reed, ed.), 2nd Ed., pp. 255–300. Academic Press, New York.
Minagawa, T., and Hamaishi, T. (1962). Enzyme production. U.S. Patent 3,031,380.
Mudgett, R. E. (1986). Solid state fermentations. In "Manual of Industrial Microbiology and Biotechnology" (A. L. Demain and N. A. Solomon, eds.), pp. 66–83. Am. Soc. Microbiol., Washington, D.C.
Nishio, N., Tai, K., and Nagai, S. (1979). Hydrolase production by *Aspergillus niger* in solid state cultivation. *Eur. J. Appl. Microbiol. Biotechnol.* **8**, 263–270.
Nomura, N., Maruo, B., and Akabori, S. (1956). Studies on amylase production by *Bacillus subtilis*. I. Effect of high concentration of polyethylene glycol on amylase formation by *Bacillus subtilis*. *J. Biochem. (Tokyo)* **43**, 143–152.
Norman, B. E. (1981). New developments in starch syrup technology. In "Enzymes and Food Processing" (G. G. Birch, N. Blakebrough, and K. J. Parker, eds.), pp. 15–50. Appl. Sci., London.
Novotel'nov, N. V., Gorbatova, K. K., and Avdonina, K. P. (1964). Antibiotic and amylase regulator isolated from wheat bran. *Izv. Vyssh. Ucheb. Zaved. Pishch. Tekhnol.* pp. 54–56.
Nyiri, L. K., and Charles, M. (1977). Economic status of fermentation processes. In "Annual Reports on Fermentation Processes" (D. Perlman, ed.), Vol. 1, pp. 365–381. Academic Press, New York.
Oji Paper Co., Ltd. (1961). Purified calcium enzymes. Jpn. Patent 2 873.
Park, Y. K., and Rivera, B. C. (1982). Alcohol production from various enzyme-converted starches with or without cooking. *Biotechnol. Bioeng.* **24**, 495–500.
Pazlarova, J., Baig, M. A., and Vortruba, J. (1984). Kinetics of alpha-amylase production in a batch and fed-batch culture of *Bacillus subtilis* with caseinate as nitrogen source and starch as carbon source. *Appl. Microbiol. Biotechnol.* **20**, 331–334.
Peltier, G. L., and Beckord, L. D. (1945). Sources of amylase-producing bacteria. *J. Bacteriol.* **50**, 711–714.
Peppler, H. J. (1983). Fermented foods and feed supplements. In "Biotechnology. Vol. 5: Food and Feed Production with Microorganisms" (G. Reed, ed.), pp. 599–615. Verlag Chemie, Weinheim.
Peppler, H. J., and Reed, G. (1987). Enzymes in food and feed processing. In "Biotechnology. Vol. 7a: Enzyme Technology" (J. F. Kennedy, ed.), pp. 547–603. VCH Verlagsges., Weinheim.
Porter, F. E., and Nack, H. (1960). Surface culture in the mass growth of bacteria. *J. Biochem. Microbiol. Technol. Eng.* **2**, 177–186.
Priest, F. G. (1977). Extracellular enzyme synthesis in the genus *Bacillus*. *Bacteriol. Rev.* **41**, 711–753.
Priest, F. G., and Thirunavukkarasu, M. (1985). Synthesis and localization of α-amylase and α-glucosidase in *Bacillus licheniformis* grown in batch and continuous culture. *J. Appl. Bacteriol.* **58**, 381–390.
Qadeer, M. A., Anjum, J. I., and Akhtar, R. (1980). Biosynthesis of enzymes by solid substrate fermentation. Part II. Production of alpha-amylase by *Bacillus subtilis*. *Pak. J. Sci. Ind. Res.* **23**, 25–29.
Quittenton, R. C. (1969). Perspectives in brewing. *Wallerstein Lab. Commun.* **32**, 77–79.

Raimbault, M., and Alazard, D. (1980). Culture method to study fungal growth in solid fermentation. *Eur. J. Appl. Microbiol. Biotechnol.* **9,** 199–209.

Ralph, B. J. (1976). Solid substrate fermentations. *Food Technol. Aust.* **28,** 247–251.

Ramakrishna, S. V., Suseela, T., Ghildyal, N. P., Jaleel, S. A., Prema, P., Lonsane, B. K., and Ahmed, S. Y. (1982). Recovery of amyloglucosidase from moldy bran. *Indian J. Technol.* **20,** 476–480.

Ramesh, M. V., and Lonsane, B. K. (1986). Inability of some *Bacillus* strains to produce thermostable α-amylase under solid state fermentation. *Annu. Conf., 27th, Assoc. Microbiologists India, Nagpur* Poster.

Ramesh, M. V., and Lonsane, B. K. (1987a). Solid state fermentation for production of α-amylase by *Bacillus megaterium* 16M. *Biotechnol. Lett.* **9,** 323–328.

Ramesh, M. V., and Lonsane, B. K. (1987b). A novel bacterial thermostable alpha-amylase system produced under solid state fermentation. *Biotechnol. Lett.* **9,** 501–504.

Ramesh, M. V., and Lonsane, B. K. (1988). Factors affecting recovery of thermostable alpha-amylase from bacterial bran produced under solid state fermentation. *Chem. Mikrobiol. Technol. Lebensm.* **11,** 155–159.

Ramesh, V., and Singh, C. (1981). Immobilization of *Bacillus subtilis* α-amylase on zirconia-coated alkylamine glass with glutaraldehyde. *Enzyme Microb. Technol.* **3,** 246–248.

Rao, M. R. R., Bindal, A. N., and Sreenivasaya, M. (1947). Studies on industrial enzymes. Part V. Extraction, concentration and purification of fungal diastases. *J. Sci. Ind. Res., Sect. B* **6,** 97–100.

Rosen, C.-G., and Datar, R. (1983). Primary separation steps in fermentation processes. *Biotechnol. 1983, Proc. Int. Conf. Commer. Appl. Implic. Biotechnol.,* 1st pp. 201–224. Online Publ., Northwood, England.

Rosendal, P., Nielsen, B. H., and Lange, N. K. (1979). Stability of bacterial alpha-amylase in the starch liquefaction process. *Starch/Staerke* **31,** 368–372.

Rutter, P. R. (1984). Mechanism of adhesion (group report). In "Microbial Adhesion and Aggregation—Dahlem Conference" (K. C. Marshall, ed.), pp. 5–19. Springer-Verlag, Berlin.

Rutter, P. R., and Vincent, B. (1984). Physiochemical interactions of the substratum, microorganisms and the fluid phase. In "Microbial Adhesion and Aggregation—Dahlem Conference" (K. C. Marshall, ed.), pp. 21–38. Springer-Verlag, Berlin.

Saito, N. (1973). A thermophilic extracellular α-amylase from *Bacillus licheniformis*. *Arch. Biochem. Biophys.* **155,** 290–298.

Saito, N., and Yamamoto, K. (1975). Regulatory factors affecting α-amylase production in *Bacillus licheniformis*. *J. Bacteriol.* **121,** 848–856.

Schultz, A. C., Atkin, L., and Frey, C. N. (1931). Enzymic material suitable for starch liquefaction. U.S. Patent 2,159,678.

Scott, D., Hammer, F. E., and Szalkucki, T. J. (1987). Bioconversions: Enzyme technology. In "Food Biotechnology" (D. Knorr, ed.), pp. 413–440. Dekker, New York.

Scott, R. (1981). "Cheesemaking Practice," pp. 191–205. Appl. Sci., London.

Scott, W. J. (1957). Water relations of food spoilage microorganisms. *Adv. Food Res.* **7,** 83–127.

Sekiguchi, J., Takada, N., and Okada, H. (1975). Genes affecting the productivity of α-amylase in *Bacillus subtilis* Marburg. *J. Bacteriol.* **121,** 688–694.

Shales, S. W. (1985). Cell adhesion—An introduction. *Ind. Biotechnol.* **5,** 69–72.

Shales, S. W., and Kumarasingham, S. (1987). Bacterial transport through porous solids: Interactions between *Micrococcus luteus* cells and sand particles. *J. Ind. Microbiol.* **2,** 219–227.

Shinmyo, A., Kimura, H., and Okoda, H. (1982). Physiology of α-amylase production by immobilized Bacillus amyloliquefaciens. Eur. J. Appl. Microbiol. Biotechnol. **14,** 7–12.
Slaughter, J. C. (1985). Enzymes in brewing industry. In "Alcoholic Beverages" (G. G. Birch and M. G. Lindley, eds.), pp. 15–27. Elsevier, London.
Smily, K. L. (1971). Continuous conversion of starch to glucose with immobilised glucoamylase. Biotechnol. Bioeng. **13,** 309–317.
Smith, G. (1969). "An Introduction to Industrial Mycology," 6th Ed., pp. 310–326. Arnold, London.
Smith, J. A., and Frankiewicz, J. R. (1976). Glucoamylase—Culture medium for its production and its use for the hydrolysis of starches. Germ. Patent 2,554,850.
Smythe, C. V., Blake, B. B., and Neubeck, C. E. (1950). Diastase and protease. U.S. Patent 2,530,210.
Sreekantiah, K. R. (1976). Studies on Pectin Degrading Strains of Fungi. Ph.D. Thesis, Univ. of Mysore, Mysore, India.
Srikanta, S., Jaleel, S. A., and Sreekantiah, K. R. (1987). Production of ethanol from tapioca (Manihot esculenta Crantz). Starch/Staerke, **39,** 132–135.
Srivastava, R. A. K., and Mathur, S. N. (1984). Regulation of amylase biosynthesis in growing and non-growing cells of Bacillus stearothermophilus. J. Appl. Bacteriol. **57,** 147–151.
Stamer, J. R. (1983). Lactic acid fermentation of cabbage and cucumbers. In "Biotechnology. Vol. 5: Food and Feed Production with Microorganisms" (G. Reed, ed.), pp. 367–397. Verlag Chemie, Wienheim.
Stein, E. A., and Fischer, E. H. (1958). The resistance of α-amylases towards proteolytic attack. J. Biol. Chem. **232,** 867–879.
Steinkraus, K. H. (1984). Acta Biotechnol. **4,** 83–88. Cited in Hahn-Hägerdal (1986).
Stewart, G. G., and Russel, I. (1985). Modern brewing biotechnology. In "Comprehensive Biotechnology. Vol. 3: The Practice of Biotechnology—Current Commodity Products" (H. W. Blanch, S. Drew, and D. I. C. Wang, eds.), pp. 335–381. Pergamon, Oxford.
Surovaya, A. V. (1944). The use of superbiolase for desizing cotton fabrics. Tekst. Prom. (Moscow) **4,** 14–18.
Takamine, J. (1894). Preparing and making fermented alcoholic liquors. U.S. Patent 525,819.
Takamine, J. (1911). Amylolytic enzyme. U.S. Patent 991,561.
Takamine, J. (1914). Enzymes of Aspergillus oryzae and the application of its amyloclastic enzyme to the fermentation industry. Ind. Eng. Chem. **6,** 824–828.
Tilden, E. B., and Hudson, C. S. (1942). Preparation and properties of the amylases produced by Bacillus macerans and Bacillus polymyxa. J. Bacteriol. **43,** 527–544.
Tobey, J. F., and Yousten, A. A. (1976). Factors affecting the production of amylase by Bacillus thuringiensis. Dev. Ind. Microbiol. **18,** 499–510.
Troller, J. A., and Christian, J. H. B. (1978). "Water Activity and Food," pp. 86–102. Academic Press, London.
Ueda, S. (1957). Studies on the amylolytic system of black-koji moulds. II. Raw starch digestibility of the saccharogenic amylase fraction and its interaction with the dextrogenic amylase fraction. Bull. Agric. Chem. Soc. Jpn. **21,** 284–290.
Ueda, S., and Kano, S. (1975). Multiple forms of glucoamylase of Rhizopus species. Starch/Staerke **27,** 123–128.
Ueda, S., and Koba, Y. (1980). Alcoholic fermentation of raw starch without cooking by using black-koji amylase. J. Ferment. Technol. **58,** 237–242.

Ueda, S., and Marshall, J. J. (1980). On the ability of pullulanase to stimulate the enzymic digestion of raw starch. *Carbohydr. Res.* **84,** 196–198.
Ueda, S., and Ohba, R. (1976). Pullulanase responsible for digesting raw starch. *Starch/Staerke* **28,** 20–22.
Ueda, S., Ohba, R., and Kano, S. (1974). Fractionation of the glucoamylase system from black-koji mould and the effects of adding isoamylase on amylosis by the glucoamylase fractions. *Starch/Staerke* **26,** 374–378.
Ueda, S., Saha, B. C., and Koba, Y. (1984). Direct hydrolysis of raw starch. *Microbiol. Sci.* **1,** 21–24.
Underkofler, L. A. (1976). Microbial enzymes. In "Industrial Microbiology" (B. M. Miller and W. Litskey, eds.), pp. 128–164. McGraw-Hill, New York.
Underkofler, L. A., Fulmer, E. I., and Schone, L. (1939). Saccharification of starchy grain mashes for the alcoholic fermentation industry. *Ind. Eng. Chem.* **31,** 734–738.
Underkofler, L. A., Barton, R. R., and Rennert, S. S. (1958). Microbiological process report, production of microbial enzymes and their applications. *J. Appl. Microbiol.* **6,** 212–221.
Vincent, W. A., and Priestley, G. (1975). Large scale production of enzymes: Techniques in fermentation. In "Handbook of Enzyme Biotechnology" (A. Wiseman, ed.), pp. 27–57. Horwood, Chichester, England.
Vortruba, J., Emanuilova, E., Kaymakchiev, A., and Pazlarova, J. (1984). Kinetics of α-amylase production in a continuous culture of *Bacillus licheniformis. Folia Microbiol.* **29,** 19–22.
Waksman, S. A., and Allen, M. C. (1933). Decomposition of polyuronides by fungi and bacteria. I. Decomposition of pectin and pectic acid by fungi and formation of proteolytic enzymes. *J. Am. Chem. Soc.* **55,** 3408–3418.
Wallerstein, L. (1939). Enzyme preparations from microorganisms. Commercial production and industrial application. *Ind. Eng. Chem.* **31,** 1218–1224.
Walton, H. H., and Eastman, J. E. (1973). Insolubilized amylases. *Biotechnol. Bioeng.* **15,** 951–962.
Wardell, J. N., Brown, C. M., and Flannigan, B. (1983). Microbes and surfaces. In "Microbes in Their Natural Environments" (J. H. Slater, R. Whittenbury, and J. W. T. Wimpenny, eds.), Symp. Soc. Gen. Microbiol., Vol. 34, pp. 351–378. Cambridge Univ. Press, London.
Weig, A. J., Hollo, J., and Verga, P. (1969). Brewing beer with enzymes. *Process. Biochem.* **12**(5), 33–38.
Welker, N. E., and Campbell, L. L. (1963a). Induced biosynthesis of α-amylase by growing cultures of *Bacillus stearothermophilus. J. Bacteriol.* **86,** 1196–1201.
Welker, N. E., and Campbell, L. L. (1963b). Induction of α-amylase of *Bacillus stearothermophilus* by maltodextrins. *J. Bacteriol.* **86,** 687–691.
Weller, C. L., Steinberg, M. P., and Rodda, E. D. (1984). Fuel ethanol from raw corn. *Trans. ASAE* **27,** 1911–1916.
Windish, W. W., and Mhatre, N. S. (1965). Microbial amylases. *Adv. Appl. Microbiol.* **7,** 273–304.
Wiseman, A. (1975). Industrial practice with enzymes. In "Handbook of Enzyme Biotechnology" (A. Wiseman, ed.), pp. 243–272. Horwood, Chichester, England.
Yabuchi, S., and Okita, H. (1985). Biologically active substance manufacture by fermentation. Jpn. Patent 60 196 190.
Yamaguchi, K., Matsuzaki, H., and Maruo, B. (1969). Participation of a regulator gene in the α-amylase production of *Bacillus subtilis. J. Gen. Appl. Microbiol.* **15,** 97–107.

Yasunobu, K. T., and McConn, T. (1970). Bacillus subtilis neutral protease. Methods Enzymol. **19,** 569–575.

Yuki, S. (1967). Genetic studies on amylase of different electrophoretic mobility produced by strains of Bacillus subtilis. Jpn. J. Genet. **42,** 251–261.

Yuki, S. (1968). On the gene controlling the rate of amylase production in Bacillus subtilis. Biochem. Biophys. Res. Commun. **31,** 182–187.

Zhang, Q., Tsukagoshi, N., Miyashiro, S., and Udaka, S. (1983). Increased production of α-amylase by Bacillus amyloliquefaciens in the presence of glycine. Appl. Environ. Microbiol. **46,** 293–295.

Zo Bell, C. E., and Anderson, D. O. (1936). Observations on the multiplication of bacteria in different volumes of stored sea water and the influence of oxygen tension and solid surfaces. Biol. Bull. (Woods Hole, Mass.) **71,** 324–342.

Methods for Studying Bacterial Gene Transfer in Soil by Conjugation and Transduction

G. STOTZKY, MONICA A. DEVANAS,[1] AND LAWRENCE R. ZEPH[2]

Laboratory of Microbial Ecology
Department of Biology
New York University
New York, New York 10003

I. Introduction
    A. Microhabitats in Soil
    B. Gene Transfer in Soil
    C. Biological Factors
    D. Effects of Biological and Physicochemical Environmental Factors on Gene Transfer in Soil
    E. Conclusions
II. Literature Review of Conjugation and Transduction in Soil
    A. Conjugation
    B. Transduction
III. Terrestrial Microcosms
    A. Introduction
    B. Simple Microcosms
    C. Complex Microcosms
IV. Selection, Preparation, Maintenance, and Storage of Soils
    A. Introduction
    B. Selection of Soils
    C. Preparation of Soils
    D. Maintenance of Soils
    E. Sampling of Soils
    F. Storage of Soils
V. Methods for Studying Conjugation
    A. In Vitro
    B. In Soil
    C. Stepwise Summary of Procedures for Studying Conjugation in Soil
VI. Methods for Studying Transduction
    A. In Vitro
    B. In Soil
    C. Stepwise Summary of Procedures for Studying Transduction in Soil
VII. Identification, Characterization, and Confirmation of Recombinants
    A. Introduction
    B. Selective Media

---

[1] Present address: Department of Biological Sciences, Rutgers University, New Brunswick, New Jersey 08903.

[2] Present address: U. S. Environmental Protection Agency, Office of Toxic Substances, Washington, D.C. 20460.

C. "Breakthrough" of Indigenous Microbes
D. Viable but Nonculturable Bacteria
E. Gene Transfer on Recovery Media versus *in Situ*
F. DNA "Fingerprinting" and Plasmid Profiles
G. DNA Probes
H. Serological Techniques
I. Heat Induction of Prophages
VIII. Quality Assurance and Quality Control
A. Sample Representativeness and Custody
B. Sampling Procedures
C. Comparability
D. Calibration Procedures and Frequency
E. Analytical Procedures
F. Experimental Design and Statistical Analyses
G. Data Analysis and Reporting
H. Internal Quality Control Checks
I. Preventive Maintenance
J. Corrective Action
IX. Appendix
A. Media Composition
B. Antimicrobial Agents Commonly Used in Selective Media
References

I. Introduction

A. Microhabitats in Soil

Soil is unique among microbial habitats in that it is a structured environment with a high solid to water ratio (Stotzky, 1974, 1986). Inasmuch as all microbes are aquatic creatures, their metabolism in soil is restricted to microhabitats wherein there is a continual supply of available water. Hence, their distribution is essentially restricted to microhabitats that contain clay minerals, because sand and silt do not retain water long against gravitational pull. Clay minerals, because of their surface activity, retain water against this pull, as the water adjacent to these active surfaces and coordinated with charge-compensating ions on the clays becomes sufficiently ordered to form a quasi-crystalline structure (i.e., the strong attraction of water molecules to the negatively and positively charged surfaces of clay minerals and their charge-compensating ions enhances the hydrogen bonding of adjacent water molecules). The ordering of this clay-associated water decreases with distance from the clay surface until a distance is reached at which the water is no longer under the attraction of the clay and is susceptible to gravity.

Clay minerals do not exist free in soil but are present as coatings, or cutans, on larger sand and silt particles or as oriented clusters, or domains, between these particles. The clay-coated particles and domains cluster, primarily as the result of electrostatic attraction between the net negatively charged faces and the net positively charged edges of clays, into microaggregates; these microaggregates, in turn, cluster to form aggregates that can range from 0.5 to 5 mm in diameter and are stabilized by organic matter and precipitated inorganic materials. These aggregates retain water, and the thickness and permanence of this water depend on the type and amount of clay and organic matter within the aggregates; this water may form bridges with the water of closely adjacent aggregates. These aggregates or clusters of aggregates, with their adjacent water, constitute the microhabitats in soil wherein microbes function. The space between the microhabitats constitutes the pore space, which is filled with air and other gases and volatiles.

As the result of the discreteness of microhabitats in soil, the probability of genetic exchange may be less than the probability in ecosystems wherein water is continuous. Except for periods when soil is saturated with water (e.g., after a heavy rain or snowmelt or when irrigated), individual microhabitats are isolated by the surrounding pore space, and movement of bacteria, transducing bacteriophages, and transforming DNA among the microhabitats is limited to areas where water bridges between microhabitats may occur. Even when the pore space is saturated, movement between microhabitats may be restricted, because the surface tension of the ordered water around aggregates may be too great to allow passive movement of bacteria or even active movement by flagellated cells. (There is no convincing evidence that bacteria are flagellated in soil, even though they may have the genetic capability to produce flagella when isolated from soil and cultured in liquid media or on agar.) However, filamentous fungi appear to be able to bridge pore spaces between microhabitats, even when the pore spaces are not filled with water, as these fungi grow apically from mycelia that have a food and water base in a microhabitat and, therefore, are independent of the nutrient and water conditions surrounding the growing mycelia. Moreover, the extending mycelia probably have a surrounding water film in which bacteria or bacteriophages may be transported from one microhabitat to another.

These conditions in soil differ markedly from those in sediments of aquatic systems. Although clay minerals in sediments also occur as domains and as cutans on larger particles and in aggregates, water-dependent microhabitats do not exist as they do in soil, because the water in sediments is essentially continuous from one aggregate to the next. Moreover, microbes in sediments appear to colonize primarily sand and silt particles, rather than clays, because water surrounds these particles, and the need to overcome the electrokinetic repulsion between net negatively charged clays and microbes is eliminated. Hence, care must be exercised in extrapolating observations on gene transfer in sediments to soil and vice versa.

B. Gene Transfer in Soil

The majority of studies on the transfer of genetic information in soil, especially with genetically engineered microorganisms, or GEMs (i.e., microorganisms in which the genetic information has been deliberately altered by recombinant DNA techniques), have been conducted with bacteria, wherein DNA can be transferred *in situ* by conjugation (cell-to-cell contact), transduction (via a bacteriophage), and transformation (uptake of "naked" DNA by an intact cell). Although these phenomena have been demonstrated in a wide spectrum of Gram-positive and Gram-negative bacteria in the laboratory (i.e., in pure culture), there is sparse information on their occurrence in soil and other natural habitats (Levy and Miller, 1989; Stotzky, 1989; Stotzky and Babich, 1986; Trevors *et al.*, 1987).

The first studies on the transfer of genetic information in soil were conducted only in sterile soil (Graham and Istock, 1978; Weinberg and Stotzky, 1972), as methods for conducting such studies in nonsterile soil had not yet been developed. Studies conducted in sterile soil have little relevance to what occurs in nonsterile soil. Hence, extrapolations from results obtained in sterile soil to what presumably occurs in nonsterile soil should be viewed with skepticism, as should studies conducted in soil extracts, wherein microhabitats have been disrupted and physicochemical characteristics altered. Nevertheless, studies in sterile soil can be valuable, when coupled with parallel studies in nonsterile soil, as techniques for subsequent use in nonsterile soil can be developed, and the effects of the indigenous microbiota and of surfaces and some other physicochemical characteristics on survival, establishment, growth, and gene transfer can be estimated.

Many of the early studies on gene transfer in soil were conducted with *Escherichia coli*, which is not an autochthonous member of the microbiota of soil. There were numerous plausible reasons why *E. coli* was used as the model bacterium in these studies. For example, (1) the genetics of *E. coli* were better defined than those of other bacteria; (2) numerous strains with a spectrum of chromosomal alterations or containing different plasmids were readily available; (3) there had been extensive genetic engineering of *E. coli* strains for use in a variety of industrial applications; (4) because of the successful experience with genetic engineering for industrial applications, it appeared reasonable to assume that strains of *E. coli* engineered to perform specific functions would be used initially for releases to the environment; (5) *E. coli* has been shown to transfer plasmid-borne genetic information to >40 genera of Gram-negative bacteria and even to some Gram-positive bacteria; and (6) though primarily an inhabitant of the gastrointestinal tract of many, but not all, animals, *E. coli* is increasingly found in fresh and estuarine waters and in soils in urban and agricultural areas, probably as the result, in large part, of the presence of human beings. The tendency now is to conduct studies on survival, growth, establishment, and gene transfer with bacterial species that are autochthonous in soil, especially as the genetics of these species become more defined.

A major impetus for these studies has been the concern about the release of GEMs to the environment, especially to soil, for agricultural (e.g., enhanced nitrogen fixation, biological pest control, reduction in frost damage) and other (e.g., degradation of recalcitrant organic pollutants) purposes. The concern is that these released GEMs will not just perform the functions for which they have been engineered but that they will somehow perturb the homeostasis of soil and cause unanticipated and undesirable changes in the activity, ecology, and population dynamics of the soil microbiota.

The survival, perpetuation, and efficacy of recombinant DNA in microbes in soil or in any natural environment depend on (1) the nature of the microbial host and the vector (e.g., a plasmid or a bacteriophage), if involved, carrying the DNA; (2) the survival, establishment, and growth of the host–vector system; (3) the maintenance, replication, and segregation of the DNA; (4) the frequency of transfer of the DNA to other microbes; (5) the adaptability of the host–vector systems, either the original or the subsequent new ones, to their ecological "niches"; and (6) the selective advantage or disadvantage conferred on the original or subsequent hosts by the DNA. These phenomena are affected by the physicochemical and biological characteristics of soil and other natural habitats (Table I).

The relative importance of individual environmental characteristics varies with the specific habitat (e.g., electromagnetic radiation is probably relatively unimportant in soil, whereas it is extremely important in aquatic habitats), and their effects are usually greater on introduced than on indigenous microbes. Moreover, none of these characteristics exerts its influence individually but, rather, in concert with other characteristics. Although the influence of one or a few characteristics may predominate in a specific habitat, these influences have indirect, but cascading, effects on other characteristics. Consequently, an alteration in one environmental factor may result in simultaneous or subsequent changes in other characteristics and, ultimately, in the habitat, and

TABLE I

FACTORS AFFECTING THE ACTIVITY, ECOLOGY, AND POPULATION
DYNAMICS OF MICROORGANISMS IN NATURAL HABITATS

Carbon and energy sources	Electromagnetic radiation
Mineral nutrients	pH
Growth factors	Oxidation–reduction potential
Ionic composition	Surfaces
Available water	Spatial relations
Temperature	Genetics of the microorganisms
Pressure	Interactions between microorganisms
Atmospheric composition	

therefore, in the ability of both introduced and indigenous microbes to survive, establish, grow, and transfer genetic information (Stotzky, 1974, 1986; Stotzky and Krasovsky, 1981). Because the possible permutations of interactions between these environmental characteristics are vast, the relative success of microbes containing new genetic information to transfer this information in soil cannot be easily predicted.

C. BIOLOGICAL FACTORS

For detectable gene transfer to occur in soil or other natural environments, the population densities of the donors and recipients of the gene(s) must be sufficiently high. This is especially true in soil, where the spatial separation of microhabitats reduces the probability of transfer. Consequently, for a bacterium containing a novel gene to transfer this genetic information at a detectable frequency, it must be able to establish (i.e., colonize) and grow to sufficiently high densities within the microhabitat, especially if it is a recombinant that has resulted naturally from transfer *in situ*. If the recombinant bacterium is introduced into soil, even at high densities, it will have difficulty colonizing and growing, as the microhabitats are already filled with organisms that are highly adapted to these microhabitats after eons of selection. Even in the gastrointestinal tract—an environment that is probably less complex than soil, especially in terms of the number of species present—the introduction of billions of cells of *E. coli* K12 containing small nonconjugative plasmids, such as pBR322 and pBR325, did not result in their colonization of the tract unless the indigenous microbiota, of which *E. coli* is a major component, was suppressed by antibiotics or starvation or the tract was essentially devoid of indigenous microbes, as in gnotobiotic or neonatal animals (see Stotzky and Babich, 1986, for references).

It must be emphasized that "gene transfer" in soil and other natural environments implies not only the transfer of genes, but also their expression, as most methods used initially to detect transfer rely on expression of the genes (i.e., phenotypic characteristics). Hence, the transfer of genetic information in soil may be more frequent than is actually detected.

The apparent lack of high frequencies of detectable gene transfer in soil may be the result not only of insufficient population densities of appropriate donors and recipients, but also of a reduction in the fitness and competitiveness of the recombinants in some microhabitats and of barriers that reduce the successful transfer and expression of genes. In pure culture, the yield of plasmid-containing transconjugants is apparently decreased by approximately the square of parental densities $<10^8$ cells/ml; for example, the number of transconjugants is reduced by approximately four orders of magnitude if the parental densities are only 10^6 cells/ml (Curtiss, 1976). However, this dependence on parental densities was not observed in sterile fresh water (O'Morchoe et al., 1988). Comparable studies have not been conducted in soil, but plasmid transfer is greatly reduced when the survival of added parentals, especially of the donors, decreases to levels below $\sim 10^4$ cells/g soil (Devanas and Stotzky, 1988b). Moreover, if the foreign DNA does not confer some selective advantage to the recombinant, the energy and precursors necessary for the replication of the additional DNA and for the synthesis of the products specified by this DNA, whether functional or not, may reduce the competitiveness of the recombinant (Curtiss, 1976), especially in soil microhabitats, which are usually oligotrophic.

Among the numerous cellular barriers that can prevent or reduce the transfer and expression of genetic information are (1) restriction–modification systems, in which restriction endonucleases of the recipient cleave incoming exogenous DNA; (2) incompatibility of the exogenous DNA with replicons in the recipient, including insufficient sequence homology for recombination; (3) inability of the recipient to provide the proteins, including the RecA protein, necessary for the establishment and replication of non-self-sufficient exogenous DNA; (4) absence in the recipient of factors necessary for transcription of the exogenous DNA; (5) absence in the recipient of appropriate enzymes for posttranslational modification of the gene product into a functional entity; and (6) fertility inhibition, in which the exogenous DNA prevents the expression of transfer genes, especially those that code for the production of pili (Miller, 1988). Despite these numerous barriers, which usually do not occur simultaneously in a recipient, none is probably sufficiently absolute to prevent gene transfer and expression in soil if the number of donors containing the novel DNA is high enough (which it probably will be in the case of planned releases) and the residence time of the novel DNA is long enough (which it may be if the DNA survives). Moreover, many of these barriers can be surmounted by alternate strategies of DNA transfer and recombination and if the recipients are stressed (Miller, 1988), which they undoubtedly will be in soil.

It would seem, therefore, that the combination of physical separation of donors and recipients, poor survival (which will reduce the population levels of donors and recipients), cellular barriers to gene transfer and expression, and decreased fitness and competitiveness of the recombinant would reduce greatly the probability of successful transfer and expression of genetic information in bacteria in soil. However, even an extremely low probability of transfer would eventually result, and probably has, in the successful establishment of recombinant bacteria in soil *in situ*. Bacteria have existed in soil for millions of years, and some transferred DNA undoubtedly conferred some selective advantage to and, thereby, increased the fitness and competitiveness of the recipient. Moreover, periodic fluctuations in the microhabitats, such as the introtion of nutrients and changes in pH, E_h, and other physicochemical characteristics, may reduce the levels of the adapted indigenous microbiota sufficiently to allow a recombinant bacterium to colonize and increase its population density before the microhabitat returns to "normal" and again enables the autochthonous microbiota to proliferate. Such periodic alterations in the environmental conditions within microhabitats appear to be involved in the establishment in soil of fungal pathogens of plants and humans introduced naturally into soils in which they were originally absent (Stotzky, 1974, 1986). Furthermore, the presence in soil of an antimicrobial agent to which a recombinant bacterium has gained resistance (e.g., antibiotics, heavy metals) or of a compound that only it can metabolize (e.g., a recalcitrant xenobiotic) will increase its survival, establishment, and growth and, hence, the probability of gene transfer. The balance between the factors that mitigate against and those that encourage the transfer and expression of genetic information in soil will not be known until extensive studies are conducted with a spectrum of recombinant bacteria introduced into a wide variety of soils in different geographic locations.

D. Effects of Biological and Physicochemical Environmental Factors on Gene Transfer in Soil

The fate of recombinant DNA in soil is ultimately dependent on the survival, establishment, and growth in soil of the microbial hosts that house the genetic material. The survival, establishment, and growth of the hosts are, in turn, dependent on their genetic constitution and on the biological and physicochemical characteristics of the recipient soil (Table I). Detectable transfer of genetic information, regardless of the mechanism of transfer, usually requires sufficiently high populations of donors (whether a bacterium, a transducing bacteriophage, or transforming DNA) and recipients. Moreover, the ability to predict the fate of introduced GEMs and the potential transfer of their novel genetic information to indigenous microorganisms in various types of soil would be enhanced if the knowledge of the effects of different biological and physicochemical characteristics could be related to the actual characteristics of the recipient soil. Unfortunately, insufficient studies have been conducted on the effects of biological and physicochemical factors on the activity, ecology, and population dynamics of microbes and on their ability to transfer genetic information in soil and other natural habitats (Curtiss, 1976; Freter, 1984; Stotzky, 1974, 1986, 1989; Stotzky and Babich, 1986; Stotzky and Krasovsky, 1981).

1. *Microbial Competition*

On the basis of numerous studies conducted *in vitro*, *in vivo*, and *in situ*, it is apparent that the survival, establishment, and growth of introduced microorganisms, whether containing recombinant DNA or not, are usually reduced when other species of microorganisms are present. This is especially true in natural habitats, such as soil, wherein the indigenous microbial populations are usually not only better adapted to the habitat, but exert competitive, amensalistic, parasitic, and predatory pressures on the introduced organisms. For example, the survival and growth of bacteria that were not genetically engineered were significantly reduced when they were inoculated into nonsterile environments (e.g., soil, water, sewage) in which they were not natural residents, whereas they survived, and even grew, in the same environments when these were sterilized (e.g., the numbers of *Salmonella typhimurium*, *Agrobacterium tumefaciens*, and *Klebsiella pneumoniae* increased by one to two orders of magnitude in sterile soil, whereas there was a reduction in numbers when inoculated into nonsterile soil) (Liang et al., 1982). There are numerous other examples in the literature of the apparent lower survival and growth of introduced microbes in nonsterile than in sterile environments (Stotzky and Babich, 1986), and this relationship appears intuitively to be valid. However, most of these studies did not consider the possibility of the "viable but nonculturable" phenomenon (see Section VII,D). Both genetically engineered and nonengineered bacteria introduced into soil and other natural habitats could become so debilitated or otherwise altered that they would not be recovered from these habitats, especially on selective media, even though they could actually be surviving and growing in their new environment. Because of the implications of such alteration to monitoring the fate of GEMs introduced into soil and other natural habitats, it must be clearly established that the apparent lack of survival and growth of an introduced GEM was not an artifact of the experimental procedures.

2. Energy Sources

The presence of available energy sources appears to enhance the transfer of genetic information, because these sources apparently increase the population densities of the donor and recipient and the subsequent growth of the recombinant bacterium. This effect is considerably less pronounced in nonsterile than in sterile soil, as the indigenous microbiota quickly mineralizes added carbonaceous substrates. If the donor or recombinant organism contains a gene that enables the metabolism of a carbonaceous compound that cannot be metabolized by other members of the indigenous microbiota (e.g., a recalcitrant pesticide), the organism may have a temporary selective advantage, as the result of its ability to utilize the compound (e.g., Chatterjee et al., 1981). Nevertheless, this advantage is probably lost when the specific compound is mineralized or transformed into an unavailable form. Moreover, the concentration of such recalcitrant compounds in soil is usually low and less than the concentration of natural organic materials that are available to essentially all the indigenous microbes. Nevertheless, the addition to soil of a compound that only an introduced GEM can utilize should enhance the survival of that GEM in soil, and no further addition of the compound after it has been degraded should reduce the level of the GEM. This may be an efficient method with which to restrict and control the population densities of some GEMs in soil.

3. Temperature

Temperatures near the optimum growth temperature appear to be necessary for the efficient in vitro transfer of genetic information in laboratory strains of bacteria, although this is not always the case (Stotzky and Babich, 1986). For example, maximal conjugal transfer of the R-plasmid, R1drd-19, in strains of E. coli decreased progressively as the temperature was decreased from 37° to 17°C, and no transfer was detected at 15°C (Singleton and Anson, 1981). Transfer of pRD1, derived from a clinical isolate of Pseudomonas and maintained in the laboratory, from an E. coli K12 donor to an E. coli K12 recipient decreased progressively as the temperature was reduced from 37° to 15°C; when

Erwinia herbicola was the recipient, transfer occurred even at 12.5°C (Kelly and Reanney, 1984). Plasmids pWK1 and pWK2, which conferred resistance to antibiotics and mercury (Hg), were transferred optimally *in vitro* from a species of *Citrobacter* and a species of *Enterobacter* isolated from soil to *E. coli* at 28°C, and frequencies were greatly reduced 15° and 37°C. In contrast, the transfer of a plasmid conferring resistance to kanamycin (Km) from a strain of *Proteus vulgaris*, isolated from the human urinary tract, to *E. coli* was about five orders of magnitude higher at 25° than at 37°C (Terawaki et al., 1967). The *in vitro* transfer frequency of a plasmid conferring resistance to streptomycin (Sm) and tetracycline (Tc) from an *E. coli* isolated from sewage to an *E. coli* isolated from creek water was highest at 25°C and lowest at 35°C, with frequencies at 15°, 20°, and 30°C being intermediate. The frequency of transfer in raw sewage *in situ* was also higher at 22.5° than at 29.5°C (Altherr and Kasweck, 1982).

Plasmids present in enterobacteria isolated from human beings, fecally polluted rivers, and sewage treatment plants were differentiated on the basis of their thermosensitivity: "thermotolerant" plasmids were transferred equally well at 22°–37°C, whereas "thermosensitive" plasmids were transferred at high frequencies at 22° or 28°C but at low frequencies at 37°C. Only 3.1% of 775 conjugative antibiotic-resistance plasmids evaluated were thermosensitive (Smith et al., 1978).

The maintenance of some plasmids by their host cells *in vitro* is also dependent on temperature, and many plasmids are lost above the optimum growth temperature (Stotzky and Babich, 1986). For example, the Ti plasmid of *A. tumefaciens* (Watson et al., 1975) and the nodulation plasmid, pW22, of *Rhizobium trifolii* (Zurkowski and Lorkiewicz, 1979) were lost when the host cells were grown at 37°C. In contrast, the survival of *E. coli* J5(RP4) and JC5466(pRD1) was reduced to a greater extent in soil maintained at 20°C than at 4°C (Schilf and Klingmüller, 1983).

There have been few controlled studies on the effects of temperature on the transfer of genetic information in soil. It appears, however, that conjugal transfer of both plasmid- and chromosome-borne genes occurs in both sterile and nonsterile soil at temperatures lower than those necessary for optimal transfer *in vitro*; for example, transfer in *E. coli* in soil occurred at temperatures between 15° and 27°C, considerably below the requisite temperatures *in vitro* (e.g., Krasovsky and Stotzky, 1987; Trevors, 1987b; Weinberg and Stotzky, 1972), and in *Bacillus subtilis* at 15°C as well as at 27°C (van Elsas et al., 1987). Transduction of *E. coli* occurred in soil at 25°–27°C (Germida and Khachatourians, 1988; Zeph et al., 1988). Transformation of *B. subtilis* in sterile soil occurred at 37°C

(Graham and Istock, 1978), in a simulated sterile marine system containing sea sand at 23°C (Aardema et al., 1983; Lorenz et al., 1988), and in suspensions of montmorillonite at 33°C (G. Stotzky and A. Golard, unpublished observations) and at 25°C (G. Stozky and M. Khanna, unpublished observations). Unfortunately, no studies have been conducted to determine the effects of temperature on transduction and transformation in soil.

Temperature in soil cannot be conveniently controlled in situ. However, the range in fluctuations in soil temperature may be an important consideration when constructing a GEM for introduction into soil.

4. pH

The hydrogen ion concentration is an important physicochemical characteristic of soil that is amenable to control in situ but whose effect on gene transfer in soil has been insufficiently studied. Conjugal transfer of plasmids in E. coli was restricted in vitro to pH 6–8.5 (Curtiss, 1976). In soil, both intraspecific and interspecific plasmid transfer was not detected until the bulk pH was adjusted to 6.8 with $CaCO_3$ (M. A. Devanas and G. Stotzky, unpublished observations), and conjugal transfer of chromosomal genes was also detected only at pH values near neutrality (Krasovsky and Stotzky, 1987; Weinberg and Stotzky, 1972) (pH values above neutrality were not evaluated in these studies). Transduction of E. coli by phage P1 was higher in a soil with a pH of 7.9 than in a soil with a pH of 6.8 (Germida and Khachatourians, 1988). However, as these soils also differed in texture, organic matter content, and other physicochemical characteristics, differences in the transduction frequencies cannot be attributed solely to these differences in pH. The effect of pH on transformation in soil has apparently not been studied.

The pH of soil can affect gene transfer both directly (e.g., survival and growth of the parentals) and indirectly (e.g., growth of the competitive and amensalistic indigenous microbiota; adsorption of transducing phages, transforming DNA, and DNase on soil particles). For example, the adsorption of bacteriophages on clay is pH-dependent (Lipson and Stotzky, 1987; Stotzky et al., 1981), and the adsorption of DNA on sea sand increased as the pH was increased from 5 to 9 (Lorenz and Wackernagel, 1987). In contrast, the adsorption of transforming DNA from B. subtilis on montmorillonite decreased as the pH was increased from 3 to 9, but the frequency of transformation increased with increases in pH (M. Khanna and G. Stotzky, unpublished observations). The pH determines the sign of the net surface charge of amphoteric materials (e.g., enzymes, bacteria, bacteriophages) and the negative charge of ionizable nonamphoteric materials (e.g., DNA). Therefore, the adsorption of these materials on soil particles, especially on most types of clay minerals, which have a pH-independent negative charge, will be influenced by the pH of soil, which, in turn, will influence the persistence and activity of these materials and the transfer of genetic information (Stotzky, 1986). Consequently, controlled studies on the effects of pH on genetic transfer in soil should be conducted.

5. Water Content

The few data that are available on the effects of water content on the transfer of genetic information in soil indicate that transfer frequencies are higher when the soil water tension is near or at the optimum for microbial growth (i.e., the −33-kPa tension). For example, plasmid transfer and survival of the donors [E. coli J5(RP4) and JC5466(pRD1)] and the exconjugants [Enterobacteriaceae strain 1(pRD1) and *Pseudomonas fluorescens* (pRD1)] were higher in soil maintained at 16% water than when allowed to dry to 4% (Schilf and Klingmüller, 1983). Similarly, plasmid transfer in sterile soil was greater between strains of B. subtilis at a water content of 20–22% (equivalent to the −33-kPa water tension) than at a water content of 8% (van Elsas et al., 1987), between strains of E. coli in sterile soil at 80% than at 20% of the water-holding capacity (WHC) (Trevors, 1987a), between E. coli and other enterobacteria and *Pseudomonas aeruginosa* in nonsterile soils at their −33-kPa tension (24–26% water) than at 16% water (M. A. Devanas and G. Stotzky, unpublished observations). No studies appear to have been conducted on the effects of water content on transduction and transformation in soil. Inasmuch as microbial growth in soil is optimal at the −33-kPa water tension, genetic transfer, regardless of the mechanism of transfer, is probably maximal at this tension, which is easily controlled by irrigation.

6. Oxygen and E_h

At soil water tensions above −33 kPa, oxygen will become progressively more limiting, and the E_h will be reduced (Stotzky, 1974). The effects of oxygen tension and E_h on the transfer of genetic information in soil has apparently not been studied, and results from the few studies conducted *in vitro* are contradictory: for example, the conjugal transfer of chromosomal genes between E. coli was similar under aerobic and anaerobic conditions (Stallions and Curtiss, 1972); the expression of antibiotic resistance by 45 different plasmids in E. coli was the same under aerobic and anaerobic conditions, although the formation of sex pili was reduced under anaerobic conditions (Burman, 1977); the frequency of conjugal transfer of R-plasmids by two donor strains of E. coli isolated from human feces was reduced 10- to 1000-fold under anaerobic conditions (Moodie and Woods, 1973).

7. Ionic Composition

The effects of the ionic composition of the soil solution, which affects the activity of water (a_w), as well as surface interactions among and between soil particles, especially clay minerals, and bacteria, bacteriophages, DNA, and proteins (Stotzky, 1986), on genetic transfer require study. The transfer, in vitro, of plasmid R1drd-19 in E. coli was apparently stimulated by concentrations of NaCl that are present in estuaries (Singleton, 1983). The survival of antibiotic-sensitive fecal coliforms was reduced in seawater, whereas that of E. coli strains containing R-plasmids was unaffected (Smith et al., 1974).

8. Electromagnetic Radiation

Although light probably affects only microbes residing on the surface of soils, this physicochemical factor can be important in arid and semiarid soils, where photosynthesis in algal crusts, both prokaryotic and eukaryotic, may be the major source of primary productivity (e.g., Skujins, 1984). Hence, the transfer of genes conferring resistance to ultraviolet (UV) radiation may enhance the survival of microbial surface dwellers (Marsh and Smith, 1969). The importance of electromagnetic radiation on gene transfer in soil has apparently not been studied.

9. Surfaces

The effects of surfaces, especially those of clay minerals, on transfer of genetic information in soil appear to have been studied to a greater extent, albeit also insufficiently, than those of other physicochemical characteristics of soil. Montmorillonite appears to enhance conjugal transfer of both plasmid- and chromosome-borne genes in both sterile and nonsterile soil, whereas kaolinite appears to have essentially no effect (e.g., Devanas and Stotzky, 1988a; Krasovsky and Stotzky, 1987; van Elsas et al., 1987; Weinberg and Stotzky, 1972). In contrast, colloidal montmorillonite reduced the in vitro transfer of plasmid R1drd-19 in E. coli by several orders of magnitude (Singleton, 1983). This apparent paradox between in vitro and in vivo effects of clay minerals may be the result of free clay particles in vitro blocking sites on the bacterial surface necessary for gene transfer, whereas few free clay particles exist in soil, as they are relatively stably maintained in cutans and domains.

Surfaces, especially those of clay minerals, can exert both direct and indirect effects on microbial activities, including transfer of genetic information, in soil. However, the mechanisms by which some clays enhance conjugal gene transfer in soil are not clear. Montmorillonite has been shown to stimulate the growth of bacteria, in part by maintaining a suitable pH for sustained growth in the microhabitats, and to reduce the growth of fungi, in part by complexing siderophores necessary for iron transport (Stotzky, 1986). Van Elsas et al. (1987) suggested that the enhanced transfer of a plasmid in B. subtilis in soil amended with montmorillonite was not the result of an effect of the clay on pH but of "a modification of the physicochemical soil environment, possibly modifying cellular physiology or promoting cell-to-cell contact" or because the "clay apparently protected the recipient population." Unfortunately, no details were presented on the mechanisms of such modifications and protection. Moreover, it is also not clear which physicochemical properties of clays (e.g., cation and anion exchange capacity, specific surface area, surface charge density, nature of the charge-compensating cations) are responsible for their effects on conjugation and survival of parentals and exconjugants in soil.

The frequency of transduction in soil was not affected by montmorillonite, although the survival of the transducing bacteriophage, P1, was increased (Zeph et al., 1988). Transduction by phage P1 was greater in a sandy (8% clay) than in a silty clay loam (21% clay) soil (Germida and Khachatourians, 1988), but as these soils also differed in pH, organic matter content, and other characteristics, and the types of clay present were not described, the differences in transduction frequencies cannot be attributed solely to differences in clay content in these two soils. The effects of surfaces on transformation in soil have not been studied. However, both sea sand (Aardema et al., 1983; Lorenz et al., 1988) and montmorillonite (G. Stotzky and A. Golard, unpublished observations) reduced the in vitro frequency of transformation in B. subtilis.

10. Interactions between Factors

The value of studies conducted on the effects on gene transfer of one physicochemical factor of soil at a time may be limited, as a change in one factor can result in changes in numerous other factors, and several factors can interact to affect gene transfer. For example, the in vitro conjugal transfer of plasmid R1drd-19 in E. coli was inhibited more by deviations in pH from the optimum of 6.9 when the temperature was simultaneously decreased from 37° to 17°C (Singleton and Anson, 1983). Studies on the effects of interactions between multiple physicochemical characteristics of soil are needed not only with respect to the transfer of genetic information, but on all aspects of microbial ecology in soil. Nevertheless, even the few studies that have been conducted on the effects of these characteristics individually on gene transfer in vitro and in soil and other natural habitats indicate that in vitro studies of conjugation, transduction, and transformation, which are usually conducted under standardized and optimal growth conditions, are not always adequate predictors of gene transfer in natural habitats, wherein these characteristics fluctuate continually and in concert.

E. Conclusions

There is insufficient information on the frequency of transfer, whether by conjugation, transduction, or transformation, and on the survival and activity of recombinant bacteria in soil and other natural environments that contain high numbers of other microorganisms not involved in the transfer. The relatively few studies that have been conducted in soil—and it must be emphasized that even these studies have been conducted primarily in the laboratory in microcosms of varying degrees of complexity or under greenhouse conditions, as few field releases have been authorized—indicate that transfer can occur in soil. However, insufficient information is available on how transfer and survival are affected by the physicochemical and biological characteristics of soil, on the numbers of donors and recipients necessary for transfer, on the numbers of recombinant bacteria that can be accurately detected, on the probability that low levels of recombinant bacteria (especially below the level of detection) can multiply sufficiently to become a significant portion of the soil microbial population, and on numerous related questions.

The major lack of knowledge, however, is in the area of the potential effects that recombinant bacteria could have on the structure and function of soil and other natural habitats; for example, what kinds of genes need to be transferred and how many recombinant bacteria need to be functioning per unit volume of soil to result in detectable changes in the activity, ecology, and population dynamics of microorganisms in soil? Even if a GEM introduced into soil survives and transfers its novel genetic information to indigenous microbes, there should be little cause for concern unless the novel genetic information, either in the introduced GEM or in an autochthonous recipient to which it has been transferred, results in some unexpected and untoward impacts. Unfortunately, it is difficult with existing knowledge and methodology to study and predict the occurrence, extent, and severity of such impacts.

Preliminary studies on the effects of adding high concentrations (e.g., 10^8 cells/g soil) of various strains of *E. coli*, *Enterobacter cloacae*, *Ps. aeruginosa*, and *Pseudomonas putida*, with and without plasmids carrying antibiotic-resistance genes, to soil have not shown any consistent and lasting effects on the gross metabolic activity (as measured by CO_2 evolution), transformations of fixed nitrogen, activity of soil enzymes (phosphatases, arylsulfatases, dehydrogenases), and species diversity of the soil microbiota (Doyle et al., 1988; Jones et al., 1988). However, in other preliminary studies, the introduction into soil of a strain of *Streptomyces lividans* that contained a plasmid carrying a lignin peroxidase gene from the chromosome of *Streptomyces viridosporus* enhanced the rate of mineralization of soil carbon (as measured by CO_2 evolution) during the 30-day incubation, especially when the soil was amended with lignocellulose (Wang and Crawford, 1988). More such studies are needed, especially with GEMs that have been engineered to perform specific enzymatic functions in soil.

It is obvious, based on the few studies that have been conducted on the transfer of genetic information in soil, that more research is necessary to determine the frequency, the location, and the effects of physicochemical and biological factors on gene transfer in soil and other natural habitats. The attainment of this knowledge will be of immense academic interest not only to microbial and macrobial ecology and evolutionary theory, but also to risk assessment and regulation of the release of GEMs to soil and other natural habitats.

II. Literature Review of Conjugation and Transduction in Soil

A. CONJUGATION

1. *Introduction*

Conjugation is the transfer of DNA from one bacterium to another by direct cell-to-cell contact and is mediated by transfer (*tra*) genes. For example, the self-transmissible fertility (F)-factor in the Gram-negative bacterium *E. coli* is a plasmid that has 15–25 *tra* genes that encode its transfer, that is, enzymes and the appropriate F-pili. The F-factor may exist autonomously in the cell, or it can be integrated into the bacterial chromosome, where it is replicated and segregated to progeny cells with the other genes on the chromosome, similar to a prophage in lysogenic bacteria. In *E. coli*, which has been more extensively studied than other Gram-negative and Gram-positive bacteria, the DNA moves from the donor F^+ cell to the recipient F^- cell through a special surface pilus called an F-pilus, the synthesis of whose structural proteins are directed by the *tra* genes. The F-pilus of the donor attaches to the recipient cell at specific surface receptors and retracts until the surfaces of the donor and recipient are touching. At the point of contact on the recipient, a channel is made through its cell envelope, and the DNA is transferred, presumably through the pilus, into the recipient cell. A bacterial cell in which the F-factor is integrated into the chromosome is called a "high frequency of recombination" (Hfr) cell. The integrated F-factor can mediate transfer of part of the F-factor and part of the donor chomosome (an Hfr conjugation) (Freifelder, 1987; Lewin, 1977; Vandemark and Batzing, 1987).

Gram-positive bacteria can also transfer DNA by conjugation, but no sex pili have been demonstrated. The recipient cells synthesize specific diffusible proteins called pheromones. When these pheromones contact a donor cell, it synthesizes surface proteins with adhesive properties that cause the donor and recipient cells to form aggregates. The sex pheromones then induce the transfer of the plasmid to the recipient, but the exact mechanism of transfer is not known (Clewell, 1981; Freifelder, 1987).

2. Transfer of Chromosomal Genes

Transfer of the bacterial chromosome from an Hfr cell to an F⁻ cell begins at the 3'-end of the origin of transfer (*oriT*) site within the integrated F-factor, where a *tra*-encoded endonuclease nicks and opens one of the two DNA strands. The free 5'-end of the open single-stranded DNA attaches to a site on the cytoplasmic membrane, and the 3'-end acts as a primer for chain elongation, using the closed, intact, circular strand as a template. The open DNA strand, after its release from the cytoplasmic membrane, serves as a template for the synthesis of its complementary DNA and moves, 5'-end first, into the recipient through a conjugation bridge (presumably the F-pilus in *E. coli*). This is an example of the unidirectional "rolling-circle" mechanism of replication. The entire single-stranded DNA from the donor chromosome and the newly synthesized complementary DNA strand must be moved to the recipient for complete transfer of all the genes on the chromosome, including the distal end of the F-factor, and to produce a complete functional integrated F-factor upon reannealing and ligation in the recipient. This process is estimated to require ~100 minutes in *E. coli*. This is a relatively long time for bacteria to stay physically attached, especially in soil and other natural habitats, and interruption of conjugation usually occurs after only 10–20% of the chromosome has been transferred (Freifelder, 1987; Lewin, 1977). Therefore, the entire F-factor is rarely transferred from an Hfr cell, as the remaining F-factor genes are at the opposite free 3'-end of the nicked chromosome. Genes that show a high frequency of recombination are those nearest the *oriT* site, and the frequency of transfer of genes decreases proportionately with their distance from this site.

An F-factor that has been integrated into the chromosome may excise itself, by expressing the appropriate enzymes coded on its DNA, by a reverse recombinational event and return to an autonomous, extrachromosomal state. Precise excision—that is, nicking at the exact sites that were ligated to the chromosomal DNA on integration—will restore the original autonomous F-factor. Imprecise excision may result in an autonomous F-factor that includes chromosomal genes (possibly introduced recombinant genes) from one or both sides of the integration site of the F-factor. This modified F-factor, sometimes referred to as an F'-plasmid, can transfer these genes to another cell.

3. *Conjugative Plasmids*

Autonomous F-factors, F'-plasmids, and R-plasmids (conjugative plasmids similar to an F-factor that carry genes coding for antibiotic resistance) are self-transmissible. Hence, they may direct their transfer to other bacteria by expressing their *tra* genes. Transfer of conjugative plasmid DNA also begins at the *oriT* site within the *tra* genes, as it does in Hfr strains. However, because the DNA sequences (i.e., the number of genes) in a plasmid are significantly shorter (1–10%) than in a bacterial chromosome, the entire plasmid is usually transferred. As in chromosomal transfer, the plasmid DNA is replicated during transfer (i.e., a copy remains in the donor cell), and the recipient cell then expresses the *tra* genes and becomes a donor cell. Consequently, in the presence of appropriate selection pressure, a plasmid may spread rapidly through a bacterial population, even though plasmid-containing cells may originally have been an insignificant part of the population. This rapid conversion of plasmid-free recipient cells to plasmid-containing donor cells appears to be responsible for the rapid increase in antibiotic-resistant nosocomial infections.

4. *Nonconjugative Plasmids*

Autonomous, extrachromosomal, and self-replicating plasmids that do not have *tra* genes are nonconjugative. Consequently, these plasmids, which may code for a variety of functions (e.g., resistance to antibiotics and other antimicrobial agents, catabolic enzymes, toxin production), cannot direct their own transfer. However, nonconjugative plasmids may be dispersed to appropriate recipient cells by coinhabiting conjugative plasmids that can "move" them (mobilization) to other cells, or there is recombination and integration into a plasmid that is self-transmissible (cointegrate formation).

a. *Mobilization.* If a plasmid is deficient in DNA sequences that code for mobility (*mob*) (e.g., for endonucleases that nick the *oriT* region) or transfer (e.g., F-pilus production) functions, the plasmid will not be transferred, even if all other functions are present, and it will persist as an autonomous, self-replicating element. However, such nonconjugative plasmids may be transferred by conjugation if the missing functions are supplied gratuitously by another plasmid present in the same cell. For example, the nonconjugative plasmid, ColE1, can be mobilized and transferred when an F-factor present in the same cell supplies the missing pilus and transfer apparatus (Freifelder, 1987).

The mobilization of nonconjugative plasmids under environmental conditions has been reported. A nonconjugative plasmid, pHSV106, coding for resistance to ampicillin (Ap) and thymidine kinase from herpes simplex virus, was mobilized by plasmids present in both laboratory and indigenous wastewater strains of bacteria in a laboratory-scale wastewater treatment facility (Mancini et al., 1987). In sterile soil, triparental matings (i.e., mating of a cell carrying a nonconjugative plasmid with a cell carrying a conjugative plasmid and then transfer of either the nonconjugative or both plasmids to a third cell) mobilized nonconjugative plasmids between strains of Streptomyces (Rafii and Crawford, 1988). Transfer of a 2.8-MDa plasmid, pFT30, from Bacillus cereus to B. subtilis occurred in sterile and nonsterile soil when a 29.5-MDa plasmid was present in B. cereus (van Elsas et al., 1987).

b. *Cointegrate Formation.* When two plasmids coexist in a cell and there is a region of homology between them (e.g., a transposon or an insertion sequence), the two plasmids may fuse to form a single plasmid termed a cointegrate plasmid. If one of the plasmids is conjugative, the cointegrate plasmid, containing genes from both plasmids, can be transferred. Once in the recipient, the cointegrate plasmid may separate into the original two plasmids. Components of an infectious R-plasmid (i.e., the RTF region, containing transfer and replication genes, and the R-determinant, containing drug-resistance genes) have been shown to form cointegrates with themselves and other R-plasmids and then to separate into new species of plasmids (segregants) of various sizes in the recipient (Freifelder, 1987; Lewin, 1977).

5. *Mobilizing Elements*

Transfer of DNA that is not self-transferable can also be facilitated by mobilizing elements such as insertion sequences (IS) and transposons (Tn). These transposable elements are specific segments of DNA that have the ability to move autonomously as discrete units to other sites in DNA. The IS are a special type of transposable elements containing 800–1400 base pairs with inverted-repeat sequences that are 16–41 base pairs long at each end. The IS encode only those genes necessary for their own transposition, including a gene for an enzyme called transposase, which mediates recombination at special sites in the terminal inverted-repeat sequences. Several distinct IS have been detected and characterized in prokaryotic and eukaryotic genomes.

The Tn are composite units with terminal IS or short inverted-repeat sequences that flank antibiotic-resistance or other genes, as well as the gene for transposase, and do not require homologous recombination genes for movement. These mobilizing elements serve as regions of homology for recombinational events that can (1) produce cointegrates of transferable and nontransferable plasmids, (2) transfer copies of genes from sites in the chromosome to sites on other replicons such as plasmids, (3) cause insertional mutations within genes, and (4) provide a detectable marker after incorporation into the genome (Freifelder, 1987).

6. Spectrum of Conjugative Bacteria

Some genera of Gram-negative bacteria that have been shown to be capable of participating in the transfer of chromosome- and/or plasmid-borne genes by conjugation are *Enterobacter, Escherichia, Salmonella, Shigella, Providencia, Vibrio, Proteus, Klebsiella, Pseudomonas, Serratia, Rhizobium, Bradyrhizobium, Bordetella, Neisseria, Rhodopseudomonas, Azotobacter, Caulobacter, Agrobacterium, Erwinia, Chromobacterium, Acinetobacter, Rhodospirillum,* and *Flavobacterium.* The Gram-positive genera that are known to transfer genes by conjugation include *Staphylococcus, Clostridium, Bacillus, Streptococcus, Lactobacillus, Nocardia, Actinomyces,* and *Streptomyces.*

Conjugal transfer of plasmids occurs in both Gram-negative and Gram-positive bacteria, and plasmid transfer from *E. coli* to various Gram-positive bacteria has been demonstrated *in vitro* (Trieu-Cuot et al., 1987).

7. Survival of Introduced Genes in Situ

When the heterologous DNA is present in a plasmid, the survival and stability of the DNA is dependent not only on the survival of the bacterial host, but also on the maintenance and faithful replication of the plasmid during the cell cycle and the partitioning of the plasmid to the progeny. Low copy number plasmids (one to two per chromosome) are usually stably maintained and stably inherited by a partitioning mechanism, whereas high copy number plasmids (more than five per chromosome) may be inherited by a random-distribution mechanism of stochastic segregation, although some form of active partitioning may also occur with high copy number plasmids (Sherratt, 1982). For example, when sterile soil was inoculated with *E. coli* K12 strain χ1666, the number of colony-forming units (CFU) increased rapidly to a maximum population level, the increase being greater when inoculated in Luria broth (LB) than in saline (Section IX). When the cells of χ1666 contained plasmid pES019, which is a 4-MDa construct of the multicopy, Tc-resistant plasmid, pBR328 (Covarrubias et al., 1981), engineered to contain resistance markers for gentamicin (Gn), chloramphenicol (Cm), and streptomycin (Sm) (W. Watkins, personal communication; Levin, 1982), the numbers of CFU also increased in sterile soil. However, segregation of the plasmid apparently occurred, as the numbers of CFU recovered on the nonselective medium, MacConkey agar (MAC), were significantly higher than the numbers recovered on the medium selective for pES019 (MAC + 25 µg/ml Cm). This difference in numbers between the plasmidless and the plasmid-containing populations was not the result of an artifact (e.g., a "viable but nonculturable" phenomenon), as the same difference occurred when the CFU recovered on MAC were subsequently transferred to MAC + Cm. The segregation occurred shortly after the addition of the host–vector system to soil, and the plasmid-free population then outgrew the plasmid-containing population. Nevertheless, despite the segregation, the population of χ1666(pES019) increased by one and three orders of magnitude when inoculated in saline or LB, respectively, and persisted near these levels throughout the 27-day incubation (Devanas et al., 1986; Devanas and Stotzky, 1988a).

When *E. coli* χ1666 and χ1666(pES019) were inoculated in saline into soil containing a natural microbial community (i.e., nonsterile soil), the numbers of indigenous total bacteria and total Gram-negative bacteria remained essentially the same throughout the 20-day study. However, χ1666 declined rapidly and was not detected 6 days after inoculation, whereas χ1666(pES019) declined more gradually to undetectable levels after 13 days. In contrast to the results obtained in sterile soil, there was no faulty segregation of the plasmid to produce a plasmid-free population, and the presence of the plasmid enhanced the survival of the host–vector system. The lack of partitioning of the plasmid in nonsterile soil was apparently the result of little or no growth of the host–vector system, because when nutrients (LB) were added on days 1 and 3, both growth and segregation occurred. The possible partitioning of plasmids should be considered if GEM are added to soils containing sources of available carbon and energy (e.g., fresh plant residues or degradable organic pollutants), as growth of the GEM may result in the eventual loss of the plasmid-containing population.

Although it has been suggested that the size of the resident plasmid can affect its maintenance (e.g., Hakkaart et al., 1985), and large plasmids appear to prolong the generation time of hosts in vitro (Godwin and Slater, 1979; Zund and Lebek, 1980), the host strain and not the size of the plasmid appeared to be important in the persistence of E. coli host–vector systems in nonsterile soil. There was little difference in the rates of decline in nonsterile soil of hosts that contained small, multicopy, nonconjugative plasmids (2.6 and 8.0 MDa) or large, low copy number, conjugative plasmids that ranged in size from 26 to 64 MDa, regardless of whether they were added in LB or saline (Devanas et al., 1986).

Moreover, the survival of E. coli strains χ1666 and PRC487 was greater when they contained plasmids [pES019 and pACYC175 (8 MDa), respectively] (Devanas et al., 1986). The survival of χ1666(pBR322) in the human gastrointestinal tract was also greater than that of the plasmidless parental (Levy et al., 1980), as was that of the highly debilitated E. coli strain χ1776 when it contained pBR322 (Levy and Marshall, 1979). Furthermore, the survival of plasmid-containing strains of E. coli was similar to that of homologous plasmidless strains in fresh water (Grabow et al., 1975), conventional sewage treatment systems (Sturtevant and Feary, 1969), and seawater (Smith et al., 1974), even in the absence of the selection pressure of the antibiotics to which the plasmids conferred resistance. Although most in vitro studies, usually conducted in chemostats under nutrient-limiting conditions, indicate that plasmid-free cells have a more rapid growth rate and quickly displace plasmid-containing cells of the same species, this is not always the case, as these studies show. Even in chemostat studies, the growth of E. coli that contained either the cos DNA fragment from phage λ (Edlin et al., 1984), the transposable element, IS50 (Hartl et al., 1983), derivatives of pBR322 that expressed resistance to Tc (Lee and Edlin, 1985), or were lysogenic for phage λ (Edlin et al., 1975) was faster than that of the homologous hosts not carrying these DNA sequences. The reasons for these observations are not known. Although it is tempting, in the case of the greater survival of plasmid-containing bacteria in soil, to attribute this to the antibiotic resistance genes carried by the plasmids, there is no convincing evidence that antibiotics are produced in soil, at least not at detectable levels (Stotzky, 1974, 1986). Regardless of the reasons involved, these observations indicate that the presence of plasmids, even if they confer no known advantage to their hosts, is not always detrimental to the competitiveness of the bacteria and that results obtained in vitro, despite their apparent logic, cannot always be extrapolated to soil in situ.

There appeared to be a more rapid decrease in the numbers of some plasmid-containing *E. coli* strains when recovered from nonsterile soil on selective media [e.g., MAC + 25 μg/ml Tc + 5 × 10^{-5} M Hg as $HgCl_2$] than on nonselective media (e.g., MAC), suggesting a loss of the plasmids. However, this was an artifact, and there was no loss of the plasmids: when cells from colonies recovered on nonselective media were transferred to selective media, the numbers of CFU that subsequently developed on the selective media were the same as those recovered on the nonselective media (Devanas et al., 1986). The cells of some host–plasmid systems were apparently sufficiently debilitated in soil to prevent their recovery directly on selective media with the added stress of an antibiotic and/or a heavy metal, and resuscitation on less stressful nonselective media was necessary for subsequent growth on selective media.

This phenomenon of "viable but nonculturable" has also been reported in other environments with bacteria not containing recombinant DNA, for example, *E. coli*, *Vibrio cholerae* (Xu et al., 1982), and *Salmonella enteritidis* (Roszak et al., 1984) in natural waters, and *E. coli* and *Streptococcus faecalis* in chlorinated water (Bissonnette et al., 1975; Camper and McFeters, 1979; Zaske et al., 1980; see also Roszak and Colwell, 1987). The mechanisms responsible for this phenomenon have not been definitively established. It has been suggested that the cell membrane may be damaged (Zaske et al., 1980) and, hence, not be capable of using an electrochemical gradient to generate high-energy intermediates for sustained survival (Sjogren and Gibson, 1981), or that the cells may be restricted in obtaining and retaining energy sources for endogenous metabolism and maintenance (Morita, 1982). Regardless of the mechanisms involved, the reduced viability or recoverability of host–plasmid systems introduced into stressful environments, such as soil, may result in erroneous data on the survival of GEMs in such environments.

Even DNA foreign to bacteria appears to be maintained in soil when introduced into bacteria in a plasmid. For example, studies on the survival and maintenance of a 0.9-MDa cDNA fragment that codes for a yolk protein in *Drosophila grimshawii* and inserted into plasmid pBR322 showed that the presence of the chimeric plasmid (C357) had little effect on the survival of the host (*E. coli* HB101) in sterile soil (Devanas and Stotzky, 1986). When *E. coli* HB101(pBR322) and HB101(C357) were added in LB to nonsterile soil, the presence of the insert of *Drosophila* DNA reduced somewhat the stability of the plasmid during the 27-day incubation. When the host–plasmid systems were added in saline, the rate of decrease of the system containing the chimeric plasmid was similar to that of the system containing only pBR322, and the decrease of both was greater than when added in LB. The presence of foreign genetic material has been shown to reduce the stability of plasmids when the host–vector systems were grown in a chemostat (Warnes and Stephenson, 1986), and other studies have indicated that pBR322 is unstable in nutrient-limited chemostat cultures (Jones et al., 1980; Jones and Melling, 1984). However, when E. coli HB101(C357) was added in either LB or saline to nonsterile soil, there was no selective loss of the chimeric plasmid, as determined by the isolation of HB101(C357) on a selective medium and colony hybridization with a ^{32}P-labeled DNA probe specific for the cDNA (Devanas and Stotzky, 1986). Nevertheless, HB101(C357) appeared to be less able to cope with conditions of starvation and competition in nonsterile soil, as the addition of LB on day 16 to soil that received the inoculum in saline did not elicit growth similar to that of HB101(pBR322). Although E. coli HB101 is used extensively for cloning, it is a poor choice for studies in soil, as it is auxotrophic and lactose-negative. Hence, an important phenotypic marker (i.e., pigment production) is not available, and it is difficult to distinguish E. coli HB101 from Gram-negative indigenous soil bacteria, which are usually also lactose-negative, on MAC.

The survival of Ps. fluorescens, resistant to Km and Sm, and B. subtilis, resistant to Tc, in two soils of different texture and planted with wheat was studied in situ (van Elsas et al., 1986). During 120 days, Ps. fluorescens decreased more slowly in the silt loam soil (a reduction of approximately two orders of magnitude) than in the coarser loamy sand (a reduction of approximately five orders of magnitude), whereas B. subtilis declined rapidly in both soils within the first 20 days and then stabilized, with the population being present primarily as spores. These results were similar in both rhizosphere and root-free soil. Although this is one of the few studies with engineered bacteria that has been conducted in soil in situ, it should not be concluded that the survival of engineered Gram-negative bacteria is greater than that of Gram-positive bacteria in soil, as the bacteria were altered by two very different mechanisms (i.e., Ps. fluorescens by Tn5 mutagenesis of the chromosome and B. subtilis by protoplast transformation with plasmid pFT30).

In a current in situ study, the survival of root-colonizing Pseudomonas aureofaciens 3732RNL11, which contains the lacZY genes from E. coli K12 (encoding β-galactosidase and β-galactoside permease, respectively) and genes conferring resistance to nalidixic acid (Nx) and rifampicin (Rf) on the chromosome (Drahos et al., 1986), in the rhizosphere is being investigated (Drahos et al., 1988; E. L. Kline, H. Skipper, and E. J. Brandt, personal communication). The genetically engineered bacterium, in aqueous suspension, was applied to seeds of winter wheat planted at a depth of ~2.4–3.8 cm, and its survival was monitored by periodically plating dilutions of rhizosphere soil on a minimal medium containing lactose, Nx, and Rf, checking for fluorescence under UV light, and enumerating colonies that showed β-galactosidase activity (i.e., they turned blue when they cleaved the X-gal reagent, 5-bromo-4-chloro-3-indolyl-β-D-galactoside). The level of detection is ~1–10 CFU/g soil or root. The numbers of bacteria decreased (after colonization of the emerging roots) from ~10^6 to 10^2 CFU/g root during the 31-week growth period (seeds were planted in November, and the crop was harvested in June). The survival kinetics of the host strain that did not contain the lacZY genes were similar to those of the recombinant bacterium. After harvest of the wheat crop, the levels of the bacteria fluctuated between 10^2 CFU/g root and undetectable levels on soybeans that were subsequently planted in the same site without tillage. Survival of the bacteria was not detected in root-free soil. No apparent transfer of the lacZY genes to indigenous fluorescent pseudomonads occurred, as determined by dot blot hybridization of a ^{32}P-labeled DNA probe (a 502-base pair nucleotide sequence of the Tn7 used to insert the genes) with DNA extracted from >5000 fluorescent bacteria isolated from the rhizosphere (third wash of roots) and nonrhizosphere soil.

The survival of engineered genes, introduced in GEMs, in soil and other natural habitats may be increased by their transfer to indigenous bacteria that are more fit for survival in these habitats (Stotzky, 1989; Stotzky and Babich, 1986).

8. Conjugation in Situ

a. Transfer of Chromosomal Genes. Conjugal transfer of chromosomal genes from prototrophic to auxotrophic strains of E. coli occurred in both sterile (Weinberg and Stotzky, 1972) and nonsterile soils (Krasovsky and Stotzky, 1987; Stotzky and Krasovsky, 1981) inoculated with these strains. The frequency of transfer was significantly higher in sterile than in nonsterile soils, indicating that the presence of the indigenous microbiota interfered with conjugal transfer.

The physicochemical characteristics of the soil affected the frequency of transfer of chromosomal genes. For example, the addition of the clay mineral, montmorillonite, with a high cation exchange capacity and surface area, enhanced the frequency, whereas the addition of the clay mineral, kaolinite, with a low cation exchange capacity and surface area, had no effect. The enhancement by montmorillonite was apparently the result of the increased growth of the donors and recipients in the microhabitats, which probably resulted in increased conjugation because of the larger number of donors and recipients. Montmorillonite has been shown to increase the growth and metabolic activity of bacteria, primarily by maintaining the pH of microhabitats at levels conducive to sustained growth (Stotzky, 1974, 1986). The effect of pH on the transfer of chromosomal genes was further demonstrated in studies that showed that both the survival of the donors and recipients and the frequency of gene transfer increased as the pH of the soil was raised to neutrality (Krasovsky and Stotzky, 1987; Weinberg and Stotzky, 1972).

b. *Transfer of Plasmid Genes.* Numerous studies have demonstrated the transfer of plasmids from bacteria isolated from soils and waters to recipient strains of the same and different species in the laboratory (Stotzky and Babich, 1986). However, few studies have demonstrated such transfer in soil and other natural environments. For example, when a strain of *Enterobacter cloacae*, which had been isolated from the rhizosphere of *Festuca heterophylla*, did not fix nitrogen, and was sensitive to antibiotics, was mated *in vitro* with *E. coli* containing plasmid pRD1, which carried genes for nitrogen fixation and resistance to three antibiotics, the exconjugants were able to fix nitrogen and were resistant to the antibiotics. However, many of the exconjugants contained plasmids that were smaller than pRD1 or contained no plasmids, indicating that all or part of the pRD1 had become incorporated into the chromosome of *E. cloacae* (Kleeberger and Klingmüller, 1980). When *E. coli* JC5466 containing the broad-host-range plasmid pRD1 was added to nonsterile soil, the frequency of transfer of the plasmid to indigenous bacteria was $\sim 1 \times 10^{-9}$ exconjugants per recipient, whereas the *in vitro* frequency of transfer to mixed bacterial populations isolated from soil ranged from 7.1×10^{-8} to 4.5×10^{-6} (Schilf and Klingmüller, 1983). The survival of *E. coli* strains J5(RP4) and JC5466(pRD1), Enterobacteriaceae strain 1(pRD1), Enterobacteriaceae strain 2(RP4), and *Ps. fluorescens* pRD1 (the latter three being exconjugants of soil isolates obtained by *in vitro* matings) was also reduced in nonsterile soil, at either 4° or 20°C with a soil water content of 16% or at 20°C with a water content of 4%, presumably as the result of either a loss of the plasmids or a selective decrease in the host–plasmid systems when they were exposed to nonselective conditions in soil. In contrast, the survival of a chromosomal transconjugant that resulted from conjugation in soil of *E. coli* strains $\chi 493$ (Hfr, prototrophic) and $\chi 696$ (F^-, auxotrophic) was greater in nonsterile soil, when reinoculated after isolation from soil and growth on laboratory medium, than that of the parentals (Krasovsky and Stotzky, 1987).

Pertsova *et al.* (1984) reported the transfer in nonsterile soil of plasmid-borne genes that coded for the degradation of 3-chlorobenzoate from *Ps. aeruginosa* and *Ps. putida* to indigenous strains of *Pseudomonas* that were taxonomically different from the donors. Transfer of antibiotic-resistance plasmids among strains of *Klebsiella* in a sandy loam soil sown with radish seeds (the seeds were inoculated with plasmidless strains and the soil with strains containing the resistance plasmids) was detected on ~24% of the radish seedlings after 1 week, but not thereafter, at frequencies of $10^{-7}-10^{-6}$, whereas in an aqueous suspension of redwood sawdust, transfer was detected in ~30% of the donor–recipient combinations at frequencies of $10^{-8}-10^{-5}$, and in broth, transfer was detected in ~60% of the combinations at frequencies of $10^{-6}-10^{-3}$ (Talbot et al., 1980).

Plasmid transfer also appears to occur among Gram-positive bacteria in soil, although this has not been studied as extensively as with Gram-negative bacteria. Both conjugative and nonconjugative plasmids were transferred (the latter by triparental matings) between strains of *Streptomyces* in sterile soil, though less frequently than on agar (Rafii and Crawford, 1988). The transfer of a 2.8-MDa plasmid, pFT30, carrying a Tc-resistance gene, from *B. cereus* to *B. subtilis* occurred in sterile soil at a frequency of 7×10^{-8} (van Elsas et al., 1987). Transfer was increased to a frequency of 1.6×10^{-6} by the addition of nutrients and bentonite (montmorillonite), but it was lower at 15°C and a WHC of ~20% than at 27°C and a WHC of ~60%. Detectable transfer of plasmid pFT30 (frequency of 9×10^{-8}) occurred in nonsterile soil only when it was amended with bentonite, and the clay also enhanced the survival of the recipients. Though not stated, the transfer of this 2.8-MDa plasmid was probably facilitated by the presence of a 29.5-MDa plasmid in the donor. Bacilli isolated from soil samples collected in hog and cattle feedlots throughout the United States contained plasmids that showed a high degree of homology with plasmids from clinical isolates of *Staphylococcus aureus*, suggesting the occurrence of natural intergeneric plasmid transfer (Polak and Novick, 1982).

The transfer of nonconjugative plasmids by mobilization, either directly or by cointegrate formation, by conjugative plasmids has been studied extensively in pure culture, but there have been few studies on the frequency of such transfer in soil and other environments. Transfer of nonconjugative plasmids by triparental mating has been demonstrated in *E. coli* in a laboratory simulation of a sewage treatment plant (Gealt et al., 1985; Mancini et al., 1987; McPherson and Gealt, 1986) and in *Streptomyces lividans* in sterile soil (Rafii and Crawford, 1988). Nonconjugative plasmids, as a result of their smaller size and, hence, lower "burden" on the host, especially if present in low copy number, may be retained longer than larger conjugative plasmids in bacteria in soil, and GEMs that will be released to soil will undoubtedly carry their heterologous DNA on nonconjugative plasmids, if not on the chromosome. Consequently, extensive studies should be conducted on the mechanisms and frequency of transfer of nonconjugative plasmids in soil.

In sterile soil, the conjugative plasmids, pDU202 (64 MDa) and pRR226 (26 MDa), which were derepressed for transfer, were transferred between strains of E. coli (Fig. 1) only when nutrients (LB) were added, and the donors and recipients increased to high numbers, and when the soil had an optimum water content (i.e., −33 kPa-water tension; no detectable transfer occurred and survival of the donors and recipients was reduced at lower water tensions) (Devanas and Stotzky, 1988a). Similar results were obtained in sterile soil with E. coli containing a 60-MDa plasmid (Trevors, 1987a; Trevors and Oddie, 1986). However, in nonsterile soil at the optimum water tension, no transfer of plasmids pDU202 and pRR226 was detected between the same strains of E. coli, even when nutrients were added initially and during the incubation, which resulted in the growth of the donors and recipients (Fig. 2) (Devanas and Stotzky, 1988a). The soil microbiota apparently interfered in some manner (probably biochemically and not physically) with the transfer of the plasmids. In contrast, Trevors (1987b) reported that a 60-MDa plasmid was transferred between strains of E. coli in nonsterile soil at 22°C and at 20 or 80% of the WHC, albeit at low frequencies, when the donor–recipient ratios were 1:1 (frequencies of 5×10^{-9} and 3.2×10^{-8} at 20 and 80% of the WHC, respectively) or 1:10 (5×10^{-9} at 80% of the WHC), but not when they were 1:100 (at 80% of the WHC).

However, the broad-host-range plasmid RP4 was transferred from E. coli to not only another strain of E. coli and related members of the Enterobacteriaceae [e.g., Enterobacter aerogenes, K. pneumoniae—both autochthonous soil bacteria—and P. vulgaris (Fig. 3)], but also to Ps. aeruginosa in both sterile and nonsterile soils (Figs. 4 and 5) (Devanas and Stotzky, 1988a,b). In sterile soil, there was growth of the donor and recipients during the first few days after inoculation, especially when inoculated in LB, and then the numbers remained essentially constant. Exconjugants were usually detected 2 hours after inoculation (the first time measured) and throughout the incubation at levels ranging from 10^4 to 10^6 CFU/g soil (oven-dry basis).

In nonsterile soil, the numbers of some recipients increased during the first few days after inoculation and then either remained relatively constant or decreased by one to two orders of magnitude to 10^5–10^7 CFU/g soil, depending on the recipient (Figs. 3 and 5). In contrast, the numbers of the donor decreased, sometimes after an initial increase, to either undetectable (<10^2 CFU/g soil) or low (e.g., 10^4 CFU/g soil) levels. The numbers of exconjugants detected generally depended on whether the donor and recipients were inoculated in saline or with

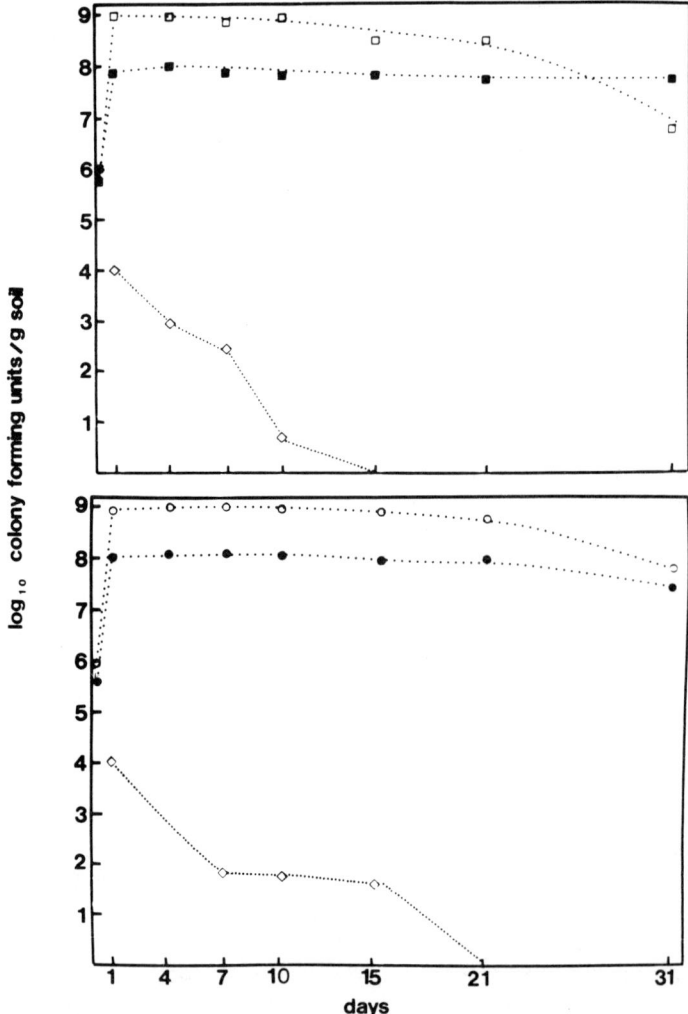

FIG. 1. Survival and transfer of plasmids pDU202 (□) and pRR226 (○) from *Escherichia coli* DU1040 to *E. coli* PRC487 (■, ●) in sterile soil. Cells were inoculated in Luria broth (LB). Total numbers of bacteria were enumerated on MacConkey agar (MAC): strain PRC487 was lactose-positive and strain DU1040 was lactose-negative. Transconjugant (◇) cells containing plasmid pDU202 were enumerated on MAC containing 25 μg/ml chloramphenicol (Cm) and those containing plasmid pRR226 on MAC containing 25 μg/ml tetracycline (Tc). Means ± SEMs, which are shown if larger than the dimensions of the symbols (Devanas and Stotzky, 1988a).

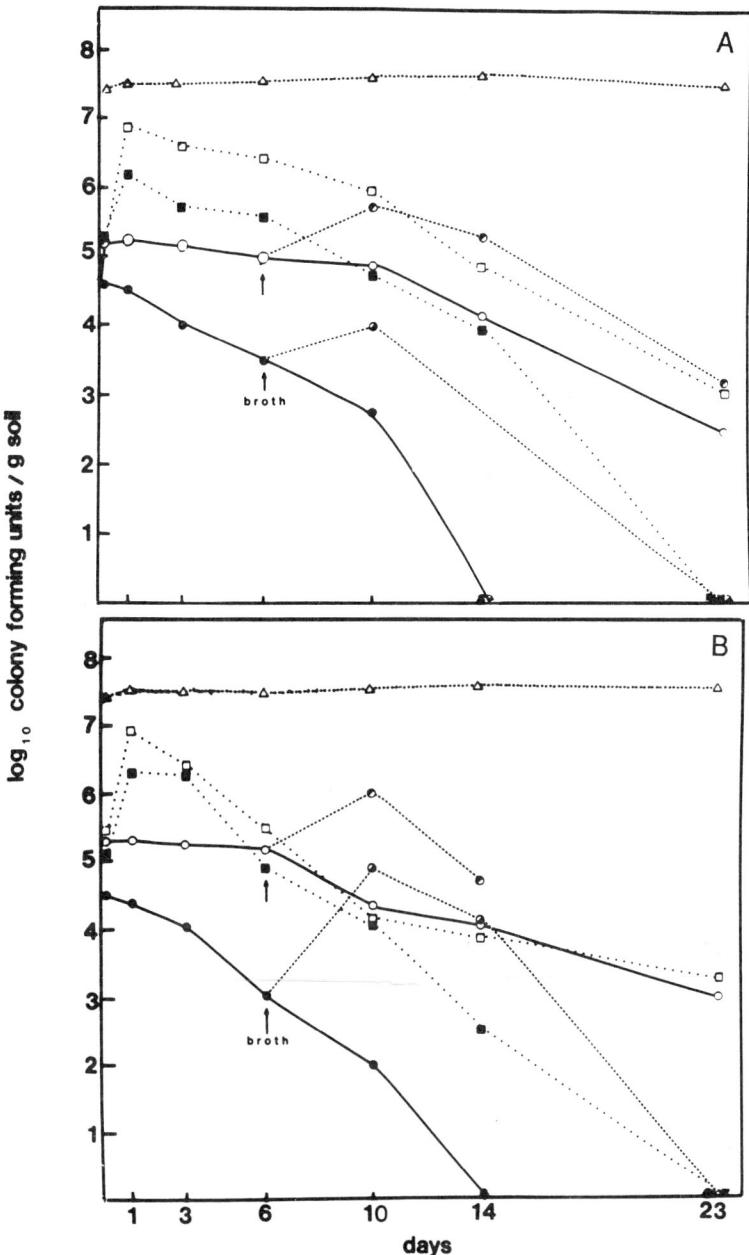

FIG. 2. Survival and transfer of plasmids (A) pDU202 and (B) pRR226 from *Escherichia coli* DU1040 (donors, open symbols) to *E. coli* PRC487 (recipient, filled symbols) in nonsterile soil. Cells were inoculated in Luria broth (□, ■) or saline (○, ●). Total bacteria (△) were enumerated on MAC, bacteria containing pDU202 on MAC containing Cm, and bacteria containing pRR226 on MAC containing Tc (see Fig. 1 for details) (Devanas and Stotzky, 1988a).

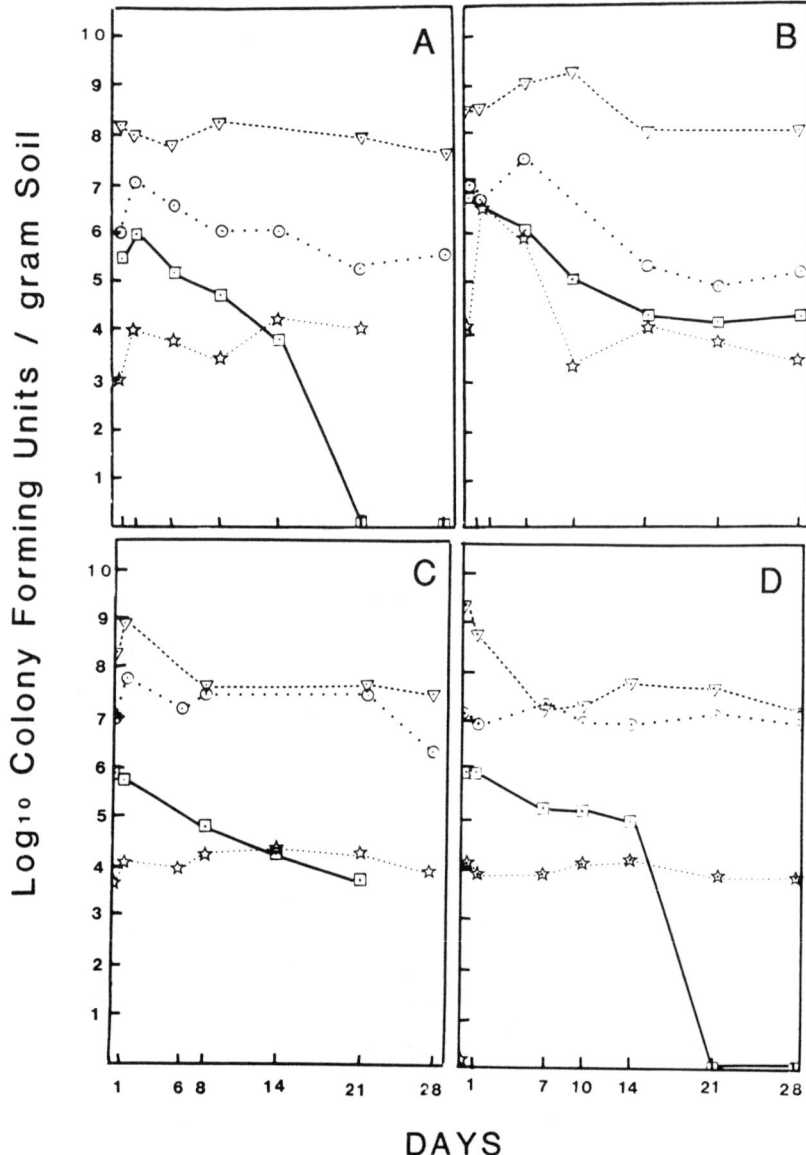

FIG. 3. Survival and transfer of plasmid RP4 from *Escherichia coli* J53(RP4) (□) to (A) *Enterobacter aerogenes*, (B) *E. coli* DU1040, (C) *Klebsiella pneumoniae*, and (D) *Proteus vulgaris* in nonsterile soil (○). Cells were inoculated in saline. Total bacteria (∇) were enumerated on Luria agar (LA); *E. coli* J53(RP4) on MAC containing Tc; *E. aerogenes*, *E. coli* DU1040, *K. pneumoniae*, and *P. vulgaris* on LA containing streptomycin (Sm); and *E. aerogenes*, *E. coli* DU1040, *K. pneumoniae* and *P. vulgaris*, carrying RP4 (☆) on LA containing Sm and Tc (see Fig. 1 for details) (Devanas and Stotzky, 1988a).

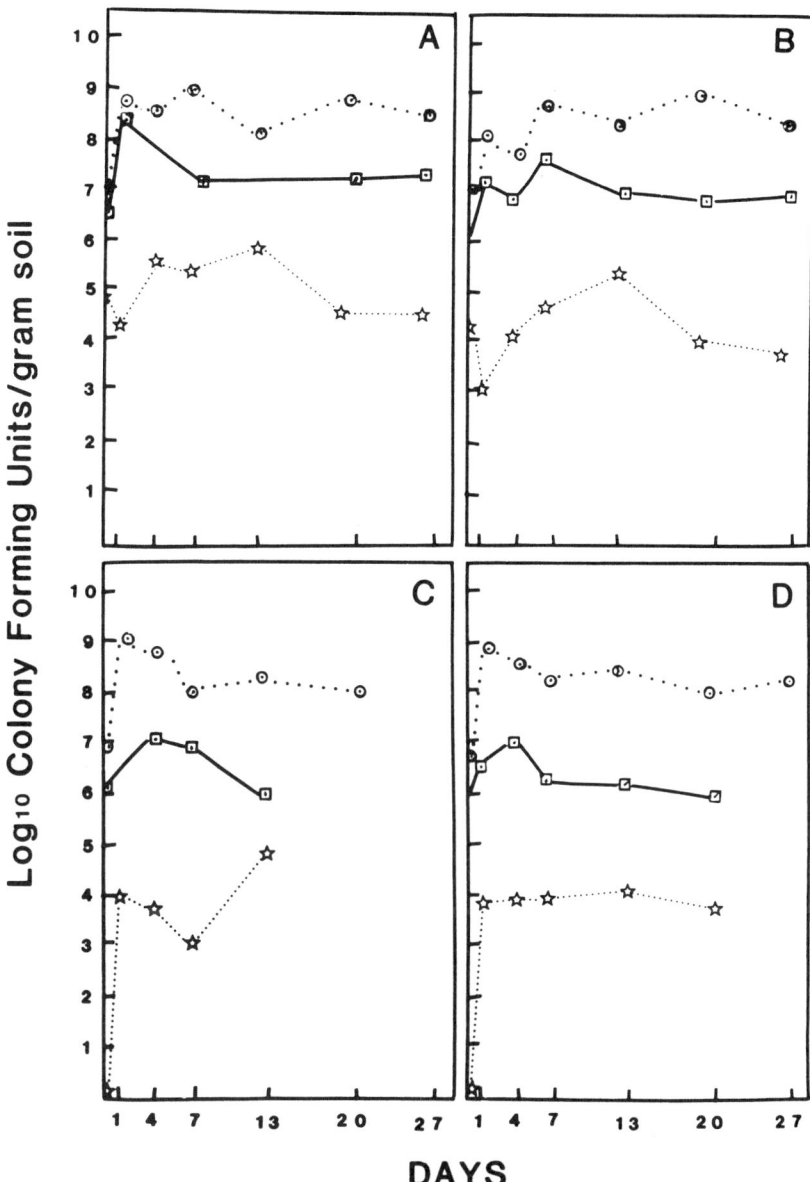

FIG. 4. Effect of clay minerals on survival and transfer of plasmid RP4 from *Escherichia coli* J53(RP4) (□) to *Pseudomonas aeruginosa* PAO1 (○) in sterile soil. Soil (Kitchawan, K) was amended with 3 or 12% (vol/vol) montmorillonite (M) or 12% (vol/vol) kaolinite (K). (A, B) K12M soil; (C) K3M soil; (D) K12K soil. Cells were inoculated in (A) LB or (B–D) saline. J53(RP4) was enumerated on MAC containing Tc, PAO1 on *Pseudomonas* isolation agar (PIA), and PAO1(RP4) (☆) on PIA containing Tc (see Figs. 1 and 3 for details) (Devanas and Stotzky, 1988a).

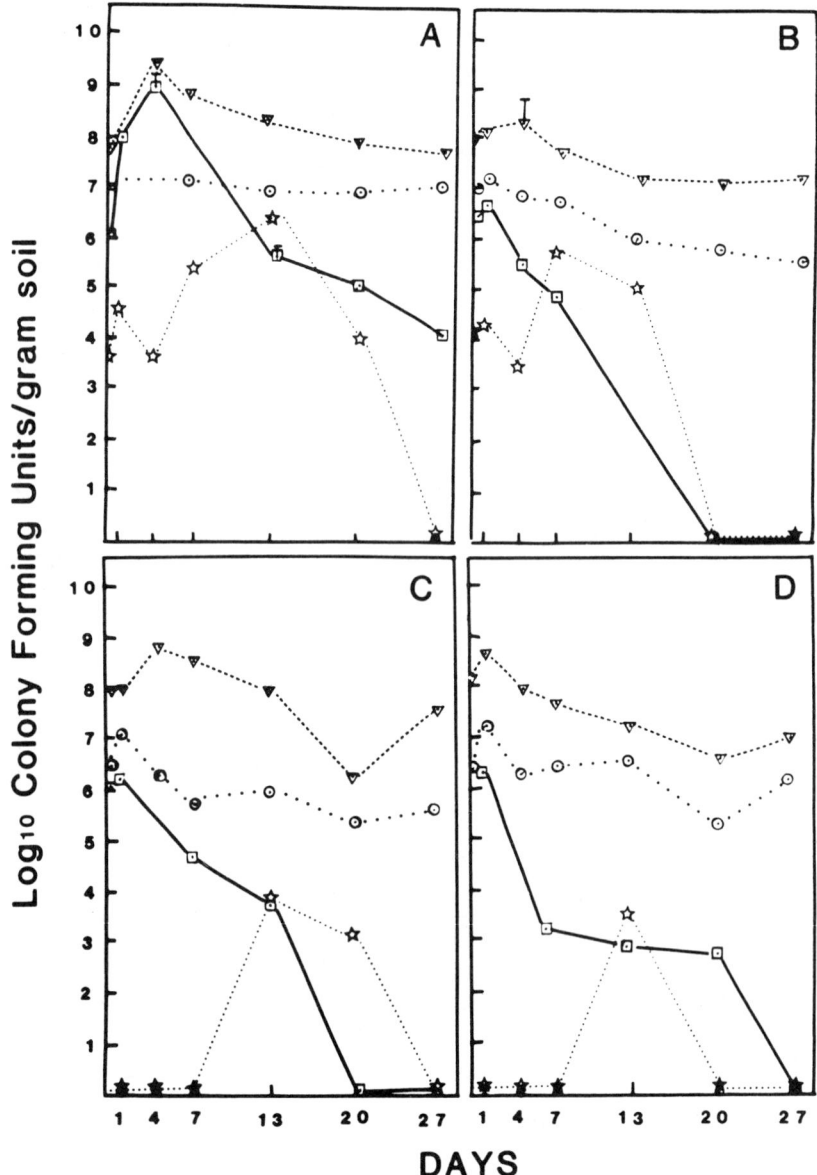

FIG. 5. Effect of clay minerals on survival and transfer of plasmid RP4 from *Escherichia coli* J53(RP4) to *Pseudomonas aeruginosa* PAO1 in nonsterile soil. ▽, Total bacteria. Other symbols, soil types, inoculation media, and enumeration conditions are as in Fig. 4 (Devanas and Stotzky, 1988a).

nutrients (LB), on the type and concentration of clay minerals with which the soil was amended, and, in part, on the species of the recipient and the survival of the donor. For example, the numbers of exconjugants of the Enterobacteriaceae remained at $\sim 10^4$ CFU/g soil throughout a 21- to 28-day incubation, even though the donor and recipients were inoculated in saline and the numbers of the donor decreased by several orders of magnitude or, with E. aerogenes and P. vulgaris as the recipients, to undetectable levels (Fig. 3). In contrast, when Ps. aeruginosa, as the recipient, was inoculated in saline, its numbers decreased slowly by about one order of magnitude, but the numbers of the E. coli donor, also inoculated in saline, decreased to an undetectable level after 20–27 days (Fig. 5). The numbers of exconjugants [Ps. aeruginosa(RP4)] detected differed with the type and amount of clay present in the soil, but in all soils, the numbers of exconjugants decreased to undetectable levels after 20–27 days, similar to the decrease in the donor. When the donor and recipients were added in LB, the numbers of recipients remained essentially constant, total bacteria increased initially and then slowly decreased, the donor increased by about three orders of magnitude during the first few days after inoculation and then decreased rapidly by about five orders of magnitude, and the exconjugants, which were initially detected 2 hours after inoculation of the parentals, increased to $\sim 5 \times 10^6$ CFU/g soil, followed by a rapid decrease to undetectable levels (Fig. 5).

9. Conclusions

The results of these studies again emphasize the importance of the type of bacteria and plasmids involved and of the physicochemical characteristics of the recipient environment to the survival of, and genetic transfer by, GEMs. For example, plasmids pDU202 and pRR226 were not transferred even between strains of E. coli in nonsterile soil, whereas the broad-host-range plasmid RP4 was transferred both intragenerically and intergenerically in nonsterile soil. Pseudomonas aeruginosa survived better than E. coli in nonsterile but not in sterile soil, reflecting the relative competitiveness of these species, the former being an autochthonous soil inhabitant. Similar survival data for these species in soil (Klein and Casida, 1967; Zechman and Casida, 1982) and in soil extracts (Walter et al., 1987) have been reported. The presence of nutrients generally enhanced survival and gene transfer in nonsterile soil but had essentially no effect in sterile soil, and suboptimal water tensions inhibited transfer. Montmorillonite, especially at high concentrations, apparently enhanced the frequency of transfer of both plasmid- and chromosome-borne genes.

It is not known whether the fluctuating survival patterns of the recombinant bacteria reflected continuous conjugation or the growth and survival of recombinants formed early in the incubation, as the recombinants were not inoculated but resulted from gene transfer in soil. Consequently, the duration of gene transfer in soil may be more apparent than real, because it is not possible to distinguish between continual gene transfer and cell division of recombinant cells formed shortly after introduction of the parentals. Nevertheless, the detection of exconjugants at levels ranging from 10^4 to 10^6 CFU/g soil for at least 20 days after inoculation of the parentals indicated that the novel genes survived. Therefore, there is a possibility that, depending on the host–plasmid system involved, the introduction of GEMs into soil and other natural environments could alter the homeostasis of these habitats.

These results also indicate that more research is needed on the transfer of genetic information by conjugation from GEMs and between indigenous bacteria in soil and other natural habitats *in situ*. Furthermore, the differences in survival and genetic transfer between the various host–vector systems studied indicate that the survival of, and genetic transfer by, GEMs that may be released to soil or other natural environments must, at this state of knowledge, be evaluated on a "case-by-case" basis.

One aspect that is clear from these studies is that the frequency of transfer of genetic material is higher in sterile than in nonsterile soil, indicating that the presence of the indigenous microbiota interferes with transfer. What is not clear is whether this interference is the result of (1) a reduction in the population densities of donors and recipients caused by competition for nutrients, water, oxygen, space, and so on, or by amensalistic, parasitic, and predatory activities of the indigenous microbiota; (2) the production of molecules by the indigenous microbiota that interfere with transfer (e.g., blocking or alteration of receptor sites on the recipients); or (3) other mechanisms. The first possibility is probably not a major factor, as the frequency of transfer in nonsterile soil was reduced even when the population densities of the donors and recipients were high enough to result in transfer in sterile soil (e.g., Figs. 1 and 4). The second possibility is suggested by observations made in studies on the *in vivo* frequency of transfer of a multiple drug-resistance plasmid from *Salmonella typhosa* to *E. coli* in the bladder of healthy rabbits (Richter et al., 1973; Stotzky and Krasovsky, 1981). The frequency was reduced by greater than one order of magnitude when *Proteus mirabilis* and a nonconjugative strain of *E. coli* (exogens) were also introduced into the bladder. When polystyrene latex particles of approximately the same size (0.8 μm diameter) and at the same concentration as the exogens were introduced into the bladder, the frequency of transfer of the plasmid from *S. typhosa* to *E. coli* was not reduced, indicating that the reduction caused by the exogens was not the result of a steric interference but of a chemical interference in the conjugal process. Similar results were observed when these studies were conducted *in vitro* in urine removed from the bladder or in a liquid mating medium. Whether such interference occurs in soil is not known and requires study. However, the apparent inability of narrow-host-range plasmids (e.g., pDU202 and pRR226) to be transferred in nonsterile soil while they were transferred in sterile soil, whereas broad-host-range plasmids (e.g., RP4) were transferred in both nonsterile and sterile soil, suggests that the transfer of some plasmids may be more susceptible to chemical interference than the transfer of others.

It is also not known whether the enhancement of the transfer of genetic material by the addition of nutrients (e.g., better transfer when the parentals were added in LB than in saline) was related to this interference. Although the addition of nutrients increased the total numbers of detectable donors and recipients, the numbers initially added, either in LB or saline, were sufficiently high for transfer to have occurred. However, the nutrients may have increased the numbers of parentals in those microhabitats where transfer was occurring, or the nutrients may have increased the robustness (or reduced the debilitation) of the parentals, which facilitated transfer. This is another aspect that requires study in soil.

The studies on conjugal transfer of chromosomal genes in soil showed that auxotrophic bacteria (in these studies, genetically deficient for the synthesis of some amino acids) can coexist by cross-feeding (syntrophism) in soil and on agar media inoculated directly from soil by the soil replica-plating technique (Stotzky, 1965b) rather than by only genetic recombination (Krasovsky and Stotzky, 1987; Weinberg and Stotzky, 1972). Similarly, co-metabolism or "shared detoxification" of inhibitors can contribute to the survival in soil of toxin-sensitive microbes in the absence of genetic recombination (e.g., Daughton and Hsieh, 1977; Senior et al., 1976; Slater and Godwin, 1980; Slater and Somerville, 1979). Furthermore, the transfer of genetic information in soil may be more apparent than real, as the transfer may occur not in soil but on the agar used for the subsequent isolation and enumeration of the donors, recipients, and putative recombinants (Walter et al., 1989b). These aspects, in addition to the "viable but nonculturable" phenomenon, must be considered and controlled in studies on transfer of genetic information in soil.

The apparent higher frequency of conjugal transfer of chromosome-borne (Krasovsky and Stotzky, 1987) than of plasmid-borne (Devanas and Stotzky, 1988a,b) genes in soil suggests that conjugation may not occur long enough in soil for the transfer of an entire large, conjugative plasmid. This suggestion was supported by preliminary observations that the frequency of transfer of marker chromosomal genes was reduced as their location was more distant from the *oriT* site. Studies on how plasmid size and the location of chromosomal genes affect the frequency of gene transfer in soil should be conducted, especially in that such studies may provide clues for the construction of recombinant DNA that would have a low frequency of transfer.

Although nonconjugative plasmids are used extensively as vectors for the introduction of heterologous DNA into bacteria, the introduced DNA can be moved to the chromosome (e.g., by a Tn) and from there be transferred to another bacterium. Hence, conjugal transfer of chromosomal genes could be an important mechanism for the dispersal of genetic information, for example, from an introduced GEM to indigenous soil bacteria. However, appropriate genetic manipulation of a chromosome containing heterologous DNA can presumably reduce the probability of transfer in soil (Watrud et al., 1986).

B. Transduction

1. Introduction

Transduction is another method by which novel genes can be transferred to indigenous soil bacteria after the release of GEMs to the environment. Moreover, bacteriophages may serve as reservoirs of bacterial DNA in soil and other natural habitats, as the packaging of genetic material in a transducing bacteriophage probably represents an evolutionary survival strategy for bacterial genes (Stotzky, 1989; Stotzky and Babich, 1986; Zeph et al., 1988). The survival of a transducing phage that contains the novel DNA could be longer than that of the introduced GEM itself, and these novel genes could reappear long after the GEM can no longer be detected if the phage infects susceptible indigenous bacteria. The probability of the occurrence of this scenario is difficult to predict, as the persistence of, and transduction by, phages in soil and other natural ecosystems has been insufficiently studied.

Two mechanisms of transduction, designated generalized and specialized transduction, have been described in which virions that contain bacterial DNA can be formed after infection of a bacterium by a phage. In generalized transduction, phages, whose DNA is replicated shortly after infection of the bacterial cell (lytic cycle), can package DNA from the bacterium, whose chromosome has been cleaved into small segments by nucleases, in place of phage DNA during incorporation of

DNA into the phage capsid (Lin et al., 1984). Usually, <1% of the bacterial genome is packaged into the phage virions. Any single or group of bacterial genes may be incorporated, depending on the size of the phage capsid, and thus the designation, generalized transduction. The transducing phages that carry bacterial DNA are capable of adsorbing on and injecting their DNA into new bacterial host cells, but the DNA will be defective for replication of the phages.

In specialized transduction, the bacteriophage genome is integrated into the bacterial chromosome, usually at a specific site, and replicates faithfully with the bacterial chromosome. The integrated phage genome is termed prophage DNA, and the bacteria that contain it are termed lysogenic (Lin et al., 1984). Upon induction of the lytic cycle of phage replication, imprecise excision of the phage genome can result in the packaging of adjoining bacterial genes on the chromosome into the virion with the prophage DNA. The most extensively described specialized transducing phage is phage λ, which can transduce the genes for galactose catabolism and biotin synthesis that are located on either side of the integration site of this prophage in *E. coli*. The imprecise excision event occurs at a low frequency (e.g., in about one bacterial cell per 10^6–10^7 cells). The ability to transfer only genes that are close to the site of phage integration is termed specialized transduction. A specialized transducing phage may be defective for some of its genes and may require the activity of "helper" phages to replicate in the new host bacterium.

Two types of transducing phages, temperate and lytic, that differ in their mode of replication in the bacterial host, have been described. Temperate bacteriophages are capable of maintaining single or multiple copies of their genome in host bacterium in a prophage state; the phage DNA is either inserted into the bacterial chromosome through a recombinational event (e.g., coliphage λ or Mu-1) or remains as an independent and autonomous replicon (e.g., coliphage P1 or phage F116 of *Ps. aeruginosa*), similar to a bacterial plasmid. A temperate phage can initiate a lytic cycle of replication and virion formation (induction), although this occurs at a low frequency under normal conditions (e.g., one in 10^5–10^6 DNA replication cycles for phage P1). Exposure of lysogenic bacteria to temperatures above the optimum for growth or to mutagenic agents can result in the induction of lytic reproduction at a higher frequency. Only between 1 and 10% of the infective virions produced after induction *in vitro* will usually establish lysogeny after infection of new host cells, as most infections result in lysis of the cells.

Lytic transducing phages undergo the normal lytic cycle of infection and formation of new virions in the bacterial host, and they are not capable of establishing lysogeny. During replication of the phage DNA, genetic material from the host bacterium that has escaped degradation by phage-encoded nucleases can be randomly packaged into intact virions at a low frequency. Hence, lytic transducing phages are all generalized transducing phages. In a typical lysate, ~1% of the virions will contain only bacterial DNA (Stent and Calender, 1978). Consequently, a phage lysate of sufficient titer (i.e., at least 10^6 phages/ml) can be expected to contain at least one virion carrying a copy of each gene on the bacterial chromosome.

2. *Transduction in Situ*

Gene transfer *in situ* by transduction has been studied to only a limited extent. There have been few studies on transduction in marine or freshwater ecosystems (Baross et al., 1974; Morrison et al., 1978; Saye et al., 1987) and in terrestrial ecosystems (Germida and Khachatourians, 1988; Zeph et al., 1988; Zeph and Stotzky, 1989), even though transducing phages for a variety of bacteria have been isolated from soil. Zeph et al. (1988) studied the transduction in soil of bacterial genes that confer resistance to antimicrobial agents by coliphage P1 Cm cts, which carries a resistance gene for Cm, and a variant of this phage, P1 Cm cts :: Tn*501*, which also carries a Hg-resistance gene on the transposon, Tn*501*. A temperature-sensitive mutation (cts) results in the inactivation, at temperatures above the optimum for growth (e.g., at 42°C), of a repressor protein coded for by phage P1 that maintains lysogeny.

Lysates of phage P1 [10^5–10^6 plaque-forming units (PFU)/g soil (ovendry basis)] were added to soil with *E. coli* (10^5–10^6 CFU/g soil), and soil dilutions were periodically plated on selective media to quantitate the numbers of *E. coli*, phage P1, *E. coli* transductants (Cm- or Hg-resistant cells lysogenized by phage P1), and total soil bacteria. Significantly higher maximum numbers of *E. coli* transductants were observed in sterile (~10^8 transductants/g soil) than in nonsterile (~10^6 transductants/g soil) soil (Fig. 6). Amendment of nonsterile soil with nutrients [LB containing 10 mM $MgSO_4 \cdot 7H_2O$ and 2 mM $CaCl_2$ (LCB)] on day 0 did not significantly affect the numbers or survival of *E. coli* transductants during the 28-day experiment, even though the numbers of total *E. coli* increased by one order of magnitude in the nutrient-amended soil (Fig. 7). Weekly amendment of the soil with LCB also did not result in significant increases in numbers of *E. coli*, phage P1, or *E. coli* transductants (Fig. 8) (Zeph and Stotzky, 1988). These results demonstrated that phage P1 was capable of infecting and transducing *E. coli* under the relatively low nutrient conditions that presumably occur in natural soil. Although the survival of phage P1 was slightly greater in soil amended

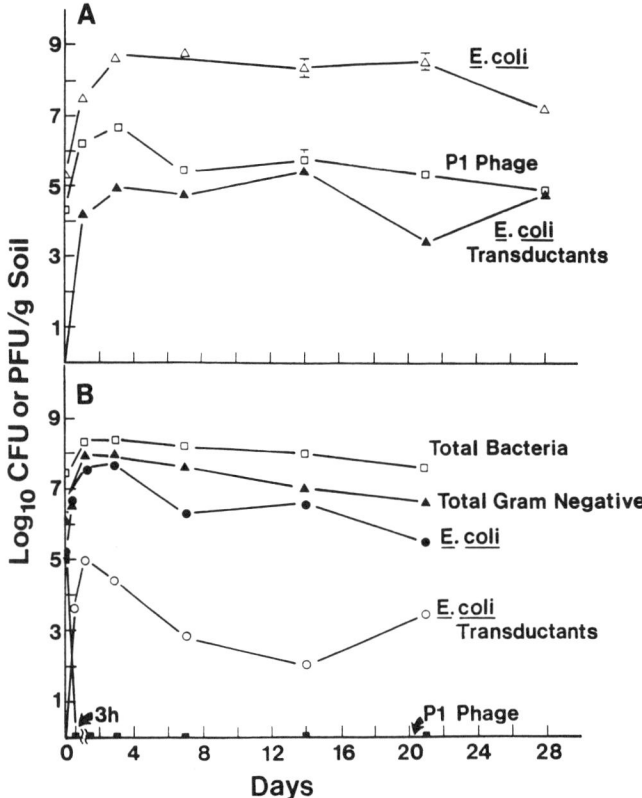

FIG. 6. Comparison of transduction of *Escherichia coli* W3110(R702) by phage P1 Cm cts in sterile (A) and nonsterile (B) soil. Bacteria and phage inocula were added in LCB (Luria broth containing 10 mM $MgSO_4 \cdot 7H_2O$ and 2 mM $CaCl_2$) on day 0. Means ± SEMs, which are shown if larger than the dimensions of the symbols (Zeph et al., 1988).

with the clay mineral, montmorillonite, than with the clay mineral, kaolinite, this improved survival did not result in significantly higher levels of transduction (Fig. 9).

When phage P1 was added to sterile soil in lysogenic *E. coli* J53(RP4)(P1 Cm cts :: Tn501) together with nonlysogenic *E. coli* W3110 (both at ~10^6 CFU/g soil) and nutrients (LCB), phage P1 was released from the lysogenic *E. coli* and infected *E. coli* W3110, which resulted in ~10^5 transductants/g soil (Fig. 10). In nonsterile soil, using *E. coli* W3110(R702) as the recipient and *E. coli* J53(P1 Cm cts) as the lysogen, the number of *E. coli* W3110(R702)(P1 Cm cts) transductants was significantly lower, as 10^2 transductants/g soil were recovered on day 1 only.

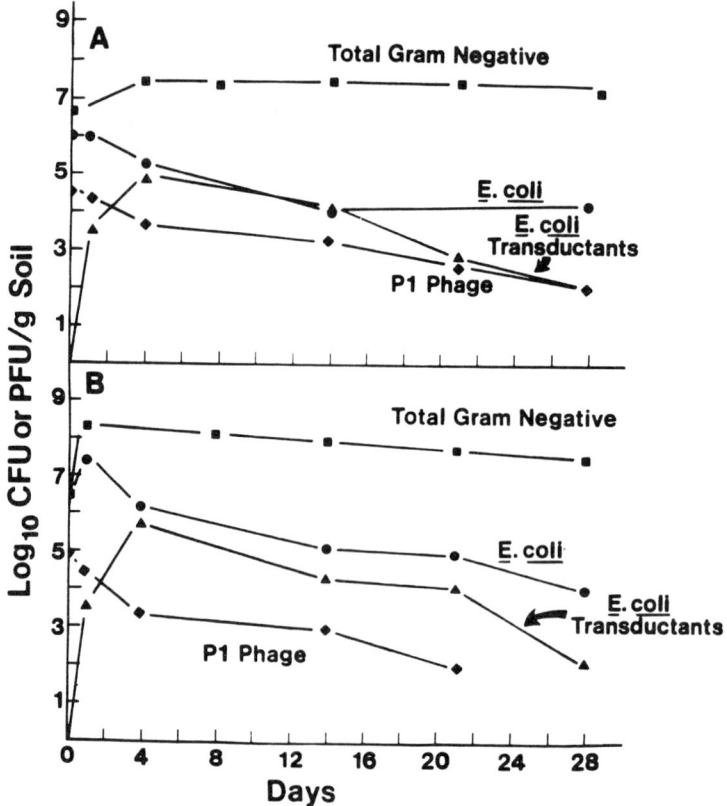

FIG. 7. Comparison of the effect of inoculating *Escherichia coli* J53(RP4) and phage P1 Cm cts::Tn501 in saline (A) or LCB (B) on transduction in nonsterile soil (see Fig. 6 for details) (Zeph et al., 1988).

The multiplication and survival of the lysogenic donor and nonlysogenic recipient *E. coli* were also significantly lower than in sterile soil (Fig. 10). Despite the lower survival and frequency of transduction in nonsterile than in sterile soil, these studies showed that phage P1 released from lysogenic *E. coli* can lysogenize susceptible bacterial cells under conditions similar to those in natural soil.

Escherichia coli transductants isolated from soil were verified by heat induction of phage P1 and with a biotinylated DNA probe, which demonstrated the presence of phage P1 DNA in the isolates (Zeph and Stotzky, 1989). Approximately 300 isolates of indigenous soil bacteria, isolated on MAC containing 75 μg/ml Cm and inoculated with a 10^{-2} dilution of soil, were negative when subjected to both verification pro-

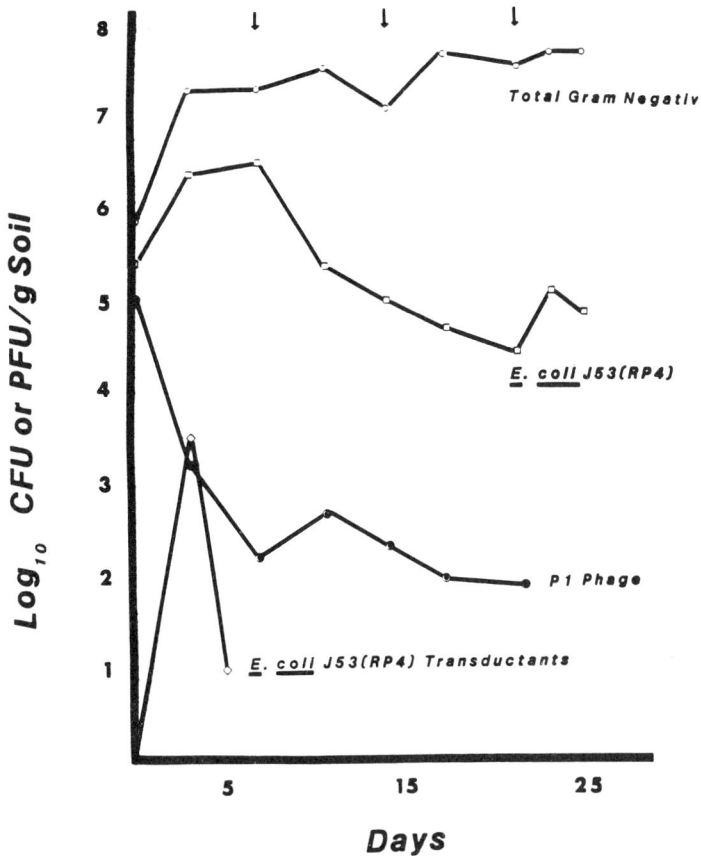

FIG. 8. Transduction of *Escherichia coli* J53(RP4) by phage P1 Cm cts::Tn501 inoculated in LCB into nonsterile soil. Sterile LCB was added on days 7, 14, and 21, as indicated by the arrows at top (see Fig. 6 for details) (Zeph and Stotzky, 1988).

cedures (Table II), even though the theoretical minimum limit of detection was 10^2 transductants/g soil, and the specificity and sensitivity of the DNA probe was satisfactory. Consequently, few, if any, indigenous bacteria in the soils studied were apparently capable of being transduced by phage P1.

Germida and Khachatourians (1988) studied the generalized transduction by phage P1 of amino acid and Tc-resistance markers on the chromosome of *E. coli* in a sandy and a silty clay loam soil. The frequency of transduction of these markers to added recipient strains of *E. coli* was 10^{-6}, which was similar to the frequency obtained *in vitro*.

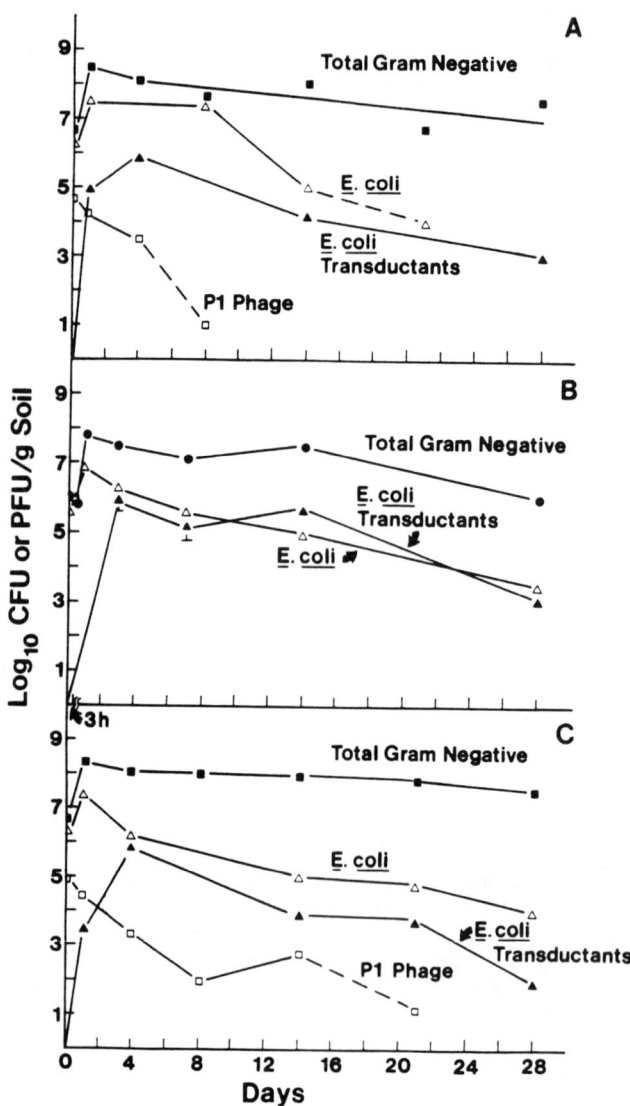

FIG. 9. Transduction of *Escherichia coli* J53(RP4) by phage P1 Cm cts::Tn501 in nonsterile soil amended with (A) 12% kaolinite, (B) 6% montmorillonite, and (C) 3% montmorillonite. Bacteria and phage inocula were added in LCB (see Fig. 6 for details) (Zeph et al., 1988).

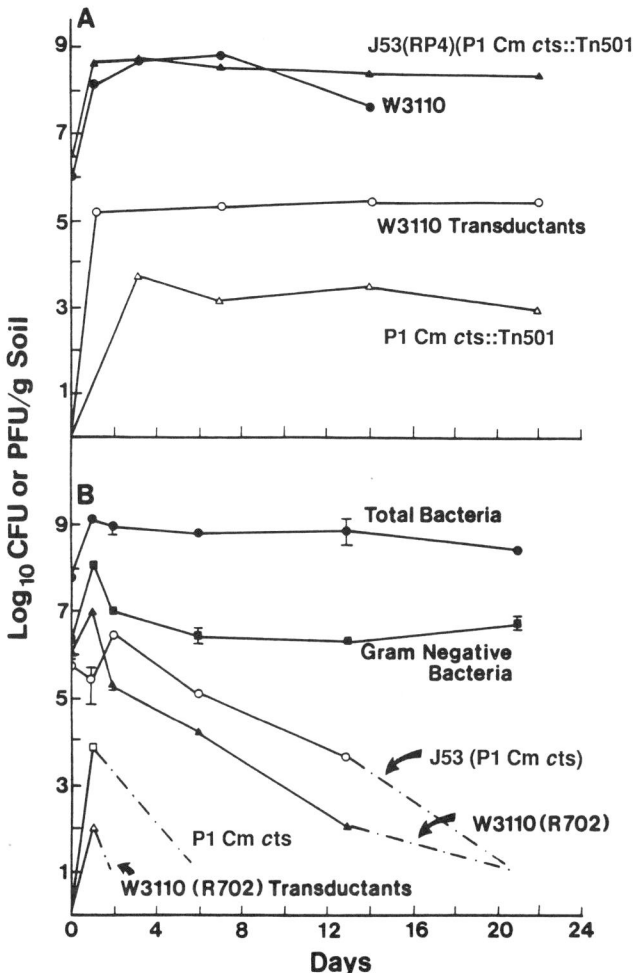

FIG. 10. Transduction in sterile (A) and nonsterile (B) soil inoculated with either *Escherichia coli* J53(RP4) lysogenic for phage P1 Cm cts::Tn501 and nonlysogenic *E. coli* W3110 (A) or *E. coli* J53 lysogenic for phage P1 Cm cts and nonlysogenic *E. coli* W3110(R702) (B). Bacterial inocula were added in LCB on day 0 (see Fig. 6 for details) (Zeph et al., 1988).

Thus, conditions in nonsterile soil were apparently not detrimental to transduction by phage P1.

A generalized transducing phage of *Ps. aeruginosa*, F116L, has been shown to transfer plasmid RP4 to *Ps. aeruginosa* PAO1 in soil (Zeph and Stotzky, 1988). When a lysate of phage F116L (10^7 PFU/g soil), raised in *Ps. aeruginosa* PAO1 carrying plasmid RP4, was added to sterile soil

TABLE II

VERIFICATION BY HEAT INDUCTION OF LYSIS AND WITH A BIOTINYLATED DNA PROBE OF PRESUMPTIVE TRANSDUCTANTS OF *Escherichia coli* AND INDIGENOUS SOIL BACTERIA ISOLATED ON MACCONKEY AGAR CONTAINING EITHER $HgCl_2$ (30 μM) or Cm (75 μg/ml)[a]

Presumed transductants	Soil	Heat induction of phage P1[b]	Hybridization with P1 DNA[c]
Indigenous	Nonsterile	0/297	0/297
E. coli W3110(R702)	Sterile	15/15	ND[d]
	Nonsterile	8/8	8/8
E. coli J53(RP4)	Sterile	13/13	ND
	Nonsterile	25/25	7/7

[a] From Zeph and Stotzky (1989).
[b] Number of isolates that were positive/number tested.
[c] Number of colonies or dot blots that were positive/number tested.
[d] Not determined.

with nonlysogenic cells of Ps. aeruginosa PAO1 (10^8 CFU/g soil), presumed Ps. aeruginosa transductants resistant to Tc were detected at ~10^3 transductants/g soil (Fig. 11). These presumptive transductants were confirmed by the demonstration that their antibiotic-resistant phenotype was the same as the antibiotic-resistance markers present on plasmid RP4 [Km, Ap, and neomycin (Nm), in addition to Tc] and by plasmid screening by gel electrophoresis, which showed the presence of plasmid DNA of the same molecular mass as plasmid RP4 (Fig. 12). No RP4 transductants of Ps. aeruginosa PAT2 or of Ps. aeruginosa PAT2 lysogenic for phage F116L were detected when these recipient bacteria were added to sterile soil with phage F116L carrying plasmid RP4.

3. Conclusions

Transduction as a mechanism of genetic transfer may be as, or more, important in soil and other natural habitats as conjugation and transformation. Bacteriophages are capable of multiplying in natural soil, which emphasizes the potential for transducing phages to transfer genes *in situ*. Furthermore, the survival of viruses, including bacteriophages, in soils and waters is enhanced by their adsorption on clay minerals, especially on montmorillonite, and probably on other particulates. Consequently, DNA in an indigenous bacterium or introduced in a GEM could be incorporated into the genetic material of a phage and could persist in soil longer than in the host bacterium itself. Such persistence would be undetected (i.e., cryptic) in the absence of a bacterial host susceptible to infection and reproduction of the phage. However, if an

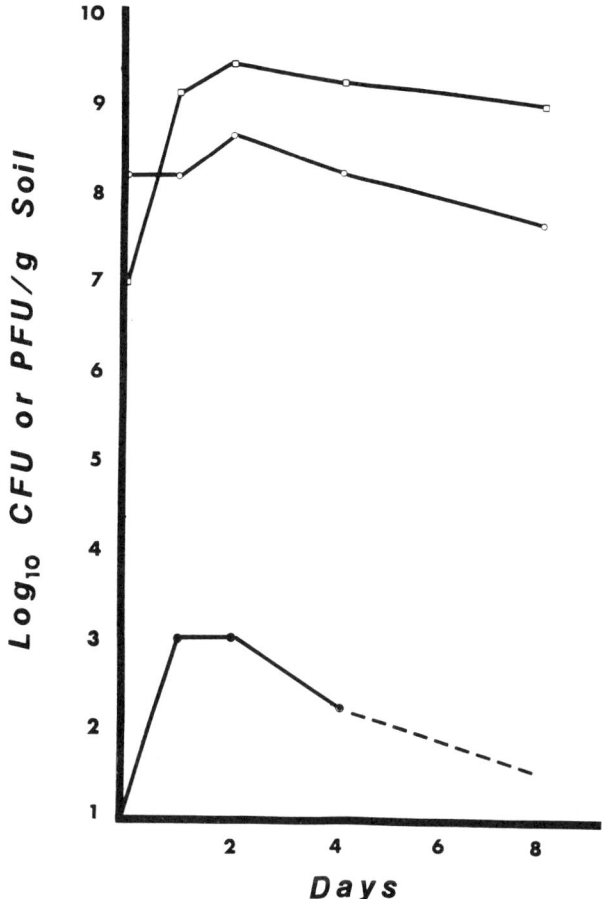

FIG. 11. Transduction of *Pseudomonas aeruginosa* PAO1 (□) by phage F116L (○) in sterile soil. The phage F116L inoculum was raised on *Ps. aeruginosa* PAO1 carrying the tetracycline-resistance plasmid, RP4 (●). The bacterium and phage were inoculated in Luria broth (Zeph and Stotzky, 1988).

appropriate host were subsequently infected, the recombinant DNA could be rapidly dispersed among susceptible bacteria; for example, a single nondefective phage that was able to accommodate the extra DNA and with a burst size of 100 could transfer the recombinant DNA to $\sim 10^8$ bacteria after only four lytic cycles. Hence, there is the potential for a novel gene to persist undetected and then unexpectedly reappear in soil and other environments if the transfer of the gene occurred by transduction.

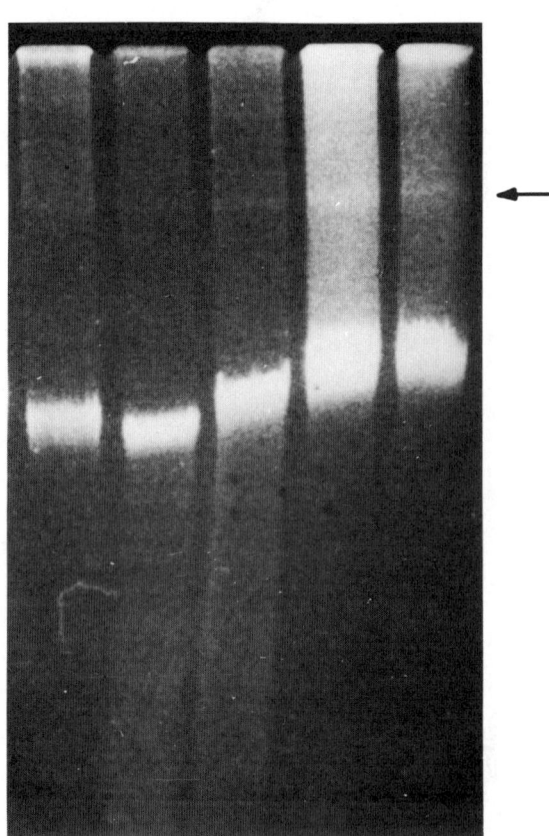

FIG. 12. Agarose gel electrophoresis of plasmid and chromosomal DNA extracts of three individual isolates of presumed transductants of *Pseudomonas aeruginosa* PAO1 isolated from sterile soil that was inoculated with phage F116L and *Ps. aeruginosa* PAO1 (see Fig. 11). Lanes 1–3, *Ps. aeruginosa* PAO1 isolates; lane 4, *Ps. aeruginosa* PAO1(RP4); lane 5, *Escherichia coli* J53(pDU202); molecular mass of pDU202 = 64 MDa; that of RP4 = 65 MDa (L. R. Zeph and G. Stotzky, unpublished observations).

Gene transfer to indigenous bacteria by conjugation can theoretically be controlled by engineering the novel genes into the chromosome and through the use of suicide plasmid vectors (Cuskey, 1988). However, placing the novel genes in the chromosome will not necessarily decrease the probability of their transfer by transduction, especially as no techniques have been developed for limiting transduction *in situ*. Few studies on the frequency of lysogeny in bacteria exist, and it is likely that even these studies underestimate the frequency, as it has been demonstrated that most strains of *Ps. aeruginosa* and several strains of *Ps. fluorescens* are lysogenic (Holloway, 1969).

To date, studies on transduction *in situ* have focused on two species of bacteria (i.e., *E. coli* and *Ps. aeruginosa*) and on a limited number of genetic markers, that is, genes encoding resistance to antimicrobial agents and amino acid auxotrophy. Studies on transduction *in situ* are needed with more bacterial species and genetic markers, particularly with GEMs that are being used or proposed for environmental releases, e.g., *Rhizobium meliloti* containing recombinant nitrogen-fixation genes.

III. Terrestrial Microcosms

A. Introduction

Terrestrial microcosms have been defined as analogs of the field, as a field within the laboratory (Pritchard, 1988), or as physical models of the natural environment in which experimental conditions in the laboratory are intended to mimic a field setting (Pritchard and Bourquin, 1984). The extent of such simulations can be as broad as isolating in the laboratory a "piece of the field" that acts ecologically similar to an identical companion "piece" in the actual field (Hicks et al., this volume) or as simple as adding sterile soil to broth in test tubes (Walter et al., 1987). The spectrum of experimental systems that have been called microcosms has resulted in numerous discussions on "what is a microcosm?" Greenberg et al. (1988) stated that "the term microcosms seems to mean all things to all men, ranging, for example, from 'synthetic' to 'natural' systems." The consensus of opinion, regardless of the style or complexity of the microcosm, was that "it is essential that the questions to be asked with the microcosm, and the experimental design, are carefully thought out" and "that those workers who use microcosms should understand the uses, limitations and bias of the particular systems and ensure that people using their results are also aware of these facts." Bull (1980) pointed out that one benefit of the use of microcosms is their replicability in statistical trials; however, they are not exact reproductions of the field but, rather, analytical tools that can be used as the basis for environmental studies.

B. Simple Microcosms

The simplest terrestrial microcosm consists of soil in a container. Sterile soil provides a gnotobiotic environment in which to evaluate GEMs in the absence of the indigenous microbiota. Soil may be amended with nutrients and clay minerals, adjusted to optimum pH values and water contents, and added to test tubes, flasks, or other types of containers. Such simple microcosms have been used to evaluate the survival, growth, gene transfer, and effects of GEMs in soil in the laboratory (Devanas et al., 1986; Doyle et al., 1988; Jones et al., 1988; Krasovsky and Stotzky, 1987; Stotzky, 1965a, 1989; Stotzky and Babich, 1986; Weinberg and Stotzky, 1972) and changes in plasmid frequencies in natural soil microbial populations (Wickham and Atlas, 1988).

Quantities of soil between 2 and 50 g are recommended for simple microcosms, as this facilitates handling of soil samples in the laboratory, enables the initial dilution of the entire soil sample in the same container, and ensures good mixing of the diluted soil sample. Sufficient dilution and mixing of the soil are necessary to prevent aggregation and settling of the soil in pipettes during dilution, which would introduce error and reduce the reproducibility of replicate samples. The soil microcosm used as an example in this document consists of a screw-cap test tube (18 × 150 mm), loosely capped and maintained in a high-humidity chamber, into which ~2 g of soil at the -33-kPa water tension (approximately equivalent to the water content at field capacity) is placed. The container should be large enough to ensure that the surface–volume ratio of the soil sample is sufficiently high to maximize gas exchange between the soil and the headspace of the container (e.g., for large samples of soil, an Erlenmeyer flask is better than a test tube). The selection, preparation, maintenance, and storage of soils is discussed in Section IV.

1. Master Jar Studies

The overall metabolic activity of microbes in soil can be determined with respirometric techniques that monitor either CO_2 evolution or O_2 uptake. These methods, especially when CO_2 evolution is used, probably provide the best and most easily measured index of the gross metabolic activity of mixed microbial populations in soil (see Stotzky, 1960, 1965a, 1974; Anderson, 1982, for details of the techniques). The "master-jar" technique (Stotzky, 1965a) (Fig. 13) enables removing subsamples of soil during an extended incubation for various analyses (e.g., transformation of substrates, species diversity, enzyme activities, survival of introduced bacteria, including GEM and their genes), in addition to continuous measurement of CO_2 evolution, without disturbing the remainder of the soil and, thereby, eliminates artifactual peaks in CO_2 evolution that can result from the physical disturbance of the soil. The soils are incubated under controlled temperatures and maintained at their -33-kPa water tension by continuous aeration with water-

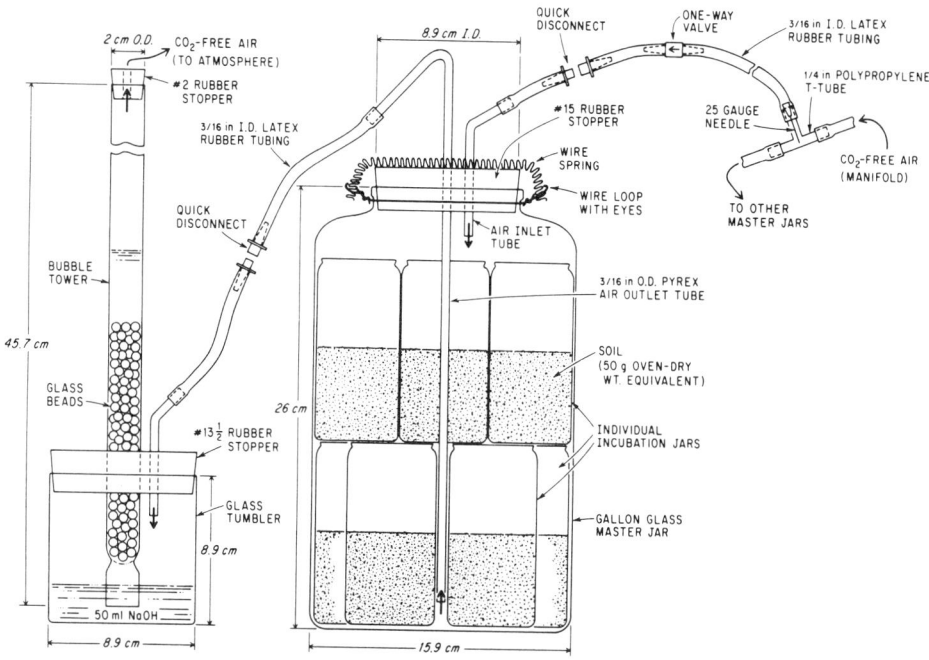

FIG. 13. Incubation unit for measuring CO_2 evolution from soil. The unit shown is used when subsamples of soil are removed during incubation. When this is not required, soil is placed directly in the master jar (Stotzky, 1965a).

saturated, CO_2-free air. The amount of CO_2 trapped in NaOH collectors is determined, after precipitation of the CO_2 with $BaCl_2$, by automatic potentiometric titration with HCl.

The gross metabolic activity of the heterotrophic soil microbiota is measured by the addition of a nonspecific substrate (e.g., glucose), and the activity of specific components of the microbiota is evaluated by the addition of specific substrates whose mineralization is dependent on the ability of these components to synthesize appropriate enzymes (e.g., celluloses, starches, lipids, lignins, aldehydes, proteins). At various times after the start of the incubation, subsamples of soil, in their own containers, are removed from the master jars and subjected to a variety of microbiological, chemical, and enzymatic analyses. Species diversity is determined by inoculating decade serial dilutions of the soil onto or into selective media, usually at least in triplicate. The soils can also be amended with the specific substrate (e.g., toluene, xylenes, 2,4-D) on which the gene product of a novel gene in an introduced GEM functions, to determine whether this provides an ecological advantage to the GEM and how this affects both nonspecific and specific metabolic activities, as well as other processes (Doyle et al., 1988; Jones et al., 1988).

2. *Soil Replica Plating*

The soil replica-plating technique (Krasovsky and Stotzky, 1987; Stotzky, 1965b, 1974; Weinberg and Stotzky, 1972) can be used to determine the growth rates of GEMs and their homologous hosts (i.e., bacteria without the novel genes), the ability of the GEMs to compete with representative soil microbes, and the transfer of genetic information. For example, the GEMs or the hosts are inoculated into the center of petri dishes containing sterile soil, and representatives of the indigenous soil microbiota (bacteria, including actinomycetes, and fungi) are inoculated into equidistant sites around the GEMs or the hosts. The plates are incubated in a high-humidity chamber, and replications from the soil plates to selective media are made periodically with a replicator constructed with stainless-steel nails and acrylic plastic and sterilized with alcohol and flaming. The design of the replicator permits numerous replications from the same soil plate without significant disturbance of the soil. The growth of all the inoculated organisms is recorded on maps of the soil plates, and growth rates (in mm/day) are calculated. The technique can also be used with nonsterile soil, but highly selective media must be used to prevent overgrowth of the GEMs by the indigenous soil microbiota (see Section VII).

C. COMPLEX MICROCOSMS

Undisturbed soil microcosms consist of soil cores of varying size that are brought into the laboratory with minimum disturbance of the structure and biotic composition of the soil. The soil sample is removed intact from the coring device, and thereby, the soil system is relatively undisturbed. Disturbed microcosms are those in which the soil ecosystem is reconstituted in the laboratory, for example, the placement of excavated soil into containers in which seeds or seedlings are planted.

Undisturbed soil cores have been utilized to study, for example, pesticide degradation (Johnen and Drew, 1977; Wingfield et al., 1977), microbial community responses to environmental perturbations (Elliot et al., 1986), soil denitrifier populations under anaerobic and aerobic conditions (Martin et al., 1988), the fate and ecological effects of GEMs (Bentjen et al., 1989), and the effects and mobility of xenobiotics in soil containing plant root systems (Hicks et al., this volume; see also Fig. 14). In the latter study, uniform and realistic distributions of the xenobiotics were attained with simulated rain events, and the data on environmen-

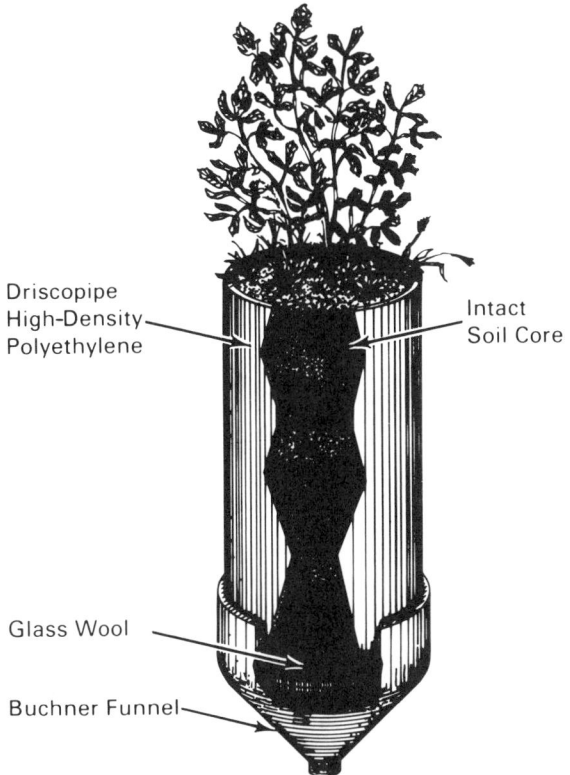

FIG. 14. Terrestrial soil core microcosm test system (Hicks et al., this volume).

tal perturbations were consistent with those obtained in parallel field studies. Other studies with intact soil cores have also adequately predicted events subsequently observed in field studies (Pritchard and Bourquin, 1984). The soil cores can be sampled for the enumeration of microbial populations at different depths through sampling ports inserted in the walls of the soil core cylinder (Ardakani et al., 1973). This also enables the soil sample to be replaced with fresh soil. Examples of various types of microcosms (Atlas and Bartha, 1981; Gillett, 1988; Johnson and Curl, 1972) and guidelines for the use of soil core microcosms, with descriptions of various soil core designs, sampling procedures, and statistical analyses, have been published (Anonymous, 1987).

With disturbed soil microcosms, it is assumed that biotic and abiotic processes and interactions representative of the *in situ* situation will be reestablished after suitable incubation or treatment (Pritchard and Bourquin, 1984). Small nonsterile soil cores (5 × 2.5 cm deep) were used to study changes in microbial populations and respiration after organic amendment (Elliot et al., 1986). Holben et al. (1988) used a soil–vermiculite microcosm (700 g of this mixture in plastic pots) to study the survival of a model GEM. A coring device was inserted through the side of the plastic pot, a sample of the mixture was removed for enumeration of the GEM, and a test tube was inserted into the hole formed by the coring device. Survival and plasmid transfer have been studied with GEMs added to a rhizosphere microcosm that consists of a peat–vermiculite mixture containing radish or bean plants. The pots or flats containing the simulated soil and plants were maintained within large chambers that enable the control of relative humidity, temperature, and light–dark cycles (Fig. 15) (Armstrong et al., 1987; Gile et al., 1982; Knudsen et al., 1988).

The soil perfusion technique is a highly sensitive, easy to sample, and continuous system for measuring the kinetics of biochemical transformations in soil. Sieved soil is added to a glass column and perfused with water at a cycling rate that achieves the desired saturation and aeration of the soil column. Various amendments can be added to the soil, and the perfusate is periodically analyzed for the biochemical transformations of interest (e.g., nitrogen transformations). This technique has been used to study the impact of GEMs and various environmental factors, such as sulfur dioxide, acid precipitation, and heavy metals, on ecological processes in soil (Doyle et al., 1988; Jones et al., 1988; see also Stotzky, 1974, 1986). The enumeration of added GEMs and of the indigenous microbiota and measurement of gene transfer can be accomplished by either sampling or sacrificing individual soil perfusion columns at various times, diluting, and plating the soil dilutions on selective media.

An apparatus for growing plants with the root systems maintained sterile has been described for the collection of root exudates and for measuring root respiration and the release of other volatiles from roots (Stotzky et al., 1962a). The apparatus consists of a growth tube containing the planted seed or seedling, a root chamber containing a solid growth medium (e.g., sand, vermiculite, soil), an irrigation chamber, an irrigation reservoir, an aeration system, a root exudate-collecting system, and inoculation ports. The root system is separated from the aboveground portion of the plant by a silicone seal, which eliminates the need for enclosing the whole plant. This system could be easily utilized for studying gene transfer in the rhizosphere.

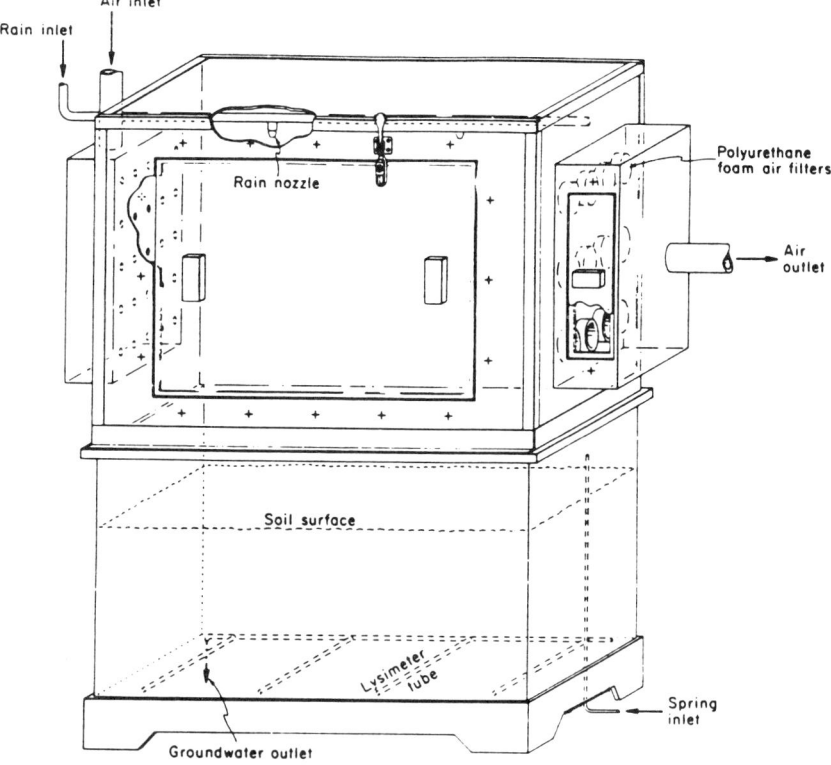

Fig. 15. A microcosm used to study the survival, fate, and genetic stability of recombinant bacteria in the rhizosphere and in herbivorous insects. Physical conditions (e.g., temperature, humidity, light–dark cycles) can be controlled in this system (Gile et al., 1982).

IV. Selection, Preparation, Maintenance, and Storage of Soils

A. Introduction

Knowledge about the structure, physicochemical composition, and microbial events in microhabitats in soil is sparse and mostly conjectural. What can be assumed with a high degree of certainty is that these microhabitats, and even sites within a single microhabitat, are heterogeneous and differ in the type and amount of clay minerals, organic and inorganic substances, and other physicochemical characteristics and, therefore, in microbial composition and activities, including the transfer of genetic information (Stotzky, 1974, 1986, 1989).

B. Selection of Soils

The selection of soils for studies on gene transfer among bacteria should be based on the relevance of the soils to projected field release sites; for example, recombinant rhizobia should be studied in representative agricultural soils on which legumes are grown commercially. A complete soil analysis should be obtained, to determine the relevance and representativeness of the soil samples to the experimental purpose and design and to correlate the survival of, and gene transfer by, the introduced GEMs with the physicochemical characteristics of the soils. The soil parameters determined should include pH, organic matter content, cation exchange capacity, water content at the -33-kPa water tension (and preferably also the WHC and the permanent wilting point, PWP), elemental analysis (including both plant nutrients and toxic heavy metals), and clay mineralogy (Babich and Stotzky, 1977). This information can most easily be obtained by sending soil samples to the soil testing laboratory of the appropriate state agricultural experiment station for analysis, or analyses can be made using the methods described in Klute (1986) and Page et al. (1982).

The baseline microbiological characteristics of a soil sample that should be obtained include numbers of total bacteria, actinomycetes, Gram-negative and Gram-positive bacteria, spore-forming bacteria, nitrifiers, denitrifiers, cellulose utilizers, fungi, and protozoa. The numbers of soil bacteria resistant to specific antimicrobial agents (e.g., antibiotics, heavy metals), especially to those agents to which the GEMs to be studied are resistant, should be evaluated. Other physiological groups of bacteria and the activity of selected enzymes (e.g., acid and alkaline phosphatases, arylsulfatases, dehydrogenases) can also be measured to provide an index of overall microbiological activity. This information is often useful, as changes in any of these parameters may provide indications of the persistence and ecological effects of the GEMs in soil.

C. Preparation of Soils

The preparation of a disturbed soil sample for use in laboratory studies on gene transfer and the ecology of GEMs begins with particle size separation. The soil is sieved (usually through a 2-mm sieve) to remove stones, gravel, roots, other debris, and large organisms. Hence, the soil particles will not vary significantly in size distribution from sample to sample, and the water content at the -33-kPa tension will be relatively constant. The sieved soil sample may then be amended in different ways, for example: with $CaCO_3$ to raise the pH, with mined clay minerals to adjust the concentration or type of clay, with water to adjust the water tension, with an inoculum of a microbiologically active soil (e.g., from the field or a flower pot) to provide an active microbiota, with nutrients (e.g., a nutrient broth or glucose plus mineral salts) to establish an active indigenous microbial population, or—usually—with some combination of these. Conditioning the soil by incubating it with a mixed-nutrient amendment (e.g., glucose plus mineral salts), in addition to sufficient water to bring the soil to its −33-kPa tension, before initiation of an experiment may be necessary if the soil has dried excessively during storage, which could result in some members of the microbial community declining in number or becoming dormant.

To determine the effects of the indigenous soil microbiota on the transfer of genetic information by the added GEMs, parallel studies should be conducted in sterile soil. The length of time for soil sterilization and the number of sterilization cycles varies with the size of the soil sample (Johnson and Curl, 1972). Sterilization with moist heat is accomplished by autoclaving the samples at 15 psi and 121°C for 30 minutes. To ensure sterilization, especially if the soil sample is >50 g, a second autoclaving should be conducted 24 hours later. This second autoclaving should kill bacterial spores, which have been heat-shocked by the first autoclaving to germinate into vegetative cells, and microorganisms protected inside soil aggregates and associated with organic matter. A sterility check should be performed by adding a nutrient medium to the soil and, after incubation for 24–48 hours, the preparation of a streak plate with a 1:10 soil–water dilution on a nonselective medium to detect the presence of micoorganisms. Sterilization with moist heat results in the formation of toxic compounds in soil, such as reduced manganese and various organic and inorganic volatiles. Therefore, the autoclaved soil samples should be maintained, preferably at room temperature and in a high-humidity chamber, for several days before inoculation of the GEMs (Stotzky, 1986). γ Irradiation is an attractive alternative technique for the sterilization of soil, as it does not result in as many changes in soil (e.g., substantial release of nutrients or toxic chemicals) as does autoclaving (McLaren et al., 1962; Stotzky and Mortensen, 1959). However, it is not as convenient as autoclaving, as it requires a source of γ rays; also, large soil samples cannot be sterilized effectively, because the penetration of γ rays into soil is poor.

D. Maintenance of Soils

The water content of the soil can be maintained at a constant −33-kPa tension during the experiment by incubating the soil samples in a high-humidity chamber or by periodically adding sterile water on a weight basis. Generally, the water content of the soil sample at the beginning of the study is adjusted to the -33-kPa water tension, as this is the water content at field capacity (i.e., the water retained by soil 24–48 hours after a rainfall or irrigation) (Cassel and Nielsen, 1986). This water tension, as well as other critical soil water contents (e.g., WHC, PWP), can be obtained from the state agricultural experiment station, as mentioned earlier.

Experiments conducted with aerobic microorganisms require adequate aeration. Hence, the soil containers should allow sufficient diffusion of air by placing closures loosely on the container, and the container should be less than one-half full to provide an adequate headspace for air above the soil sample. Alternatively, the soil can be aerated with a constant or intermittent stream of water-saturated air, which will provide both sufficient oxygen and water to maintain the soil at its -33-kPa water tension (Stotzky, 1965a).

The effects of nutrient amendment of the soil on survival and gene transfer can be studied by periodic additions of either simple (e.g., glucose) or complex (e.g., cellulose, humic materials) organic compounds or with a specific substrate utilizable only by the GEM, as with GEMs capable of degrading xenobiotics (e.g., toluene, 2,4-D). The effect of nutrient amendments on available water content, pH, and other physicochemical characteristics of the soil must be evaluated.

E. Sampling of Soils

Sampling of soil inoculated with GEMs can involve the sacrifice of individual microcosms or of multiple samplings from the same microcosm. Individual microcosms can consist of test tubes or flasks containing subsamples of the same soil that are sacrificed at different times (see Section III). Usually, triplicate or more replicates are sacrificed at each sampling time. The soil sample is commonly diluted with sterile tap water, saline (0.85% NaCl), or a phosphate buffer (pH 7) to a 1 : 2 (e.g., 10 g soil–10 ml diluent) or a 1 : 10 (e.g., 2 g soil–18 ml diluent) dilution (see Wollum, 1982, for other diluents), and the subsequent appropriate serial decade dilutions are used to inoculate either spread or pour plates containing selective media. Numerous techniques for dispersal of the microbes in the diluent have been described, and generally, each procedure is selective for a specific portion of the soil bacterial population (Casida, 1968). The efficacy of recovery of inoculated GEMs is usually dependent on the type and ionic strength of the diluent, the amount of mixing, and the temperature at which the sample is mixed (Jensen, 1968).

Methods for multiple samplings from undisturbed soil microcosms (e.g., soil cores), for studying the survival of, and gene transfer by, GEMs are presented in Section III. Multiple samplings from the same microcosms should be so performed that the soil structure of the microhabitats is not significantly disrupted, as this could affect microbial events and subsequent samplings. Microbial counts may vary with the depth in the soil core, and contamination of a soil sample from one depth with soil from other depths must be avoided.

F. Storage of Soils

Soil samples should be stored in bulk, so that repeat experiments can be conducted with the same soil samples without resampling in the field, which could result in variability in samples and microbiological responses. Bulk soil samples can be stored in open plastic bags, usually in metal or plastic garbage cans, and the water lost as a result of air drying can be replaced, along with desired nutrients, clays, or other amendments, to condition the soil before experimental use, as discussed earlier. However, the remoistening of air-dried soil can result in increases in microbial populations caused by the release of nutrients, and the soil should be allowed to incubate for at least 1 week before the initiation of an experiment. Soil samples can also be stored in thin-walled polyethylene bags that prevent the loss of moisture and allow the exchange of oxygen, carbon dioxide, and other gases. However, storage in even such bags can result in significant changes in the composition of the soil microbial community (Stotzky et al., 1962b). Storage at 4°C reduces the loss of water, but such storage probably affects the microbial composition of soil, as this temperature may not be favorable for the survival of some soil microbes and vice versa.

V. Methods for Studying Conjugation

A. In Vitro

Conjugal gene transfer in bacteria can be demonstrated in various ways, depending on the location of the genes (i.e., on the chromosome or on plasmids) that will be transferred (Curtiss, 1981). The frequency of transfer (i.e., the frequency of recombination and of expression of the genes, as this is what is actually measured) of the experimental strains should be determined in standard laboratory conjugation studies, so that the potential for gene transfer by conjugation in these strains, especially when introduced into soil, can be assessed. Moreover, this will enable meaningful comparisons between the experimental strains and well-studied and documented strains, both in vitro and in situ.

In all studies of gene transfer, both in vitro and in soil, appropriate controls must be included to determine the frequencies of spontaneous mutation, as these frequencies may differ in vitro and in soil. In vitro mutation frequencies are most easily determined during verification of the phenotypes of the donors, recipients, and plasmids. Mutation frequencies in soil are determined by inoculating the recipients and the donors, with and without the plasmids of interest, individually into soil and, after various periods of incubation, plating appropriate soil dilutions on all selective media used to enumerate donors, recipients, and recombinants. For example, a lactose-positive, Cm-sensitive recipient should be plated on MAC containing Cm at the same concentration that will allow the growth of a lactose-negative, Cm-resistant donor and of the resultant recombinants. The frequency of gene transfer must be corrected for the frequency of spontaneous mutation in the parentals.

1. Chromosomal Transfer

A model for the transfer of chromosomal genes in soil that can be used to develop studies with other bacteria is the transfer of prototrophic markers from Hfr E. coli K12 donor strains to auxotrophic F^- E. coli K12 recipient strains (Krasovsky and Stotzky, 1987; Weinberg and Stotzky, 1972). The mating type and phenotype of the E. coli strains used in these studies were: χ503, Hfr, prototrophic, Sm-sensitive (Sm^S); χ493, Hfr, prototrophic, Sm^S; and χ696, F^-, auxotrophic for leucine, proline, and arginine (Leu^-, Pro^-, Arg^-), and Sm-resistant (Sm^r) (Bachmann and Low, 1980; Curtiss and Renshaw, 1969).

a. *Media.* The type of medium (i.e., liquid or agar) used for the growth of cultures will depend on the design of the experiment and the information desired. For rapid (e.g., overnight) production of a large number of cells, as for an inoculum, shaken broth cultures are best. A solid medium should be used to distinguish metabolic traits of individual colonies, which is necessary for the enumeration and phenotypic screening of transconjugants. The preferred type of media for various procedures are described here. Commercial sources of standard media and formulations of specific media are listed in the Appendix (Section IX).

Cultures should be maintained at 4°C on agar slants [e.g., on Bacto-Penassay Agar (PA)] and transferred monthly. Some of the phenotypic markers that are commonly and most easily used to demonstrate gene transfer are unstable or revert and, therefore, can be lost from the gene pool, unless there is selection pressure for the markers. Consequently, to ensure that the resistance markers are not spontaneously lost from the cultures, the strains should be maintained on selective media (see later), or frozen suspensions should be used as the source of cells for each experiment (Lewin, 1977).

The phenotypes of all strains should be verified immediately before use in an experiment. For example, the phenotypes of the E. coli strains already described were verified and maintained on the following media: the prototrophs on minimal agar (MA); the auxotrophs on MA + 200 μg Sm, 20 μg L-leucine, 30 μg L-arginine, and 22 μg L-proline/ml; and the recombinants on MA + Sm (Curtiss, 1965; Curtiss et al., 1968; Krasovsky and Stotzky, 1987).

b. *Inoculum Preparation.* Parental strains (i.e., donors and recipients) are usually grown overnight (~18 hours) at 37°C in 125- or 250-ml Erlenmeyer flasks containing 25 or 50 ml of Bacto-Penassay Broth (PB). The period of growth for adequate production of cells depends on the generation time of the strains. The donor and recipient strains are shaken at ~120 rpm; however, the donor strain is grown in stationary culture for at least 30 minutes before mating to minimize the mechanical loss of pili. Both cultures are grown to a density of ~10^9 cells/ml, which is determined most easily by spectrophotometry at 520 nm, after correlation and standardization of absorbance units with the CFU of each strain. The cultures can be either washed (at least twice) with a suitable mating medium (e.g., minimal mating medium, 3M; LB) by centrifugation at 1000 g for 20 minutes at 4°C or just centrifuged to concentrate the cells. The cultures are then resuspended in the required amount of mating medium to yield the desired concentrations of donor and recipient cells.

c. *Mating.* An excess of recipient cells is preferable in both *in vitro* and *in situ* matings (Curtiss, 1981). Different donor–recipient ratios (ranging from 1 : 20 to 1 : 1) are obtained by varying the numbers of the recipient and donor cells added. The parental cultures can be mated in liquid or on agar media (Atlas et al., 1988; Curtiss, 1981; Krasovsky and Stotzky, 1987; Miller, 1972; Weinberg and Stotzky, 1972). Conjugation can occur with either type of medium, provided that the medium contains adequate levels of nutrients and energy to support the growth of the parentals and the transconjugants and that no growth-inhibiting substances are present. Schema for mating, for both chromosomal and plasmid transfer, in liquid and on solid surfaces are available (e.g., Curtiss, 1981; Walter et al., 1987) (Figs. 16 and 17).

The selection of either a liquid or a solid medium for the evaluation of rates of transfer of both chromosome- and plasmid-borne genes requires some knowledge of the type of pili produced by the donor. Rigid pili

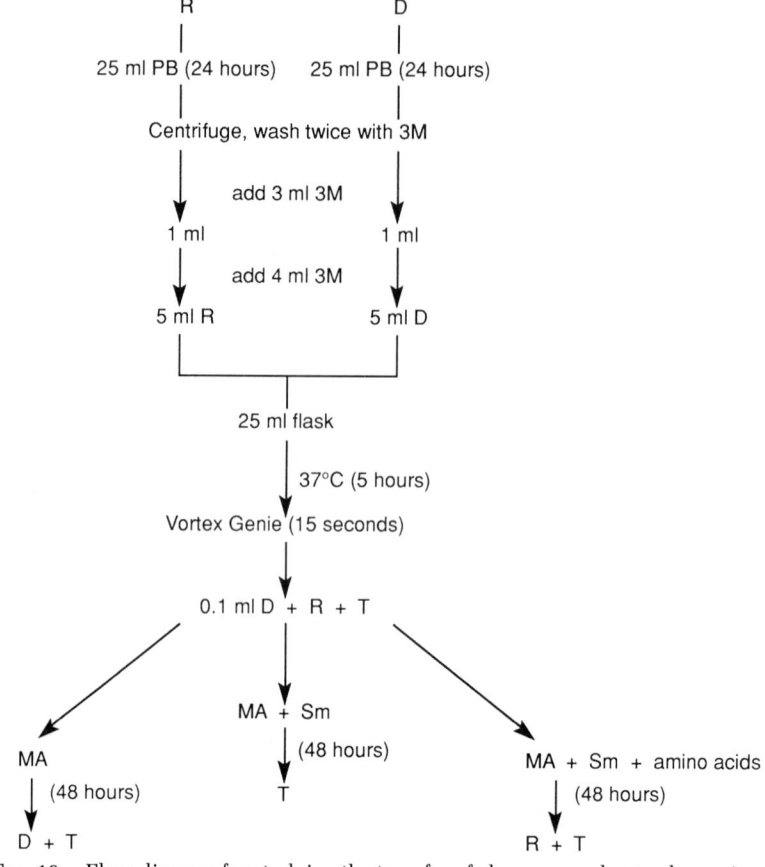

FIG. 16. Flow diagram for studying the transfer of chromosomal genes by conjugation in vitro (Krasovsky and Stotzky, 1987). R, Recipient; D, donor; T, transconjugant; PB, Bacto-Penassay broth; 3M, minimal mating medium; MA, minimal agar; Sm, streptomycin.

require a static solid environment, such as that provided by agar, whereas flexible pili usually attach to conjugative partners on a solid medium and in liquid. The pili produced by the strains of *E. coli* used herein as examples function adequately both in liquid and on solid media, and frequencies of conjugal transfer are comparable with both types of media.

For broth matings, each parental culture should be mated in at least triplicate. The environmental conditions for mating depend on the system being used; for example, an F-mediated chromosomal transfer between *E. coli* should be incubated at 37°C, without shaking, in ~10 ml of a nutrient broth contained in a 250-ml Erlenmeyer flask (Curtiss, 1981).

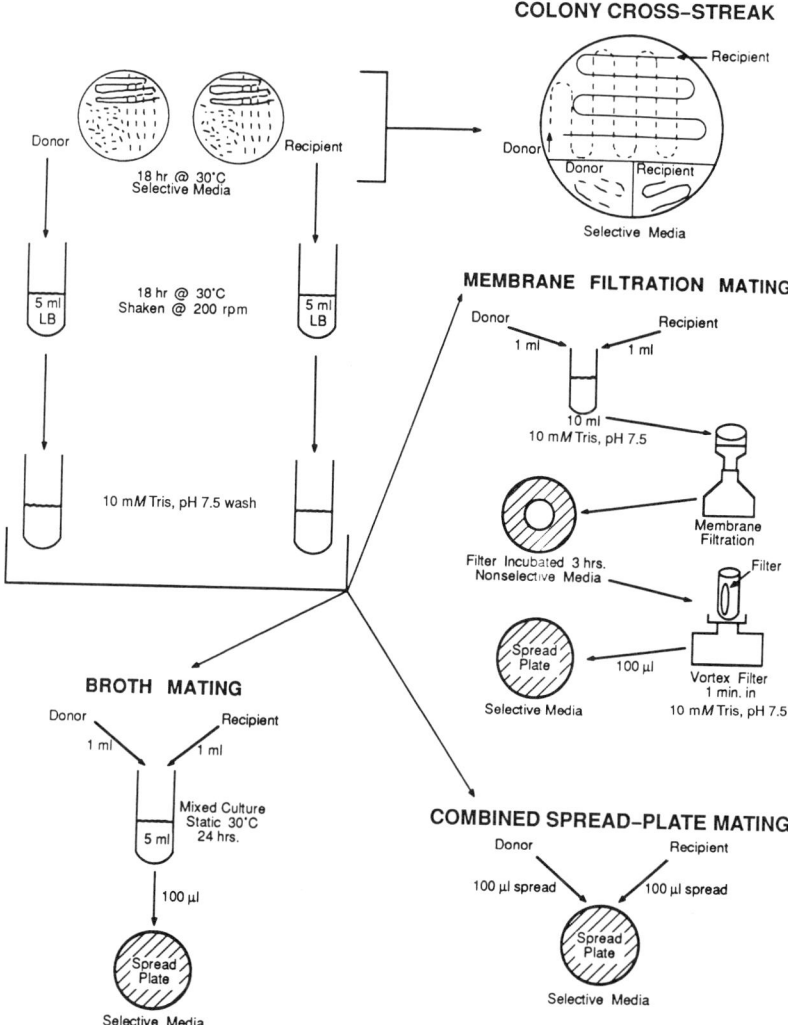

FIG. 17. Combined mating techniques for studying the transfer of genes by conjugation in vitro (Walter et al., 1987). LB, Luria broth.

The incubation time for chromosomal gene transfer can vary, and sampling at intervals for at least 100 minutes when using E. coli is recommended. Nutrient content, incubation conditions, and other experimental parameters should be optimized and standardized in vitro before experiments on in situ gene transfer are designed with specific strains.

For plate matings, parental cultures can be streaked separately and sequentially on agar or they can be mixed before placement on a solid medium, to ensure that the parental cells make contact and form conjugative pairs or aggregates (Fig. 17) (Curtiss, 1981; Walter et al., 1987). In the colony cross-streak method, the parental strains are streaked individually at right angles to each other on the surface of the solid mating medium: several streaks of one culture are streaked in one direction across the surface, and the second culture is streaked perpendicular to and crosses the streaks of the first culture. Hence, the parentals are mixed in the area where the two streaks cross. If the medium is selective for the transconjugants, neither parental will grow, and only recipient cells that conjugated and successfully received and incorporated the prototrophic markers will produce colonies. The frequency of gene transfer by this procedure is generally difficult to quantitate, as even if the numbers of each parental streaked and of the transconjugant colonies that develop are known, there is no definitive way to estimate the number of mating pairs that actually occurred in the mix of the cross-streak. If the medium is nonselective, both parentals will grow, and it will be difficult to quantitate donors, recipients, and transconjugants, as all the cells must be removed from the mating plate, diluted, and then each phenotype enumerated on selective media. Consequently, the colony-streak procedure is not recommended when quantitation of conjugation frequency is required. Conversely, it is a simple method with which to obtain transconjugants for further use in studies in soil. Moreover, it can be useful in the rapid screening of different strains for their ability to conjugate on a solid medium.

In the combined spread–plate procedure, the parentals are mixed as uniformly as possible over the surface of an agar plate. This can be achieved by mixing aliquots of each parental culture in a test tube or flask before spreading the mixture, or the parentals can be uniformly spread sequentially on the plate (Walter et al., 1987). The same limitations in the interpretation of results described for the colony cross-streak method apply to the combined spread–plate procedure.

In the membrane filter procedure, aliquots of the parental strains are mixed in a test tube or flask, and the mixture is filtered, with vacuum, through a sterile filter membrane. The parental strains are impinged on the surface of the membrane, which increases the probability of the mating pairs being in close proximity. Membranes of nitrocellulose (e.g., Millipore, Bedford, Massachusetts) or polycarbonate (e.g., Nucleopore, Pleasanton, California) with pore sizes smaller than the diameter of the test strains (e.g., 0.22 or 0.45 μm) are commonly used. The membrane is placed aseptically on a nonselective agar medium on which the parentals will grow and conjugation can occur. The membrane is then removed from the agar surface, the cells are resuspended, diluted, plated on selective media, and incubated, and the CFU of each phenotype are recorded (Curtiss, 1981). The limitations in quantitation are similar to those for the colony cross-streak and combined spread–plate methods.

d. *Incubation.* The mating mixtures should be incubated at temperatures and for durations that are optimum for conjugation of the strains. Depending on the bacteria involved, the mating mixture can be either shaken gently (30–120 rpm), to enhance aeration, or incubated without shaking in a vessel with a large surface–volume ratio to permit adequate diffusion of oxygen into the medium, for example, 10 ml of mating mixture in a 250-ml Erlenmeyer flask (Curtiss, 1981; Miller, 1972). The temperature selected for the mating depends on the optimum temperature for growth of the parental strains. For many laboratory strains, 37°C is optimum (Khalil and Gealt, 1987).

For example, for chromosomal transfer between the *E. coli* K12 strains $\chi 503$ and $\chi 696$ described earlier, the mixtures were incubated at 37°C for at least 5 hours with gentle shaking at 120 rpm (Krasovsky and Stotzky, 1987). Other strains may require different conditions of time, temperature, aeration, and nutrients. Therefore, optimal conditions for *in vitro* gene transfer should be determined for each combination of strains (Atlas *et al.*, 1988; Curtiss, 1981; Miller, 1972).

e. *Recovery of Donors, Recipients, and Transconjugants.* Conjugation can be interrupted after various periods of mating by mechanical separation of the cells, for example, on a Vortex Genie for ≥15 seconds (Miller, 1972). The time intervals at which the mating mixture is sampled for the recovery of transconjugants depend on the bacteria being studied. After the appropriate incubation period, serial decade dilutions of the mating mixture are made, and aliquots of appropriate dilutions are spread-plated on selective and nonselective agar media.

Media used for the recovery and enumeration of donor, recipient, and transconjugant populations will depend on the types of markers present in the parental strains. Auxotrophic mutants have been typically used as the recipients, and gene transfer is measured by the frequency at which a specific auxotrophic marker is replaced by the specific protrophic gene sequence in the donor. Therefore, MA, containing an agent to counterselect against the donor, can be used for isolating prototrophic transconjugants, MA can be used for isolating donors (and transconjugants), and MA supplemented with the required nutritional factors and the counterselecting agent can be used for isolating recipients (and transconjugants). For example, in the mating between *E. coli* strains $\chi 503$ and $\chi 696$ described already, MA was used for the recovery of donors and transconjugants, MA + Sm + amino acids for the recovery of recipients and transconjugants, and MA + Sm for the recovery of transconjugants (Fig. 16). Aliquots of the mating mixture were also plated on MA + Sm amended with various combinations of amino acids for which the recipient was auxotrophic, to determine the numbers of partial and complete transconjugants. Other selectable phenotypic characteristics can also be utilized (e.g., catabolic functions, resistance to heavy metals or antibiotics) for quantifying the frequency of transfer of chromosomal genes (Curtiss, 1981). The markers used for phenotype verification and strain maintenance can usually also be used for the recovery of the parental strains, but recovery of the transconjugant will require specific and unique media to detect the transferred gene(s) of interest.

f. Syntrophy. A potential artifact that can confound the enumeration of transconjugants in soil is syntrophy, or cross-feeding, a type of mutualism that enables different auxotrophic strains to grow together, but not apart, when the nutrients necessary for the growth of each auxotroph are absent or their concentrations are very low, which is not uncommon in the soil environment. Syntrophy can enable auxotrophic organisms to grow on minimal media selective for transconjugants, as the amino acids, vitamins, and other growth factors required by one auxotrophic strain may be excreted by the other strain, which can result in an exchange of nutrients under conditions that would not support the growth of either auxotroph when cultured alone. Consequently, any isolates that grow on selective minimal media must be verified to be transconjugants, rather than auxotrophic parentals that are cross-feeding on the recovery media, by streaking purified colonies of presumed transductants individually on selective minimal media or by the colony cross-streak method (Krasovsky and Stotzky, 1987; Weinberg and Stotzky, 1972).

g. Frequency of Recombination (FOR). This measure of successful conjugal transfer of chromosomal genes is calculated from the ratio of the numbers of CFU recovered on media selective for the transconjugants to the total numbers of CFU recovered on media selective for the donor (on which the transconjugants will also grow):

$$FOR = \frac{\text{CFU on TC medium}}{\text{CFU on D medium} - \text{CFU on TC medium}}$$

where TC medium is the selective medium on which only transconjugants will grow, and D medium is the selective medium on which donors and transconjugants will grow.

When an Hfr strain is mated with an F^- strain, the amount of the donor genome and the specific genes transferred depend on the duration of mating and on the location of the integration site of the F-factor (i.e., the *oriT* site), respectively. By sampling and plating cultures of mating bacteria on selective media at different times, the order of genes on the donor chromosome can be determined by the times at which they appear in the recipient (i.e., the transconjugant) and by their order and relative location (Curtiss, 1981; Miller, 1972). Genes that are located near each other on the chromosome are transferred together with a high frequency. Moreover, the frequency of chromosomal transfer of genes near the *oriT* site may be greater than that of plasmid transfer. Bacteria in soil may not conjugate long enough to transfer a large plasmid, but they probably conjugate long enough to transfer several genes located close to the *oriT* site. Conversely, if the marker genes are located too far upstream from the *oriT* site, the apparent FOR may be low, despite significant conjugation between parentals and the transfer of genes that were not detected. These principles of chromosomal gene transfer must be considered when designing studies to observe the potential for gene transfer by conjugation in the environment.

2. *Plasmid Transfer*

Many of the procedures used for the study of chromosomal transfer can also be used to study plasmid transfer by conjugation. Host strains that will maintain the plasmid(s) being studied, before and after conjugation, are required. The strains should not contain other plasmids that may interfere with conjugation, either by surface exclusion (i.e., prevention of "superinfection" by plasmids of the same type) or by competing during cell division for membrane-bound segregation systems that provide for maintenance and inheritance (i.e., incompatibility) (Freifelder, 1987; Lewin, 1977). Ideally, the strains should be devoid of plasmids other than the ones being studied, as shown by phenotypic expression and by the inability to detect other plasmid DNA by agarose gel electrophoresis after extraction of plasmid DNA by standard procedures (Curtiss, 1981; Maniatis *et al.*, 1982; Miller, 1972) (see Section VII). Selective enumeration of donors, recipients, and exconjugants is facilitated if the chromosome of the donor and recipient strains is "tagged," that is, contains antibiotic-resistance marker(s) or some other easily characterizable phenotype, such as the production of a red color by lactose fermentors on MAC (Atlas *et al.*, 1988; Curtiss, 1981; Devanas *et al.*, 1986; Miller, 1972) or a blue color on X-gal agar by cells containing the ZY genes of the lactose operon (Drahos *et al.*, 1986). The *lacZY* genes can also be used as plasmid markers.

The choice of plasmids to use experimentally will depend on the type of system to be studied and the type of information desired. The enumeration of donors and exconjugants is facilitated if the plasmid encodes catabolic pathways, resistance markers to antibiotics and/or heavy metals, or some other easily characterizable phenotype (Maniatis et al., 1982; Miller, 1972).

a. *Media.* The media used for the maintenance of the hosts, with or without plasmids, and for mating and recovery of the donors, recipients, and exconjugants depend on the nutritional requirements of the bacterial strains. Most bacteria will grow adequately on standard laboratory media, such as nutrient broth (NB), tryptone glucose yeast extract broth (TGYB), and LB, and on agars with the same nutritional composition. Environmental isolates, however, may require growth factors that are not present in standard commercial media. Such fastidious strains can usually be maintained on media supplemented with extracts of soil or with specific nutrients (see Appendix, Section IX). In contrast, many commercial media may be too nutritious for indigenous soil bacteria and, possibly, also for introduced bacteria after they have adapted to the oligotrophic conditions in soil.

Slower growth and, hence, fewer cells occur on agar than in liquid media, thereby reducing the number of mutants that develop in the cultures and extending the intervals between necessary transfers of the cultures to fresh medium. A solid medium is also better if resistance markers must be maintained by a constant selection pressure, as the destruction or alteration of the selective chemical (e.g., an antibiotic) is less than in liquid media, as the result of the slower rates of diffusion of degradative extracellular enzymes from a resistant colony on agar. Liquid culture is generally better for obtaining large numbers of bacteria in a short period of time (e.g., in 18 hours), which is important for the preparation of inocula.

Bacteria will usually not maintain extrachromosomal genetic information that is not used (i.e., that is not induced to transcribe). For example, plasmids that encode resistance to antibiotics (R-plasmids) will not be replicated, segregated, or maintained after several transfers if there is no selection pressure of the respective antibiotic(s) in the medium. Bacterial strains containing R-plasmids are best maintained on a selective agar medium that contains at least one of the antibiotics to which the plasmid encodes resistance. If the plasmid is especially unstable and the loss of plasmid genes is likely, several or all of the compounds (e.g., antibiotics, heavy metals, substrates for degradation) for which the plasmid encodes resistance or utilization should be included in the selective maintenance medium. For example, an *E. coli* strain that carries the plasmid RP4 is maintained on LA containing

25 μg/ml Tc. When the host is a strain of Pseudomonas, which are generally more resistant to many antibiotics than most members of the Enterobacteriaceae, the basal resistance level to specific antibiotics of the plasmidless host must be determined and the selective antibiotics (e.g., Tc) added at higher concentrations for both maintenance of the plasmid and selection of exconjugants. For example, Ps. aeruginosa PAO1 is resistant to Tc at concentrations >100 μg/ml, and the quantity of Tc necessary for maintenance and selection of plasmid RP4 in this strain is 200 μg/ml.

b. *Inoculum Preparation.* Strains can be grown in a nonselective broth to obtain large numbers of cells. However, if the plasmid phenotype is unstable and a significant proportion of the cells is likely to lose the plasmid when grown under nonselective conditions, then selective pressures must be maintained. In either situation, the phenotype of the donor and recipient strains must be verified before use. One method to maintain selection for the plasmid markers is to grow the strain overnight on a selective agar medium. For example, cells of E. coli containing plasmid RP4 were grown on MAC + 25 μg/ml Tc, scraped, diluted with saline to 10^4 CFU/ml, of which 0.25 ml was added to 25 ml of LB to yield a density of 10^3 CFU/ml, and grown overnight at 37°C with shaking (Devanas et al., 1986). This produced a suspension of cells that contained the plasmid, as all cells grew on the selective medium. Consequently, loss of plasmids that could result from their partitioning (segregation) during growth under nonselective conditions can be minimized. If the donor strain readily loses the plasmid under nonselective conditions, suspensions of cells scraped directly from a selective medium can be used. The concentration, washing, diluting, and inoculation procedures are as described before for chromosomal transfer.

c. *Mating.* The conditions selected for mating will vary with the type of system to be studied. For example, F-factors, their derivatives, and other plasmids derepressed for transfer are transferred readily at high frequencies (~100%); other conjugative plasmids have lower frequencies (10^{-2}–10^{-7}) (Lewin, 1977). Various proportions of donor to recipient cell numbers should be evaluated, as the optimum donor–recipient ratio for plasmid transfer depends on the species and strains used (Curtiss, 1981; Miller, 1972). This ratio should be determined in preliminary studies. In the model systems of E. coli J53(RP4) as the donor strain and Ps. aeruginosa PAO1, E. aerogenes, K. pneumoniae, and P. vulgaris as the recipient strains, the overnight broth cultures (usually 10^9 CFU/ml) were diluted with sterile saline to cell densities of ~10^6 CFU/ml and then added to the mating medium (LB) in a donor–recipient ratio of 1:5, which ensured that each potential donor was likely to encounter a suitable recipient (Devanas and Stotzky, 1988a). Successful mating and transfer may occur within 1–4 hours with plasmids derepressed for transfer. For other plasmids, longer periods of mating may be necessary. The mating mixture is then diluted, and samples are plated on appropriate selective agar media for recovery of donors, recipients, and exconjugants, as described before for transfer of chromosomal genes.

d. *Calculation of the Frequency of Conjugation (FOC).* The frequency of plasmid transfer by conjugation is calculated by dividing the number of exconjugants by the number of donors in the mating mixture (Curtiss, 1981):

$$FOC = \frac{CFU \text{ of exconjugants}}{CFU \text{ of donors}}$$

B. IN SOIL

1. *Chromosomal Transfer*

The design of studies to examine the potential for conjugal transfer of chromosomal genes in soil must consider all the variables that affect chromosomal transfer *in vitro*, as well as the physicochemical factors inherent in the soil environment. The selection of donor and recipient strains and the methods used for their maintenance, culture, and preparation as inocula are essentially the same as those employed for *in vitro* studies. Consequently, only additional aspects unique to studies of *in situ* chromosomal gene transfer are described.

a. *Inoculation.* The use of donor and recipient cells that are in log phase appears to enhance conjugal chromosomal transfer in soil (Weinberg and Stotzky, 1972). One method for the preparation of log phase cells for inocula uses cells that were grown overnight, washed by centrifugation (two to three times at 1000 g), followed by resuspension of the cells of each strain in a phosphate buffer, usually at pH 7.

The sequence of inoculation of the donor and recipient cells has been shown to affect their survival and the frequency of conjugation (Krasovsky and Stotzky, 1987; Weinberg and Stotzky, 1972). The survival of both donors and recipients was higher when the recipient cells were inoculated before the donor cells than when the donors were added before the recipients. When the donor and recipient cells were inoculated together (i.e., mixed before addition to soil), their survival and the FOR were greater than when they were inoculated sequentially (Krasovsky and Stotzky, 1987). However, mixing of the parentals before their addition to soil may result in mating-pair formation outside of the soil, thereby producing artifactually inflated frequencies in soil. Hence, mixing donors and recipients before addition to soil is not recommended for studies designed to observe the transfer of chromosomal genes in soil.

b. *Incubation.* Soils used for *in situ* mating experiments should be incubated under conditions that maintain the moisture and temperature conditions defined for the experiment. To maintain the optimum soil water tension (i.e., -33 kPa), high-humidity chambers can be used (Stotzky, 1986). Large glass containers (e.g., chromatography tanks, desiccators) containing 1–2 inches of water are covered with lids that have a vent that is loosely plugged with nonabsorbent cotton. This arrangement produces a water-saturated atmosphere that maintains the optimum soil water tension. Temperature ranges of natural soils can be simulated by placing the high-humidity chambers in incubators maintained at the desired temperatures.

c. *Sampling.* The choice of sampling regimes will depend on the type of microcosm used and on the purpose of the experiment. Simple test tube microcosms, in which the entire soil sample is "sacrificed," reduce the sampling error associated with repeated replicate samplings from larger batch volumes. However, sampling error can be introduced at the level of preparation and inoculation of the individual tubes. Hence, precise weighing of soil and uniform inoculation are necessary. Different types of soil microcosms and sampling systems are discussed in Section III.

The samples of soil are subjected to serial decade dilution with standard laboratory diluents, such as sterile distilled or tap water, saline, buffers, or 0.1% peptone. The range of dilutions suitable for donors, recipients, and recombinants will depend on the cell densities added. Dilutions from 10^{-5} to 10^{-7} should be suitable for the enumeration of the indigenous soil microbiota. The types and numbers of indigenous microbes present in the soils being studied should be evaluated before any experimental strains are introduced, so that observations made after introduction of the strains, which may alter the profile of indigenous microbes, can be compared with these baseline measurements. Aliquots of the dilutions of sample are plated on appropriate selective media, incubated, and the CFU recorded. A tared sample of the soil is dried for 24 hours at 100°C, and the CFU are expressed and compared on the basis of 1 g of oven-dried soil.

d. *Recovery of Donors, Recipients, and Transconjugants.* The appropriate selective media are based on the phenotypic markers of the donor, recipient, and transconjugant strains, as in *in vitro* studies. In studies with nonsterile soil, the media used for the isolation of donors, recipients, and transconjugants may not be selective enough to eliminate or reduce sufficiently the indigenous soil microbiota to enable isolation and distinction of the introduced strains. To suppress the indigenous populations of fungi, 50–200 μg/ml cycloheximide (Cy) should be added to the recovery media. The concentration of Cy needed will depend on the type of soil and on the numbers and types of fungi present. Indigenous populations of bacteria may be counterselected by the antibiotics used to isolate the parentals or by a combination of antibiotics to which most of the indigenous populations, but not the parentals, are sensitive (e.g., a mixture of Sm, Ap, Nx, and Rf). Media that suppress indigenous Gram-positive bacteria can be formulated based on the phenotype of the experimental Gram-negative strains, as in the case of E. coli χ503 and χ696, which are tolerant of 400 μg/ml eosin and 65 μg/ml methylene blue, whereas Gram-positive bacteria are not (Krasovsky and Stotzky, 1987). The concentrations of the various antimicrobial agents necessary to suppress the indigenous microbiota will vary with different soils and should be determined in preliminary studies.

To distinguish between transconjugants that resulted from chromosomal transfer (as well as exconjugants that resulted from plasmid transfer, see later) in soil from those that may have resulted from matings between donors and recipients on the recovery medium, Nx is incorporated into the selective medium, which will inhibit gene transfer by Nx-sensitive donors (Curtiss, 1981; Walter et al., 1989b) (see Section VII).

e. Calculation of Frequency of Recombination (FOR). The FOR is calculated by dividing the number of transconjugants by the number of donors isolated from the soil (see Section V,A,1,g).

2. Plasmid Transfer

Studies to measure conjugal transfer of plasmid genes in soil require careful design and selection of plasmids and of donor and recipient strains, as well as extensive controls for both gene transfer and environmental variables. The nutrient content of the soil, the method of inoculation of the parental strains into the soil, the recovery of donor, recipient, and exconjugants from the soil, and the transfer of the plasmid to indigenous bacteria must also be considered.

a. Inoculation. The sequential addition of donor and recipient cells may affect the survival of the parental strains and the potential for gene transfer, particularly if one mating partner is not added for some time after the addition of the other partner. For example, preincubation of the recipient, Ps. aeruginosa PAO1, in soil did not appear to have any significant effect on its survival, whereas the donor, *E. coli* J53(RP4), declined rapidly after addition to soil. Gene transfer occurred immediately after log phase donor cells were added to preincubated recipients (Fig. 18), whereas exconjugants were recovered only after 1–3 weeks of incubation of the soil when log phase recipients were added to preincubated donors (Fig. 19) (Devanas and Stotzky, 1988b).

The presence of nutrients, whether introduced with the inocula or as amendments during the incubation, affect the survival and potential for gene transfer in soil (Devanas et al., 1986). Studies on survival, recovery, and gene transfer in soil should be designed to include unamended soil as well as soil amended with nutrients.

b. *Recovery of Donors, Recipients, and Exconjugants.* The soil is sampled the same as for chromosomal transfer, and the procedures for

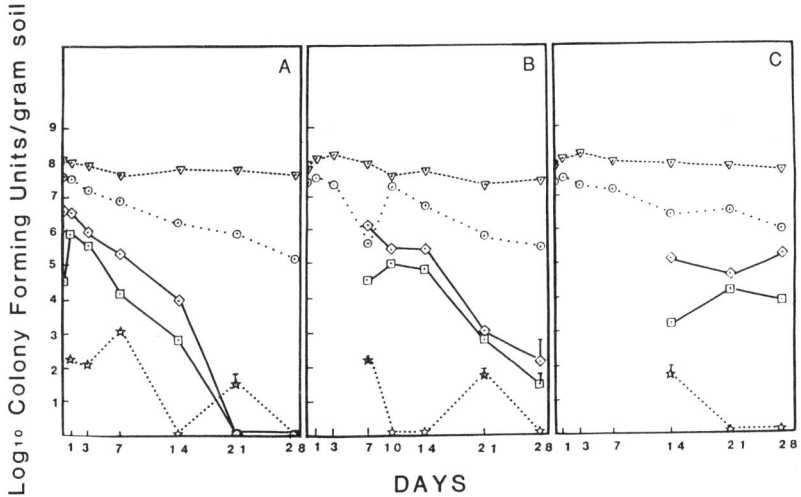

FIG. 18. Effect of preincubation of recipient *Pseudomonas aeruginosa* PAO1 cells (○) on the survival and transfer of plasmid RP4 from donor *Escherichia coli* J53(RP4) cells (◇, □) in nonsterile soil. All cells were inoculated in saline. Donor cells were added (A) immediately (i.e., on day 0) or (B, C) after the soil had been incubated with recipient cells for (B) 7 or (C) 14 days. *Escherichia coli* J53(RP4) was enumerated on MacConkey agar (MAC) containing 25 μg/ml tetracycline (Tc); *Ps. aeruginosa* PAO1 on Pseudomonas isolation agar (PIA); and PAO1(RP4) (☆) on PIA containing 200 μg/ml Tc. Total bacteria (∇). When the CFU of J53(RP4) on MAC + Tc (□) were significantly lower than the CFU of *E. coli* J53 on MAC (◇), selected colonies from the MAC plates were transferred to MAC + Tc. All transfers grew on MAC + Tc, indicating that the plasmid was not selectively lost from the J53(RP4) population but that the stress in soil produced a "viable but nonculturable" population of J53(RP4) that did not grow on MAC + Tc when isolated directly from soil (Devanas and Stotzky, 1988b).

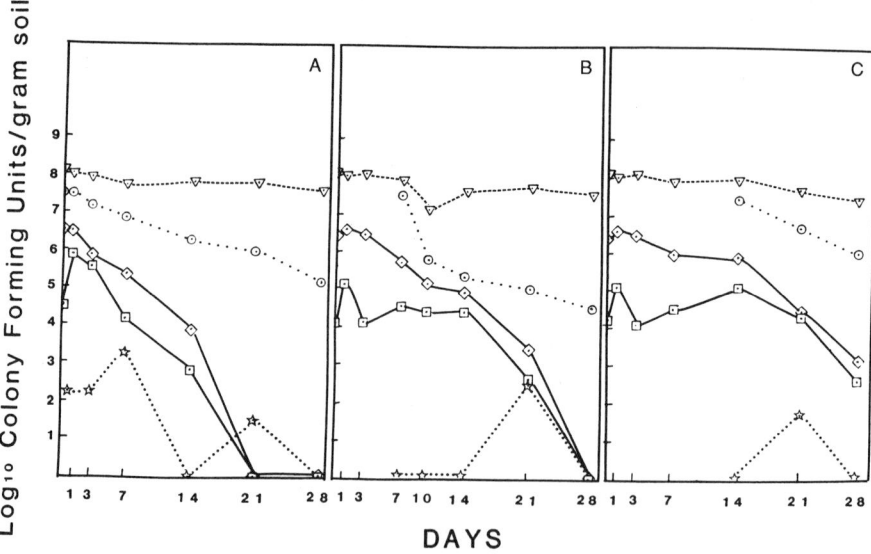

FIG. 19. Effect of preincubation of donor *Escherichia coli* J53(RP4) cells (◇, □) on the survival and transfer of plasmid RP4 to recipient *Pseudomonas aeruginosa* PAO1 cells (○) in nonsterile soil. All cells were inoculated in saline. Recipient cells were added (A) immediately (i.e., on day 0) or (B, C) after the soil had been incubated with donor cells for (B) 7 or (C) 14 days. Total bacteria (▽) and PAO1(RP4) (☆). (See Fig. 18 for details.) (Devanas and Stotzky, 1988b.)

the recovery of donors, recipients, and exconjugants are the same as those for in vitro plasmid transfer. However, the recovery of some strains from soil may sometimes be reduced, as the stress of being in soil may sufficiently debilitate these strains and affect their subsequent recovery. This phenomenon of "viable but nonculturable" has been described for bacteria in various aquatic systems (Roszak et al., 1984; Zaske et al., 1980). A similar phenomenon appears also to occur in soil, as shown by the differences between the CFU of some strains recovered on highly selective media (e.g., MAC + Tc) and on less stressful selective media (e.g., MAC) (Devanas et al., 1986). A number (50–100) of the colonies recovered on the less stressful media should be transferred to the more stringent selective media to determine if the differences in the proportion of CFU recovered by direct isolation on each medium are real or if the reduced numbers that developed on highly stressful recovery media were the result of a "viable but nonculturable" condition (see Figs. 18 and 19). The stress of being in soils may also retard the rate of growth of cells even on nonselective and nonstressful media. Consequently, repeated observation of the recovery plates, marking the colonies as they appear, and reincubating the plates until no new colonies appear may be necessary.

c. *Transfer to Indigenous Bacteria.* Gene transfer to indigenous bacteria is generally difficult to demonstrate, as selection for gene transfer to these bacteria is based solely on the presence of the transferred marker genes. All colonies that appear on media selective for exconjugants must be tested for the presence of the marker genes by phenotypic (e.g., morphology, Gram stain, biochemical characteristics) and physical (e.g., DNA extraction, agarose gel electrophoresis, DNA fingerprinting, DNA probes) methods (see Section VII). Such analyses should clarify whether the exconjugant isolated is the added recipient or an indigenous soil bacterium to which the introduced genes have been transferred. The intrinsic resistance of the indigenous soil bacteria to the antimicrobial agents used to evaluate the transfer of marker genes must be determined by plating appropriate soil dilutions of a control (i.e., uninoculated) soil sample on the various selective media used.

d. *Limits of Detection of Gene Transfer.* There are always limits to the detection of microbial populations in any natural environment. These populations may (1) not grow on any of the recovery media, (2) be present in such low numbers that unrealistically large volumes must be sampled, (3) adhere tightly to surfaces and not be removed for recovery by the diluents in serial dilution, and (4) be obscured or overgrown on the recovery media by faster-growing strains. All these problems occur in soil and make the detection of introduced GEMs and, especially, of any recombinants that may result from gene transfer difficult. The lower limit of the numbers of CFU detectable in soil samples is a function of the size of the soil sample, the ratio of the initial dilution of the soil sample (e.g., 1:10), and the volume of the dilution plated. For example, the recovery of 1 CFU when 0.1 ml of the 10^{-1} dilution of soil is plated represents a lower detection limit of 100 CFU/g soil. The limit of detection of a recombinant bacterium in a sample of soil is generally in the range of 1–20 CFU/g soil, and the level of confidence in this range is a function of the number of replicate soil samples evaluated and the number of CFU of the recombinants recovered (Devanas et al., 1986).

C. Stepwise Summary of Procedures for Studying Conjugation in Soil

Prepare soil (e.g., sieve; amend with clays, $CaCO_3$, etc.; condition; add to appropriate microcosm).
Prepare suspensions of donor and recipient cells.
Wash suspensions by dilution or centrifugation.
Adjust concentration of donor and recipient cells for appropriate conjugation ratio (1:1–1:20).

Add donor and recipient cells sequentially to soil with sufficient water (or a nutrient solution) to bring the soil to its −33-kPa water tension.

Incubate soil, preferably in a high-humidity chamber.

Sample soil periodically (either by sacrificing multiple microcosms or from a batch system).

Prepare appropriate serial decade dilutions of the soil.

Plate dilutions on selective media suitable for growth of donor, recipient, and recombinant cells, as well as of indigenous soil bacteria.

Incubate plates, and record numbers of CFU that develop on each medium.

Verify presumptive donor, recipient, and recombinant cells for phenotype and/or genotype (see Section VII).

VI. Methods for Studying Transduction

A. In Vitro

1. Preparation of Bacteriophage Lysates

Two methods for preparing phage lysates of high titer are commonly used, depending on the phage–host system that is being studied. The first method involves inoculating a liquid culture of a susceptible host bacterium with an appropriate phage or inducing (e.g., by heat, UV radiation, chemical mutagens) a culture of a lysogenic bacterium. An example of the use of a lysogenic culture is described for the preparation of lysates of phage P1. Examples of other methods for preparing phage lysates in liquid culture can be found in Adams (1959).

Phage P1 Cm cts contains a temperature-sensitive mutation that enables the induction of the lytic cycle at temperatures above (e.g., 42°C) those optimal for growth of the host bacterium, E. coli (e.g., 37°C). Escherichia coli AB1157, lysogenic for phage P1 and stored on LA slants containing 30 μg/ml Cm to maintain the temperate phage P1 DNA in the bacteria, is grown at 30°C in LCB to an optical density of 0.2 (at 600 nm) and shaken at 42°C for at least 2 hours until lysis is evident (30°C is used for growth, to reduce spontaneous induction of lytic reproduction as the result of the temperature-sensitive mutation). Chloroform (0.5 ml/100 ml of culture) is added, with mixing, for 1 minute to lyse cells and release additional phage virions, and the remaining intact cells are removed by filtration through a 0.45-μm nitrocellulose filter membrane (Millipore). The phage particles are stored in LCB at 4°C. Phage P1 titers between 10^8 and 10^{10} PFU/ml are commonly obtained.

An alternative method, which is more suitable with certain phages for the preparation of lysates of sufficient titer, is the agar-overlay method. An example of this technique, used with the generalized transducing phage, F116L, that infects Ps. aeruginosa, is described (Miller and Ku, 1978). Although phage F116L does not contain a temperature-sensitive mutation, many lysogenic bacteria can be induced to lysis by growth at temperatures above their optimum, as the repressor protein that is involved in the maintenance of lysogeny will be impaired in activity or inactivated by such higher temperatures. Phage F116L (10^5 PFU in 0.1 ml) and Ps. aeruginosa PAO1 (10^6 CFU in 0.1 ml) are mixed in 2.5 ml of melted LA (top agar), to give a multiplicity of infection (MOI; ratio of the number of phages to the number of host cells) of 0.1, and poured on the surface of LA plates (bottom agar). After incubation overnight (~18 hours) at 37°C, the top agar layer is removed in 5 ml of LB, the suspension is vortexed for 1 minute and centrifuged for 10 minutes at 5000 g to pellet the bacteria and agar, and the supernatant is filtered through a 0.45-μm nitrocellulose membrane to remove remaining bacterial cells. The phage F116L stocks (10^8–10^{12} PFU/ml) are stored at 4°C.

The lysates can be purified of bacterial cell debris and soluble cytoplasmic components by various techniques, such as differential centrifugation on cesium chloride (CsCl) gradients or polyethylene glycol (Miller, 1972). However, the removal of such contaminants is seldom necessary or recommended for studies that are intended to simulate *in situ* situations.

2. Preparation of Bacterial Cultures

The bacterial cultures used as hosts for *in vitro* transduction should be grown to log phase in liquid culture (e.g., in LCB), and the optimum MOI for the specific host–phage system, determined in preliminary experiments, should be used. For transduction to *E. coli* of resistance to Cm by phage P1 Cm cts, an MOI between 1 and 5 is typically used with 10^7–10^8 CFU/ml of *E. coli*.

3. Transduction Procedure

Transduction experiments *in vitro* should be conducted in liquid rather than on solid medium, to ensure adequate mixing of the phage and bacteria during the adsorption period. Phage P1 is allowed to adsorb on *E. coli* for 30 minutes at 30°C without shaking in LCB. The presence of Ca^{2+} and Mg^{2+} in this medium promotes the adsorption of phage P1 on the bacteria. The adsorption period is restricted to 30 minutes to prevent both the phage from completing a reproduction cycle and initiating a burst and the transduced *E. coli* cells from multiplying. After the

30-minute adsorption period, E. coli transductants are enumerated by plating serial decade dilutions of the bacterium–phage suspension on either LA or MAC containing 30 µg/ml Cm. The E. coli inoculum is also plated on this medium, to determine the number of spontaneous mutants to Cm resistance.

Quantitation of the bacterial and phage inocula should be conducted to determine the exact MOI in the transduction mixture. Dilutions of the E. coli suspension are plated on LA, and the plates are incubated at 30°C until colonies appear. Phage P1 is titered on LCA bottom and top agar at 42°C, and plaques are counted after 24 hours.

The frequency of transduction is expressed as the number of transductants divided by the number of bacteria originally present in the transduction mixture. The number of phages added can be used as the denominator instead of the number of bacteria, to give a transduction frequency related to the total number of phage particles.

B. In Soil

The design of transduction experiments in soil must consider the complex nature of soil and the fact that studies will be conducted over a period of several days or longer, thereby making it difficult to calculate the frequency of transduction. The major factors to consider are the MOI, the use of either a phage lysate or a lysogenic bacterium as the source of the transducing phage, whether the recipient bacteria should to be added to soil after washing (e.g., in saline) or with nutrients (e.g., LB), and the methods used to enumerate selectively the added bacteria, transductants, and phages.

1. Inoculation and Amendment of Soil

The optimum MOI in soil should be determined before extensive experiments are designed, as it may differ from that *in vitro*. The number of phages added should be appropriately adjusted to the number of bacteria added to the soil. In studies of transduction by lysates of phage P1 in soil, 10^5–10^6 CFU/g soil of recipient E. coli J53(RP4) or W3110(R702) are typically added, to avoid adding unrealistically high numbers of bacteria, and the MOI is ~3. *Escherichia coli* and phage P1 are added individually in 0.1 ml of LCB or saline, so as to bring the final wet weight of the soil to 2 g (Zeph et al., 1988). When lysogenic strains of E. coli are used as the source of phage P1, the same numbers of lysogenic and recipient E. coli cells are added. Control studies with soil inoculated with the same numbers of only phage P1 or only E. coli donors or recipients must be conducted, to determine both survival and mutation rates.

The addition of nutrients can be achieved most easily by diluting the host and transducing phage in fresh LCB before their addition to soil. Amendment of the soil with LB or other nutrient solutions can be made periodically, to measure the effects of nutrients on transduction and on the multiplication and survival of the phages and host cells. The potential for transduction in the absence of exogenous nutrients can be determined most easily by diluting the host and phage inocula in saline before adding them to soil. For example, inocula of $E.\ coli$ and phage P1 (10^8–10^9 CFU or PFU/ml) raised in LCB are diluted 1:1000 in saline, and a final inoculum of 10^5–10^6 CFU or PFU/g soil is added. Concentration and washing in saline of the host cells can be accomplished simultaneously by centrifugation at 5000 g for 10 minutes. However, because the viability of the inoculum could be affected by centrifugation, viable counts of the washed suspensions should be determined.

2. *Incubation and Sampling*

Soil samples containing phage and bacteria should be incubated at ~25°C (the average temperature of soils in temperate zones) and maintained at their −33-kPa water tension in a high-humidity incubator and on the basis of periodic weight measurements and the addition of sterile distilled water (sdH$_2$O) when necessary. Three to four samplings of the soil for the enumeration of recipient and donor bacteria, phage, and transductants should be made during the first week (e.g., on days 0, 1, 3, and 7), and then sampling should be at least weekly until there are no significant changes in these populations. If sufficient personnel and funds are available, the numbers of total bacteria, fungi, protozoa, and other indicators of species diversity should be evaluated concurrently.

In studies on the transduction of $E.\ coli$ by phage P1 in soil, experiments were conducted for 28 days (Zeph et al., 1988). Individual soil tubes were sacrificed at each sampling by the addition of 18 ml of sdH$_2$O to duplicate or triplicate soil tubes that contained 2 g soil (i.e., a 1:10 dilution), and subsequent serial decade dilutions were spread-plated on selective media. Plates for the enumeration of $E.\ coli$ donors, recipients, and transductants were incubated at 30°C, for phage P1 at 42°C, and for total and Gram-negative indigenous soil bacteria at 25°C.

3. *Enumeration*

The efficacy of the selective media used for the enumeration of the various groups of bacteria (i.e., transductants, donors, recipients, and indigenous) in nonsterile soil is an important parameter to establish before the start of an experiment, especially in that the numbers of bacteria transduced in soil can be relatively low. In studies on the transduction by phage P1 Cm cts::Tn501 of Cm and Hg resistance to E.

coli, the resistance markers on the phage genome were utilized to monitor the number of transduced *E. coli* in soil, and the antibiotic resistances coded on nonphage plasmids enabled following the fate of the recipient or donor (lysogenic) *E. coli* populations. *Escherichia coli* transductants were enumerated on MAC containing Cm (90 μg/ml; MAC–Cm) or Hg (20 or 30 μM; MAC–Hg), because the lactose-utilizing strains of *E. coli* form distinctive dark red colonies on MAC. Recipient *E. coli* J53(RP4) and W3110(R702) were enumerated on MAC containing Tc (25 μg/ml; MAC–Tc), resistance to which is encoded on the plasmids.

In studies in nonsterile soil with lysogenic *E. coli* donors and nonlysogenic recipients, the donor *E. coli* J53(P1 Cm cts) was enumerated on MAC–Cm, the recipient *E. coli* W3110(R702) on MAC–Tc, and the transduced *E. coli* W3110(R702)(P1 Cm cts) on MAC containing both Cm and Tc at the concentrations shown before. The lowest limit of detection of *E. coli* donors, recipients, and transductants in nonsterile soil was 10^2–10^3 CFU/g soil, as growth of indigenous soil bacteria resistant to the concentrations of Cm, Tc, and Hg used interfered with accurate detection of lower numbers of *E. coli* (Zeph et al., 1988). The problem of interference by indigenous soil bacteria resistant to antimicrobial agents is discussed later.

When lysogenic *E. coli* J53(RP4)(P1 Cm cts::Tn501) donors and nonlysogenic *E. coli* W3110 were added to sterile soil, Hg-resistant transductants of *E. coli* W3110 [i.e., W3110(P1)] were enumerated on M9 minimal agar containing 30 μM Hg (M9M). The auxotrophic *E. coli* J53(RP4)(P1) donor does not grow on minimal agar, and it was enumerated on MAC–Tc. MacConkey agar was used to enumerate total *E. coli* [*E. coli* J3(RP4)(P1), W3110, and W3110(P1)], which enabled the numbers of *E. coli* W3110 to be calculated from the difference between the total *E. coli* and the *E. coli* J53(RP4)(P1) plus W3110(P1) counts.

Auxotrophic recipients can also be utilized to study transduction in soil. For example, a Sm-resistant *E. coli* K12 recipient that was auxotrophic for threonine and leucine was used to study transduction by phage P1 in nonsterile soil (Germida and Khachatourians, 1988). Lysates of phage P1 were raised on prototrophic *E. coli*. The minimal salts agar medium (MSA), which contains Sm, Cy to suppress fungi, and eosin and methylene blue to suppress the growth of bacteria other than enterobacteria, was satisfactory for the isolation of *E. coli* transductants prototrophic for threonine and leucine and resistant to Sm.

Minimal media can also be used to counterselect an auxotrophic *E. coli* lysogen, as discussed earlier (Zeph et al., 1988). Nevertheless, the use of markers for resistance to antimicrobial agents is more efficient for monitoring transduction in soil, as many soil bacteria will often grow more readily on minimal media than on media containing antimicrobial agents. Moreover, when using a minimal medium to select for transductants, incubation of the plates for >3 or 4 days can result in the syntrophic growth of the auxotroph and the formation of background lawns on the plates. This makes it difficult to differentiate between cells that have been transduced to prototrophy and cells of the auxotrophic recipient.

4. Detection of Transductants in Nonsterile Soil

The detection of transductants in nonsterile soil is difficult, as there are no efficient methods for the enumeration of low levels of transduced bacteria. Chromosomally borne genes for resistance to Nx and other antibiotics or to heavy metals can be exploited in the isolation of a host bacterium containing either or both of these groups of genes. However, isolation of specific hosts based on resistance, either chromosomal or vector-borne, to many commonly used antibiotics (e.g., Sm, Cm, Tc) or some heavy metals (e.g., Hg) has limitations, as many bacteria in many soils now exhibit resistance to these antimicrobial agents. The limit of detection for most methods is, at best, between 10^2 and 10^3 transductants/g soil, and it is often several orders of magnitude higher. In some soils, the "background" resistance to many common antimicrobial agents can be as high as 50% of the isolatable bacteria. Increasing the level of resistance in the host bacterium through mutation and selection or concurrently using a second marker for antibiotic or heavy-metal resistance often results in reduced recovery, as some bacteria are not able to express readily the antimicrobial-resistance phenotype after exposure to the stressful conditions in soil (Devanas et al., 1986; Pettibone et al., 1987). Furthermore, resistance to antimicrobial agents may be lost if the resistance genes are located on a plasmid that may disappear through segregation during the period of the soil study. A "viable but nonculturable" physiological state, in which the cells remain viable but are so debilitated that they will not form colonies on selective media (e.g., Devanas et al., 1986; Roszak and Colwell, 1987), has been suggested as one cause of the poor recovery of some introduced bacteria from soil. However, by initially plating the soil dilutions on nonselective media, to allow "resuscitation" of the stressed bacteria, and then replica-plating to selective media, specific donors, recipients, and transductants can be enumerated (see Section VII).

5. Recovery of Phages from Soil

Difficulties can be encountered in the recovery of phages from soil, because adsorption of phages on soil particulates, especially on clay minerals, can result in apparent decreases in phage titers in soil. For example, phage P1 adsorbs strongly on soil components, probably on the clay fraction; >95% of the added phage P1 pelleted with the soil particulates after centrifugation of soil amended with this phage (L. R. Zeph and G. Stotzky, unpublished observations). The usual technique for the detection of phage particles in soil, which is filtration of appropriate soil dilutions through a 0.22- or 0.45-μm membrane filter, removes phage P1 from the dilutions, as the phage binds on soil constituents present in the dilutions and impacted on the membrane.

An alternative technique involves the use of nonfiltered soil dilutions and plaque-assay lawns of antibiotic-resistant *E. coli* grown on media containing the appropriate antibiotic at concentrations inhibitory to soil bacteria (Zeph et al., 1988). For example, 100 μg/ml Tc is added to LCA bottom and top agar, which allows the growth of confluent lawns of *E. coli* J53(RP4) and the formation of plaques by phage P1. To enumerate free phage P1 in soils incubated with *E. coli* lysogenic for the phage, chloroform should be added to a concentration of 2% (vol/vol) to the initial soil dilution for 5 minutes, with periodic vortexing, to kill the lysogens and to prevent their prophages from forming plaques on the assay lawn. Once the chloroform has settled to the bottom of the initial dilution tube (usually after ~1 minute), dilutions are prepared for plaque assay of the phage, as described before. The use of chloroform should be restricted only to chloroform-resistant phages, such as phage P1.

Because the recovery of some phages from soil can be low, specific eluents have been used to desorb various phages from soil particulates. For example, the substitution of skim milk for dH_2O as the initial diluent increases the recovery of phage P1 from <10% (dH_2O) to 60% (skim milk) of the original number of phage particles added to soil (L. R. Zeph and G. Stotzky, unpublished observations). Other common eluents used to desorb phages from soil are casein, nutrient broth, beef extract, glycine, EDTA, and egg albumin (Bitton, 1980). However, recovery of phage P1 with the eluents so far tested has been <100%.

6. Suppression of Fungal Growth on Selective Media

Fungal growth on selective media containing dilutions of nonsterile soil can be suppressed by the addition of Cy to the media. The concentration of Cy to use depends on the population level of fungi in the specific soil and the soil dilutions to be plated. A concentration of 50–200 μg/ml Cy will usually inhibit essentially all fungal growth in the 10^{-1} dilution of nonsterile soil (Tremaine and Mills, 1987; Wollum, 1982).

7. Transduction Frequency

The calculation of transduction frequencies *in situ* is complicated by the fact that studies in soil are usually conducted for periods of days or weeks. Inasmuch as transduced bacteria may multiply in soil and produce cells with the same phenotype, it is not possible to distinguish between multiplication and transduction. Consequently, estimates of transduction frequency are best obtained early in a soil study before significant multiplication of cells occurs. The transduction frequency is expressed as the number of transductants divided by the number of recipient bacteria originally added.

C. Stepwise Summary of Procedures for Studying Transduction in Soil

Prepare soil (e.g., sieve; amend with clays, $CaCO_3$, etc.; condition; add to appropriate microcosm).

Prepare suspensions of cells of recipient and lysogenic bacteria or phage lysate.

Wash cell suspensions by dilution or centrifugation.

Adjust suspensions for appropriate MOI for transduction (MOI = 1 to 5).

Add cells of recipient and lysogen or phage lysate sequentially to soil with sufficient water (or a nutrient solution) to bring the soil to its -33-kPa water tension.

Incubate soil, preferably in a high-humidity chamber.

Sample soil periodically (either by sacrificing multiple microcosms or from a batch system).

Prepare appropriate serial decade dilutions of the soil.

Plate dilutions on selective media suitable for growth of recipient, lysogen, and transductant, as well as of indigenous soil bacteria. Plaque assay the dilutions for free phage.

Incubate plates, and record numbers of CFU and PFU that develop on each medium.

Verify presumptive recipient, lysogen, and transductant cells for phenotype and/or genotype (see Section VII).

VII. Identification, Characterization, and Confirmation of Recombinants

A. Introduction

The isolation of genotypically identical bacteria, whether donors, recipients, or recombinants, for identification is most commonly and easily done by separating individual cells on a solid nutrient medium. Streaking or spreading serial dilutions of soil directly on the surface of the medium or suspending the dilutions in molten agar before making pour plates are means of separating cells so that they produce separate colonies (i.e., aggregates of cells) that presumably arise from each individual cell. However, even the formation of a single colony does not guarantee that the colony is a pure culture, and it is advisable to restreak the colony of interest until colonies are produced that are identical to one another, both macroscopically and microscopically. Even pure cultures can exhibit some variation (e.g., smooth–rough colony morphology, presence and absence of spores, and variability in the Gram stain). Microbiology laboratory manuals describe in detail the methods for obtaining and maintaining pure cultures from isolated colonies (Atlas et al., 1988; Ausubel et al., 1987; Benson, 1985).

B. Selective Media

Conventional methods of enumeration and identification depend on the growth and "phenotypic expression" of measurable traits of an organism under conditions that restrict or inhibit organisms without the traits (e.g., selective media). Many methods can be used to select or enrich for specialized, individual phenotypes, and most microbiology laboratory manuals provide descriptions of various selective and differential media and their applications. When sampling soil environments, specialized media will be necessary for the enumeration of typical groups of soil microbes (e.g., nitrifying, nitrosofying, denitrifying, sulfate-reducing, and cellulolytic bacteria; fungi; protozoa), if knowledge of the spectrum of soil microbes present is desired. A soil extract medium will be required to recover oligotrophic microbes that are not recovered on more copiotrophic standard laboratory media (Wollum, 1982) (see Appendix, Section IX).

Media for the detection and differentiation of GEMs on the bases of their metabolic and cultural requirements are useful for the detection, identification, and enumeration of novel genes in soil. The use of strains that are labeled or "tagged" with genes that enable a particular metabolic activity (e.g., toluene degradation, use of X-gal) or confer resistance to an antimicrobial agent (e.g., Tc or Hg) facilitates the detection of the GEM containing the novel genes. If the strains are to be released to the environment, it is essential to choose antibiotics that are not used for treatment of humans or other animals, because microbial resistance to clinically useful antibiotics is a major health problem.

Genes that are carried extrachromosomally may be unstable unless there is selection pressure, because the plasmids may be lost through segregation (see Section II). Therefore, to monitor the persistence of introduced novel genes, it is preferable to use genes that are located on the chromosome rather than on a transposable element that readily recombines with other sites or elements, such as transmissible plasmids. It may be beneficial to use a strain with two genetic markers, so that a spontaneous mutation will not eliminate the detectability of the phenotype when it is plated on media containing each selective agent individually, as well as on media containing both agents. The additive effect of a second antibiotic or heavy metal in the selective medium may also reduce the background level of contaminating indigenous microbes, as autochthonous microbes are not usually resistant to multiple antibiotics (see Section V).

The use of microbes tagged with genes that code for metabolic activities that can be linked to a chromogenic substrate in a selective medium is a distinct advantage in their detection and enumeration. For example, the insertion of genes for lactose fermentation into a nonenteric bacterium will facilitate its enumeration on media that are selective for coliforms [e.g., eosin methylene blue agar, MAC, Endo-agar, X-gal agar, IPTG (isopropylthio-β-D-galactoside, an analog of lactose) agar], as the chromogenic substrates aid in the visualization of colonies of tagged organisms (Jain et al., 1988). Various cloning vectors have been developed that use the lacZ (β-galactosidase) gene as a marker, such as phage M13 (Messing, 1979) and plasmid pUC (Vieira and Messing, 1982, 1987). These DNA sequences that code for lactose metabolism have been integrated into the chromosome of the host to provide a marker for monitoring the GEM in the environment (Drahos et al., 1986). The xylE gene from the TOL plasmid of Ps. aeruginosa has been cloned into other genera of bacteria, which then produce a yellow color on catechol-containing media, as the result of the activity of catechol oxygenase (Walter et al., 1989a).

Some strains of Pseudomonas (e.g., Ps. aeruginosa, Ps. fluorescens) release fluorescent water-soluble pigments that are easily visualized under UV light. Various types of Pseudomonas isolation media are commercially available.

C. "Breakthrough" of Indigenous Microbes

Many indigenous soil microbes are inherently resistant to the antibiotics and/or heavy metals routinely used in selective media for the isolation of cells containing novel genes. The use of additional antibiotics in the medium to reduce the growth of indigenous bacteria is sometimes effective. The upper limits of tolerance of the novel genes and the hosts in which they may reside to the selective antibiotics and/or heavy metals must be defined, so that the maximum levels of the antimicrobial agents can be used in the recovery medium to reduce the growth of the indigenous microbiota. Prolonged incubation of plated soil dilutions, even on media containing multiple antibiotics, may also permit the "breakthrough" growth of indigenous microbes, particularly of fungi (Krasovsky and Stotzky, 1987). Hence, shorter incubation periods and the enumeration of small colonies may be necessary to estimate the populations of recombinants before they are overgrown by indigenous microbes. Morphological differences (e.g., in colony size, shape, texture, and pigmentation) between the colonies of the GEM and those of indigenous microbes sometimes enable the distinction between introduced experimental and autochthonous strains.

The high numbers of indigenous soil bacteria (10^8–10^9 CFU/g soil) will sometimes have a "sparing" effect on the antimicrobial action of the selective agents. Under these conditions, the next higher dilution of the soil sample may reduce the numbers of indigenous microbes sufficiently to reduce or eliminate the sparing effect and, thereby, breakthrough.

Indigenous soil fungi may frequently break through the selective recovery medium and obscure or inhibit the growth of the bacterial strains of interest. The antifungal agent Cy should be incorporated into the recovery medium at concentrations from 50 to 200 μg/ml, depending on the soil, to curtail fungal growth with no effect on the growth of bacteria.

D. Viable but Nonculturable Bacteria

A phenomenon referred to as "viable but nonculturable" was first described for aquatic bacteria that could not be recovered or detected on complex synthetic laboratory media but which retained the ability to produce a virulent infection in laboratory animals (Colwell et al., 1985; Roszak and Colwell, 1987). The inability to detect some introduced GEMs directly from soil on initial isolation media has also been reported (Devanas et al., 1986). If the host–novel gene systems to be studied show a tendency toward "viable but nonculturable," it may be advisable to use initially a less selective recovery medium (e.g., MAC) rather than one that may be inhibitory (e.g., MAC + Tc) to the microbes stressed in the soil environment, even though the phenotype of the host–gene system is for resistance (e.g., Tc^r). Colonies recovered on the less selective medium are then tested on a more stringent selective medium for expression of the resistance phenotype. In other words, a "resuscitation" step is introduced for the "stressed" experimental GEMs, after which the complete resistance profile is expressed (see Section V).

Oligotrophic media may be necessary for the recovery of indigenous soil organisms that are not accustomed to the relatively high levels of nutrients contained in standard laboratory media. Various oligotrophic media have been suggested for the recovery of microorganisms from soil (Wollum, 1982).

E. Gene Transfer on Recovery Media versus in Situ

One of the purposes of developing methods for the monitoring, isolation, and identification of GEMs is to detect the transfer of their novel genes in situ. If the concentration of parentals in the soil dilutions is high, there is a possibility of cell contact between the parentals on the recovery medium. Consequently, there is the possibility that any gene transfer observed, particularly by conjugation, did not occur in soil but only on the recovery medium. Several methods have been suggested to determine whether the presence of recombinants on the recovery medium is a result of gene transfer in situ or in vitro (Curtiss, 1981). In one method, the recipient strain is selected to be resistant to Nx, which is incorporated into the medium along with other selective agents, for the recovery of the recombinant. Donor cells will be incapable of participating in gene transfer on this medium, as Nx inhibits the DNA gyrase necessary for the transfer of DNA (Curtiss, 1981; Walter et al., 1987). Cells that grow on the Nx-containing medium and have the selectable phenotype for the novel genes are assumed to be recombinants that resulted from gene transfer in soil.

F. DNA "Fingerprinting" and Plasmid Profiles

Restriction endonuclease analysis or DNA "fingerprinting" can be used to detect and monitor the fate of specific genes. Restriction endonucleases recognize specific nucleotide sequences in double-stranded DNA and cleave the DNA at a specific sequence that is usually 4–8 base pairs long. Different restriction endonucleases recognize different sequences, and depending on the frequency of the recognition sequence, a specific endonuclease will cleave the DNA numerous times, which will result in a number of different-sized DNA fragments that can then be separated according to their apparent molecular weight (M_r) by agarose gel electrophoresis. The M_r of each fragment is estimated by comparing the migration of the DNA fragments to DNA standards of known M_r.

Markers of M_r that are commonly used to produce a standard curve are restriction endonuclease digests of DNA from coliphage λ or plasmid pBR322, and many DNA fragments and plasmids of known M_r are commercially available. The distribution of different-sized fragments resulting from the cleavage by different endonucleases is unique for each chromosome or plasmid analyzed; hence, a "fingerprint" of the specific DNA is obtained (Ausubel et al., 1987).

The bacterial strains that express the specific phenotype determined by the genes of interest are isolated in pure culture, as described before. The total DNA is extracted from these cultures by any of several procedures (Ausubel et al., 1987; Crosa and Falkow, 1981; Maniatis et al., 1982). Chromosomal DNA is separated from plasmid DNA by linearization of chromosomal DNA by mechanical shearing followed by differential centrifugation techniques (e.g., CsCl–ethidium bromide density gradient centrifugation) or by agarose gel electrophoresis. The choice of procedure will depend on the information desired from the isolated DNA. The appropriate DNA containing the gene of interest (i.e., either on the chromosome or on a plasmid) is treated with a series of endonucleases, which are commercially available (e.g., International Biotechnologies Inc., New Haven, Connecticut; New England Biolabs, Beverly, Massachusetts; Bethesda Research Laboratories, Gaithersburg, Maryland) (Ausubel et al., 1987).

Crude but rapid screening of intact plasmid preparations can be used to evaluate hundreds of colonies a day for the presence of plasmid DNA by gel electrophoresis (Kado and Liu, 1981). This is a well-established method for the characterization of plasmids. Plasmid profiles (number and size per strain) have been used to identify bacterial populations and to study plasmid epidemiology in natural environments, including soil (Jain et al., 1988).

G. DNA Probes

1. Introduction

DNA–DNA hybridization can be used to detect specific gene sequences in GEMs that have been added to soil or other natural environments and in indigenous bacteria to which the genes may have been transferred. A highly specific and unique DNA sequence in the chromosome of the host or in the vector (e.g., a plasmid or a bacteriophage), preferably in the novel DNA, is isolated and purified, as described before, and labeled with either a radioactive nuclide (e.g., ^{32}P) or a chromogenic agent (e.g., biotin–streptavidin–alkaline phosphatase dyes) (Ausubel et al., 1987). DNA probes are often prepared from genes cloned in plasmid vectors so that the probe DNA can be conveniently produced in sufficient quantities. This DNA probe is hybridized with DNA from bacteria (usually those that express the specific phenotype on selective media) that have been isolated from soil and immobilized and lysed on a nitrocellulose or nylon filter membrane. Bacteria whose DNA hybridizes with the probe are then located on the filter.

One advantage of hybridization with DNA probes is the relatively high degree of specificity when compared with classical microbiological plating methods on selective media. Another advantage is that microorganisms can be detected directly in soil without prior growth on selective media. Techniques for DNA hybridization in studies of microbial ecology include Southern blot, dot or slot blot, and colony hybridization (see later).

Methods that isolate total bacterial DNA from soil and then hybridize this DNA with specific DNA probes may be valuable in (1) monitoring the survival, establishment, and growth of, and gene transfer by, GEM in soil (see Holben and Tiedje, 1988; Steffan et al., 1988); (2) the detection of groups of microorganisms capable of performing the same function (e.g., containing nitrogen fixation (nif) genes); (3) monitoring gene flow among indigenous bacteria; and (4) the detection of nonculturable microorganisms (Jain et al., 1988). These direct DNA methods, however, have a detection sensitivity of only $\sim 4 \times 10^4$ cells/g soil (Holben et al., 1988), whereas some selective isolation methods have a sensitivity approaching 2×10^1 cells/g soil (Devanas et al., 1986). Moreover, direct methods for the detection of novel genes in soil, including DNA hybridization, do not distinguish between viable and nonviable microorganisms. The choice between using DNA hybridization or classical microbiological methods will depend on the information and the sensitivity required. In all probability, both methods, in various combinations, will be used. Furthermore, DNA hybridization methods that are more sensitive, specific, and rapid, as well as less expensive, will undoubtedly be developed.

2. *Types of DNA Probes*

The probe DNA must be labeled with a tag that will enable its detection when hybridized to homologous (i.e., target) DNA sequences in the cells being probed. The first labeling technique for DNA probes was the use of radioactive nuclides (e.g., ^{32}P), which were incorporated into the nucleotides of the probe DNA (Ausubel et al., 1987; Maniatis et al., 1982). ^{32}P is preferred to ^{35}S or ^{3}H, as it has a higher specific activity and, therefore, results in the greatest sensitivity. Moreover, the activity of enzymes that act on DNA labeled with ^{32}P is not affected, as ^{32}P causes only small changes in the structure of DNA. After hybridization to its target DNA, the probe DNA is visualized by exposure of the hybridization membrane to X-ray film, which is sensitive to the γ rays emitted by ^{32}P.

An alternative to autoradiography are chromogenic techniques that use DNA labeled with nonradioactive molecules, such as biotin. The biotin-labeled DNA is visualized by the application of a streptavidin–alkaline phosphatase (Leary et al., 1983) or streptavidin–peroxidase (Syvanen et al., 1986) conjugate, followed by enzymatic conversion of a chromogenic substrate to produce a color, which identifies the target DNA with which the probe has hybridized on the filter.

3. Preparation of DNA Probes

Labeling of the probe DNA with radioactive nucleotide triphosphates can be accomplished by nick translation. DNA polymerase I adds labeled nucleotides to the 3'-OH-terminus at nicked sites along one strand of double-stranded DNA. This enzyme also has exonuclease activity and removes unlabeled nucleotides that are then replaced with radiolabeled nucleotides, which results in DNA of high specific activity, for example, $>10^8$ cpm/μg of DNA when labeled with ^{32}P (Maniatis et al., 1982). Nick translation is not the only method for radiolabeling DNA for use as probes. End-labeling of small DNA fragments with γ-^{32}P-nucleotides, using polynucleotide kinase from coliphage T4, or labeling with α-^{32}P-oligonucleotides, using random primers and the Klenow fragment of E. coli polymerase I or phage T4 polymerase, can be used to obtain radiolabeled DNA with specific activities that are often higher than those obtained by nick translation (Bentjen et al., 1989). Procedures for the isolation of DNA, probe preparation, hybridization with target DNA, and detection of the hybridization signal are detailed in Ausubel et al. (1987) and Maniatis et al. (1982).

Three different techniques are available for the preparation of nonradioactive DNA probes: nick translation, chemical labeling, and photolabeling. Commercially prepared kits are available for all three methods of DNA labeling, which result in nucleotides labeled with biotin.

Nick translation is the most frequently used method of labeling DNA with biotin (Brigati et al., 1983). Kits, in which biotin-labeled dATP or dUTP and enzymes are provided for the nick translation reaction, are available from Bethesda Research Laboratories and Enzo Biochem, Inc. (New York, New York). Procedures outlined by the manufacturer or in Ausubel et al. (1987) and Maniatis et al. (1982) can be used to label DNA with biotinylated nucleotides, as the nick translation reaction is the same as used with radionucleotides.

Chemical labeling of the DNA probe involves the insertion of an antigenic sulfone group into the cytosine moieties of the denatured DNA (Renz and Kurz, 1984). A specific monoclonal antibody is then bound to the sulfone group, followed by the application of an enzyme–antiimmunoglobulin conjugate that catalyzes a chromogenic substrate system and enables visualization of the labeled probe DNA. Kits are available from FMC Bioproducts (Rockland, Maine).

Procedures for the direct photolabeling of DNA with biotin have been described by Forster et al. (1985). This is a nonenzymatic method in which the probe DNA is labeled by irradiating photoactivated biotin with a high-intensity lamp in a mixture containing the DNA. Photolysis of an azide group on photobiotin results in covalent bonding to the base group on the nucleic acids. Visualization is achieved by the application of a streptavidin–enzyme conjugate and a chromogenic substrate, as described earlier. Highly purified DNA must be used, as the highly reactive photoactivated biotin will bind to proteins, Tris buffer salts, and polyethylene glycol, and cause nonspecific signal development. A commercial kit that supplies the photobiotin and procedures for labeling is available from Sigma Chemical Co. (St. Louis, Missouri) or Bethesda Research Laboratories.

4. *Hybridization Techniques*

a. *Southern Hybridization.* Hybridization of DNA probes to target DNA that has been separated on agarose gels by electrophoretic techniques and transferred to a nitrocellulose or nylon filter membrane is termed Southern hybridization (Southern, 1975). Details of the procedures used for Southern hybridization are presented in Ausubel et al. (1987) and Maniatis et al. (1982).

Two distinct protocols for isolating the target DNA from soil samples for Southern hybridization have been developed: (1) direct extraction of DNA from soil samples (Ogram et al., 1987) and (2) extraction of the bacterial population from soil, followed by lysis of the bacterial cells (Holben et al., 1988). Direct extraction of nucleic acids from soil is more efficient than the recovery of intact bacterial cells. However, direct extraction suffers from the limitations that a heterogeneous DNA sample is obtained, as eukaryotic and cell-free DNA are also extracted, and that humic acids, among other soil constituents, contaminate the DNA sample, which could result in decreased hybridization of the DNA probe to target DNA. Although the recovery of intact bacterial cells by differential centifugation is a more lengthy process and results in the recovery of only ~80% of the total bacteria presumably present in soil (Holben et al., 1988), the DNA sample is not contaminated with humic acids and nonbacterial DNA, and it undergoes less shearing, so that the sample contains DNA of more uniform size than that obtained by direct extraction. Treatment of the purified DNA extract with restriction enzymes before hybridization enables the comparison of the size of the labeled fragment with a known restriction map of the DNA from the GEM of interest and, thereby, easier identification of restriction fragments that contain the novel gene (Jain et al., 1988). Moreover, restriction analysis of DNA isolated from soil, either directly or indirectly, can detect gene rearrangements that may have occurred in soil (Holben et al., 1988). Gene rearrangements and deletions have also been detected in fresh water, suggesting that genetic instability may be enhanced in natural habitats (O'Morchoe et al., 1988).

b. *Slot Blot and Dot Blot Hybridization.* Slot- or dot-blotting techniques of DNA hybridization involve the placement of bacterial cells, viruses, or extracted DNA at specific locations on nitrocellulose or nylon filters. The cells or viruses are then lysed, and the DNA is immobilized on the filter and hybridized with the DNA probe. For example, dot blots are prepared by placing aliquots (5–10 μl) of target DNA directly on the filter with a micropipettor, forming a circular dot of bacteria, phage, or DNA for hybridization. Alternatively, slot blot apparatuses are available in which samples are placed in slots (30–40 samples per filter are possible) and immobilized on the hybridization filter by the application of a vacuum (Steffan and Atlas, 1988). Slot blot hybridization with radiolabeled DNA probes has been used to detect novel genes in GEM added to a nonsterile sediment microcosm (Sayler et al., 1986; Steffan and Atlas, 1988), and dot blot hybridization with a biotinylated DNA probe has been used to evaluate presumed *E. coli* transductants isolated from nonsterile soil that was inoculated with *E. coli* and phage P1 (Zeph and Stotzky, 1989).

c. *Colony Hybridization.* The hybridization of a DNA probe to target DNA released from bacterial colonies impregnated into nitrocellulose or nylon filters is termed colony hybridization (Grunstein and Hogness, 1975). Descriptions of the methods employed in colony hybridization with DNA probes can be found in Ausubel et al. (1987) and Maniatis et al. (1982). Colony hybridization techniques have been used to detect specific gene sequences naturally present in soil bacteria (Barkay et al., 1985) and novel genes introduced into bacteria that had been added to soil (Rafii and Crawford, 1988). Normally, dilutions of soil are plated on selective media, and plates with well-separated bacterial colonies are used as the source of colony DNA. The colonies are transferred to a filter membrane for lysis and hybridization. Alternatively, the screening of large numbers of bacteria for the presence of novel DNA can be accomplished if lawns of bacteria ($\sim 10^5$–10^6 cells/plate) are used in colony hybridizations with ^{32}P-labeled probes (Sayler et al., 1985), particularly if the isolation of the organism of interest from discrete colonies is not desired. However, when this procedure is used with biotin-labeled DNA probes, problems can arise with nonspecific signal development during the probe visualization step (see later).

The advantage of the colony hybridization technique is that it often enables a more rapid screening of environmental samples for the presence of microorganisms containing a novel gene(s) than Southern hybridization or slot- and dot-blotting techniques. However, the microbes must grow on the isolation media, to produce sufficient target DNA for hybridization, in contrast to the Southern hybridization method. For some purposes, such as monitoring gene transfer from an introduced GEM to the indigenous bacterial population, colony hybridization is the preferred method, because growth of the introduced GEM can be inhibited on selective media, thereby allowing hybridization of the probe with colony DNA from only the indigenous population. Jain et al. (1988) emphasized that preliminary testing of the DNA hybridization protocol is important to avoid false positive results when attempting to detect GEMs in environmental samples: a good selective medium must be used to prevent growth of indigenous bacteria that can give false positive hybridization signals, and the specificity of the DNA probe for the novel gene must be ascertained before its use in the monitoring of GEMs in the environment.

5. *Sensitivity and Specificity of DNA Probes*

The sensitivity of radiolabeled DNA probes is generally greater than that of nonradioactive probes. Zwadyk et al. (1986) reported that a ^{32}P-labeled probe was 100 times more sensitive than biotinylated probes in detecting plasmid pBR322 DNA with Southern hybridization, and a minimum of 22 ng of target DNA was necessary for detection with biotinylated plasmid pBR322 DNA probes. However, when purified target DNA was spotted on nitrocellulose filters, Zwadyk et al. (1986) detected as little as a 9.7 pg of target DNA with a ^{32}P-labeled probe and 39 pg with a biotinylated probe prepared with a kit (Bethesda Research Laboratories). Zeph and Stotzky (1989) detected 50 pg of target DNA from phage P1 using the same biotinylated probe kit. Leary et al. (1983) reported greater sensitivity with a biotinylated DNA probe made from linearized rather than from intact plasmid DNA, and they detected 3.1 pg of target DNA by Southern hybridization.

Another expression of sensitivity is the minimum number of cells containing target DNA that can be detected in a background of indigenous bacteria on enumeration plates by colony hybridization. This is useful in comparing DNA probes on the basis of their ability to detect GEMs added to natural environments that may contain high numbers of indigenous bacteria. Sayler et al. (1985) utilized ^{32}P-labeled DNA probes to detect Ps. putida added to sediment microcosms at a sensitivity of 1 colony of Ps. putida in a background of 10^6 indigenous bacterial colonies that contained nonhomologous DNA.

The minimum number of cells containing homologous DNA that can be detected is also used as a measure of the sensitivity of DNA probes. For example, Holben et al. (1988) were able to quantitate Bradyrhizobium japonicum, added to nonsterile soil, at levels of 4.3×10^4 cells/g soil, which was equivalent to 0.2 pg of target DNA per 1 μg of total purified DNA isolated from soil bacteria. Frederickson et al. (1988) used the most probable number (MPN) method, combined with DNA hybridization of DNA isolated from individual MPN tubes, to detect Ps. putida, added to nonsterile soil, at concentrations as low as 10 cells/g soil and Rhizobium leguminosarum at 100 cells/g soil. The application of DNA probe techniques to studies of microbial ecology has been reviewed by Holben and Tiedje (1988).

The relatively low sensitivity of DNA probes in detecting and enumerating bacteria containing homologous DNA in soil can be overcome, in part, through the amplification of the target DNA sequences. The polymerase chain reaction (PCR) procedure has been used to increase probe sensitivity by using the hybridized DNA probe as a primer for chain elongation beyond the target DNA sequence (Somerville et al., 1988; Saiki et al., 1988). After hybridization of the probe to target DNA, DNA synthesis is initiated from the 3'-OH-terminus of the probe DNA strand with radiolabeled nucleotides and DNA polymerase. The PCR is repeated 20–30 times by remelting the DNA strands and hybridizing with fresh probe DNA, followed by another chain elongation reaction, which results in an enhancement of the sensitivity of the DNA probe by at least three orders of magnitude. The specificity of the probe depends on the initial hybridization to target DNA, and the sensitivity is enhanced by DNA synthesis from the free 3'-OH end into adjacent genes (Colwell et al., 1988). Steffan and Atlas (1988) employed the PCR method to increase the sensitivity of a probe for the detection of catabolic genes in Pseudomonas cepacia AC1100 that was added to sediment microcosms.

The use of biotinylated DNA probes has an inherent problem in that contamination of the sample with protein will result in nonspecific color development caused by the binding of the avidin–enzyme conjugate to protein components in the cell lysate, in addition to binding to the biotin moiety on the labeled probe. Consequently, all protein must be removed from the hybridization filters when using biotinylated DNA probes, particularly when whole bacterial cells are lysed on the filter, as in slot or dot blots or colony hybridizations. Treatment of the filter with chloroform, phenol, and proteolytic enzymes will remove significant amounts of cell debris and protein (Haas and Fleming, 1986; Rafii and Crawford, 1988; Zeph and Stotzky, 1989).

In summary, radiolabeled DNA probes have the advantage of being more sensitive in many applications than biotinylated probes, although sensitivity is being improved in the latter. However, biotinylated DNA probes are less hazardous than radiolabeled DNA, have a much longer storage life (≥ 1 year compared to 2 weeks for ^{32}P-labeled DNA), and have a lower cost.

H. Serological Techniques

Serological methods, based usually on the expression of specific surface antigens, require the isolation of the bacteria of interest on media that are relatively selective (i.e., that eliminate or reduce the growth of unrelated soil bacteria). The bacteria are purified and then challenged with a spectrum of polyclonal or monoclonal antibodies raised against the bacteria of interest. The reactions between the bacteria and the antisera (usually polyclonal) are then scored as being (1) nonreactive, (2) cross-reactive with many sera without distinctive features and common to all isolates, (3) cross-reactive with distinctive features between isolates and a few antisera, or (4) a specific reaction with a single antiserum. Although this method can be highly specific in some instances, the surface antigens of some bacteria change during growth, and they may be different in soil than in the pure cultures that were used to raise the antigens. Moreover, serological data do not always agree with data obtained with DNA probes and with other characteristics of the bacteria (e.g., Schofield et al., 1987).

Although polyclonal antisera are easier to prepare, they are less specific that monoclonal antibodies. The use of labeled monoclonal antibodies provides a highly specific technique for the direct detection of GEMs in the environment. However, there are significant limitations in this methodology. For example, the presence of the novel DNA in the microbe is only presumptive and must be verified by other methods, the sensitivity is low when employed in soil studies, and it is not possible to differentiate between viable and nonviable cells (Colwell et al., 1988).

The detection of GEMs in the environment with monoclonal antibody techniques has been reviewed by Omenn (1986).

I. Heat Induction of Prophages

Verification of the transfer of novel genes by transduction is possible with some phages by induction of the lytic cycle of replication of the prophage. Zeph et al. (1988) confirmed by heat induction of lysis that presumptive *E. coli* transductants, isolated from soil inoculated with *E. coli* and phage P1 on the basis of the antimicrobial-resistance phenotype coded for on the phage P1 genome, contained the phage (Table II). The phage P1 strain used had a temperature-sensitive mutation in the $c1$ gene, which codes for a repressor protein that maintains lysogeny. The presumed *E. coli* transductants were grown in LCB at 30°C, with shaking, until log phase (optical density of 0.2 at 600 nm), followed by induction of lysis by incubation at 42°C. The presence of phage P1 virions in the culture was demonstrated by spotting aliquots of the heat-induced lysate on lawns of *E. coli*. It might be possible to use this technique to confirm the presence of any prophage DNA in presumed transductants, as incubation of lysogenic bacteria at temperatures above their optimum for growth should inactivate or impair the activity of the repressor protein that maintains lysogeny. Moreover, other agents that can induce the lytic cycle (e.g., UV radiation, mitomycin C) should be evaluated with lysogens that do not respond to elevated temperatures.

When attempting to demonstrate gene transfer to indigenous soil bacteria by transduction by phage P1, isolates of soil bacteria were grown at 25°C until they reached log phase, exposed to 42°C for 30 minutes, and then reincubated at 25°C to allow lysis of the bacteria by the phage. Continued incubation was not at 42°C, as most soil bacteria have temperature optima for growth between 20° and 30°C and may not grow at the 42°C used to induce prophage P1.

VIII. Quality Assurance and Quality Control

A. Sample Representativeness and Custody

All soil samples should be contained in separate and well-identified (e.g., labeled with waterproof ink) bags or plastic-lined garbage pails. All bacterial cultures should be maintained at 4°C in pure culture on slants in well-identified test tubes and transferred regularly. Purity and phenotype should be determined after each transfer. Soil stocks should be located in a separate locked room, and stocks of bacteria, phages, media, clays, and other materials should be in locked laboratories to which only personnel of the laboratory and appropriate security personnel have access. These stocks should be under the direct supervision of the project manager.

B. Sampling Procedures

Stocks of soils, clays, bacteria, phages, media, and other materials should be thoroughly mixed before each experiment and before subsampling to ensure maximum homogeneity. The number of subsamples will vary from experiment to experiment, depending on results of previous experiments, and should be based on significance between treatments at a probability level (p) of <0.05.

C. Comparability

Microbiological data should be reported as number of CFU or PFU/g soil, oven-dry equivalent. Frequency of recombination is expressed as a percentage of donor, recipient, or donor plus recipient, depending on the study. The denominator used to obtain the percentage must be clearly indicated. All other data (e.g., physicochemical soil characteristics; nomenclature and phenotype of bacteria, plasmids, and bacteriophages; M_r of DNA fragments) should be reported in standardized units.

D. Calibration Procedures and Frequency

All calibration activities should be thoroughly documented. Calibration records should be on permanent file in bound notebooks.

1. Instrumental Procedures

Thermometers for monitoring temperatures of refrigerators, freezers, incubators, laboratories, and so on should be verified biannually against a thermometer traceable to the National Institute of Standards and Technology. Buffer solutions for the calibration of pH meters can be obtained from commercial sources (two to three pH values that bracket the working range of pH should be evaluated by this method, in addition to electronic resistance methods). Pipettors and glass pipettes should be calibrated every 6 months by gravimetry; random samples of disposable pipettes should also be checked periodically for accuracy. Spectrophotometers should be calibrated biannually with decade dilutions of several sources or types of DNA, protein, and bacteria. Calibration of agarose gel electrophoresis is conducted with commercial DNA M_r standards (these standards should be run with all unknown preparations). Microscopes should be calibrated annually with decade dilutions of standard-sized polystyrene latex particles. For instruments used routinely, a checklist should be fastened on or near each instrument, and each user should initial and date the list after routine calibrations have been made before and after analysis of samples.

2. *Analytical Procedures*

^{32}P and other DNA probes should be verified for specificity with the DNA against which they were prepared. Similar verification can be made for probes for gene products. These verification procedures can also serve to test the reactivity of the X-ray films and the efficacy of the film developer solutions, the enzyme color development reaction for nonradioactive DNA probes, and the sensitivity of monoclonal antibodies in serological techniques.

3. *Microbiological Procedures*

The phenotype of all novel bacteria and the infectivity of the phages should be verified in pure culture before experiments are conducted in soil. Furthermore, studies in sterile soil can provide another level of verification for studies in nonsterile soil. Spectrophotometric calibration of bacterial numbers should be correlated periodically with microscopic and plate counts. The efficacy of the sterilization of soils must be determined by plating on several media, ranging from nutrient-rich to nutrient-poor.

E. ANALYTICAL PROCEDURES

Quality assurance must be established by numerous replicates per experiment and repeat experiments. When more than one genetic marker is contained in the novel DNA, direct or replica plating should be to selective media specific for the detection of each marker, either individually or in combination. Furthermore, DNA probes should be used when necessary (e.g., when there appears to be a loss or segregation of the novel DNA). When analytical procedures developed by others are used, the verification procedures should be those of the developers. It must be emphasized that the study of gene transfer in soil is a relatively new area of research, and sufficient published data are not availble for routine verification and quality assurance. Therefore, verification of gene transfer by novel bacteria by several techniques is required.

F. Experimental Design and Statistical Analyses

The experimental design and the statistical analysis to be applied to the data must be determined before the start of each study. The assistance of a statistician is helpful at this stage. The number of replicates to be used should be sufficient to detect a $p < 0.05$. This number of replicates is determined in preliminary studies, with the application of appropriate statistical theory and formulas. At the end of each experiment, the data must be statistically analyzed (e.g., means ± standard errors of the means, Student's t-test, ANOVA, regression analyses) and plotted or tabulated (Atlas and Bartha, 1981). As mentioned earlier, data should be expressed in standard terms, or if not, the dimensional units should be clearly defined.

Experimental errors are detected by abnormally high coefficients of variation (>50%) between replicates and by unanticipated and unexplainable deviations in time–response curves. Inasmuch as experiments must be not only sufficiently replicated and statistically analyzed, but also repeated, errors will be readily detected. In the event of such errors, replicated experiments should be repeated. Although statistically significant differences may be observed in some experiments, the potential ecological significance of these differences must be further evaluated (e.g., in repeat experiments or in more complex microcosms), especially if the differences are transitory.

G. Data Analysis and Reporting

Data should be recorded in ink on appropriate data sheets, in duplicate, in permanent, bound laboratory notebooks. All entries should be initialed. One copy should reside with the research technician collecting the day-to-day data, and the other with the project manager.

H. Internal Quality Control Checks

The routine quality controls of a microbiology laboratory should be followed. For example, periodic checks (at least monthly) must be conducted on laboratory water for microbiological quality ($<10^4$ CFU/ml) and conductance (<2 μS/cm). The temperature of refrigerators ($\pm 2°C$), freezers ($\pm 4°C$), incubators ($\pm 2°C$), and autoclaves (121°C) should be continually monitored (usually twice on a measurement day); pH and quality checks should be made on media lots. Sterility checks on media lots, water, soils, and other autoclaved or radiation-sterilized materials should be made by plating. A logbook of internal quality controls should be maintained. These internal quality control checks are routine in basic research laboratories.

I. Preventive Maintenance

All major pieces of laboratory equipment should be monitored regularly and serviced routinely, as individually required, through service contracts with outside contractors, as well as internally.

J. Corrective Action

If any experiments or parts of experiments have been inappropriately designed, conducted, or analyzed, the entire experiment must be repeated. It must be emphasized, however, that gene transfer among bacteria in soil is essentially an unexplored area and that the development of appropriate experimental design, conduct, and analyses is an intrinsic part of the research.

IX. Appendix

A. Media Composition

All media are prepared with dH_2O.

1. Luria Broth (LB) (Maniatis et al., 1982)

Tryptone	10 g
Yeast extract	5 g
Glucose	1 g
Sodium chloride	5 g
dH_2O	1 liter

2. Luria Broth with Calcium and Magnesium (LCB) (Zeph et al., 1988)

    LB containing:
    10 mM $MgSO_4 \cdot 7H_2O$
    2 mM Anhydrous $CaCl_2$

3. Luria Agar (LA) (Maniatis et al., 1982)

    LB containing:
    Bacto Agar   15 g/liter

4. *Luria Top Agar* (Maniatis et al., 1982)

   LB containing:
   Bacto Agar   7.5 g/liter

5. *Minimal Salt Agar (MSA)* (Germida and Khachatourians, 1988)

Glucose (or lactose)	10 g
Vitamin B_{12}	0.1 g
Tetracycline	0.1 g
Streptomycin	0.2 g
Eosin	0.4 g
Methylene blue	0.07 g
Leucine	0.02 g
Threonine	0.08 g
Bacto Agar	15 g
dH_2O	1 liter

6. *M9 Minimal Agar Containing $HgCl_2$ (M9M)* (Zeph et al., 1988)

Na_2HPO_4	6 g
KH_2PO_4	3 g
NaCl	0.5 g
NH_4Cl	1 g
Bacto Agar	15 g
dH_2O	1 liter

   After autoclaving, filter-sterilized solutions of the following are added to yield the indicated final concentrations: 0.2% glucose, 2 mM $MgSO_4 \cdot 7H_2O$, 0.1 mM $CaCl_2$, and 30 μM $HgCl_2$.

7. *Minimal Liquid (ML)* (Curtiss, 1965)

NH_4Cl	5 g
NH_4NO_3	1 g
Na_2SO_4	2 g
K_2HPO_4	9 g
KH_2PO_4	5 g
$MgSO_4 \cdot 7H_2O$	0.1 g
Glucose	5 g
dH_2O	1 liter

   Autoclave glucose separately at 121°C and 15 psi; pH 6.8.

8. *Minimal Agar (MA)* (Curtiss, 1965)

   Salts in ML:
   Glucose (autoclave separately)   5 g

Bacto Agar 15 g
dH$_2$O 1 liter
pH 6.8

9. *Minimal Mating Medium (3M)* (Curtiss et al., 1968)

NH$_4$Cl 5 g
NH$_4$NO$_3$ 1 g
Na$_2$SO$_4$ 2 g
K$_2$HPO$_4$ 4.9 g
KH$_2$PO$_4$ 6.3 g
Glucose (autoclave separately) 5 g
dH$_2$O 1 liter
pH 6.3

10. *Soil Extract Agar (SEA)* (Allen, 1957)

KH$_2$PO$_4$ 0.5 g
Glucose 1 g
Soil extract 100 ml
Tap water 900 ml
Adjust to pH 7.0

Preparation of soil extract: 1.0 kg of garden soil is heated with 1000 ml tap water in the autoclave (121°C, 15 psi) for 30 minutes. A small amount of CaCO$_3$ is added, and the soil suspension is filtered several times through a double-paper filter (Whatman No. 1) until no turbidity is evident. This solution is bottled in 100-ml quantities and autoclaved before storage.

B. ANTIMICROBIAL AGENTS COMMONLY USED IN SELECTIVE MEDIA[a]

All concentrated stock solutions are to be labeled with name of agent, concentration, date, and analyst's initials.

Agent	Abbreviation	Special instructions	Initial diluent	Concentration (mg/ml)	Approximate stability at 4°C (days)	Concentration (μg/ml)	Amount to be added to 1 liter (ml)
Ampicillin	Ap	Filter[b]	dH$_2$O[d]	25	7	200	8
Carbenicillin	Cb	Filter	dH$_2$O	50	14	100	2
Cephalothin	Ceph	Filter	0.1 M PB[e]	12.5	5	25	2
Chloramphenicol	Cm	Filter	EtOH[f]	25	30	25	1

Agent	Abbreviation	Special instructions	Initial diluent	Concentration (mg/ml)	Approximate stability at 4°C (days)	Concentration (μg/ml)	Amount to be added to 1 liter (ml)
Chlortetracycline	CTc	Use foil[c]	50% EtOH	10	5	15	1.5
Cycloheximide	Cy	Use foil	dH$_2$O	12.5	5	25	2
Gentamicin	Gn	Filter	dH$_2$O	12.5	30	25	2
Kanamycin	Km	Filter	dH$_2$O	25	30	50	2
Mercuric chloride	Hg	Filter	dH$_2$O	40	30	40	1
Nafcillin	Nf	Filter	dH$_2$O	1	5	2	2
Nalidixic acid	Nx	Filter	0.1 M NaOH	16	30	32	2
Rifampicin	Rf	Use foil	MeOH[g]	25	5	100	4
Streptomycin	Sm	Filter	dH$_2$O	12.5	30	25	2
Tetracycline	Tc	Use foil	50% EtOH	10	5	15	1.5
Tobramycin	Tb	Filter	dH$_2$O	5	30	10	2
Trimethoprim	Tp	Filter	15% 0.06 M HCl	2	30	50	25

[a] Courtesy of A. Porteous.
[b] Filter-sterilized.
[c] Store in dark covered with foil.
[d] Distilled water.
[e] Phosphate buffer, pH 6.
[f] Ethanol.
[g] Methanol.

ACKNOWLEDGMENTS

Many of the specific techniques described in this document for studying gene transfer in soil were developed by the authors during the past few years, in part with support from the United States Environmental Protection Agency, especially in the form of cooperative agreements (CR812484, CR813431, CR813650) with the Corvallis Environmental Research Laboratory (CERL). The primary motivation for these studies was the need for information on the survival of, and gene transfer by, genetically engineered bacteria that could be used in risk assessment of the release of genetically engineered microorganisms to the environment. The support, encouragement, and suggestions provided by Dr. R. J. Seidler, J. L. Armstrong, and other members of CERL are gratefully acknowledged.

REFERENCES

Aardema, B. W., Lorenz, M. G., and Krumbein, W. E. (1983). Protection of sediment-adsorbed transforming DNA against enzymatic inactivation. *Appl. Environ. Microbiol.* **46**, 417–420.

Adams, M. H. (1959). "Bacteriophages." Interscience, New York.

Allen, O. N. (1957). "Experiments in Soil Bacteriology." Burgess, Minneapolis, Minnesota.

Altherr, M. R., and Kasweck, K. L. (1982). *In situ* studies with membrane diffusion

chambers of antibiotic resistance in *Escherichia coli. Appl. Environ. Microbiol.* **44,** 838–843.
Anderson, J. P. E. (1982). Soil respiration. In "Methods of Soil Analysis" (A. L. Page, R. H. Miller, and D. R. Keeney, eds.), Part 2, pp. 831–872. Am. Soc. Agron., Madison, Wisconsin.
Anonymous (1987). "Standard Guide for Conducting a Terrestrial Soil-Core Microcosm Test." Am. Soc. Test. Mater., Philadelphia, Pennsylvania.
Ardakani, M. S., Rehbock, J. T., and McLaren, A. D. (1973). Oxidation of nitrite to nitrate in a soils column. *Soil Sci. Soc. Am. Proc.* **37,** 53–56.
Armstrong, J. L., Knudsen, G. R., and Seidler, R. J. (1987). Microcosm method to assess survival of recombinant bacteria associated with plants and herbivorous insects. *Curr. Microbiol.* **15,** 229–232.
Atlas, R. M., and Bartha, R. (1981). "Microbiol Ecology." Addison-Wesley, Reading, Massachusetts.
Atlas, R. M., Brown, A. E., Dobra, K. W., and Miller, L. (1988). "Experimental Microbiology: Fundamentals and Applications." Macmillian, New York.
Ausubel, F. M., Bent, R., Kingston, R. E., Moore, D. D., Smith, J. A., Seidman, J. G., and Struhl, K. (1987). "Current Protocols in Molecular Biology." Greene and Wiley (Interscience), New York.
Babich, H., and Stotzky, G. (1977). Effects of cadmium on fungi and on interactions between fungi and bacteria in soil: Influence of clay minerals and pH. *Appl. Environ. Microbiol.* **33,** 1059–1066.
Bachmann, B. J., and Low, K. B. (1980). Linkage map of *Escherichia coli* K-12. *Microbiol. Rev.* **44,** 1–56.
Barkay, T., Founts, D. L., and Olson, B. H. (1985). Preparation of a DNA gene probe for detection of mercury resistance genes in gram-negative bacterial communities. *Appl. Environ. Microbiol.* **49,** 686–692.
Baross, J. H., Liston, J., and Morita, R. Y. (1974). Some implications of genetic exchange among marine vibrios, including *Vibrio parahaemolyticus*, naturally occurring in the Pacific oyster. In "International Symposium on Vibrio parahaemolyticus" (T. Fujino, G. Sakaguchi, R. Sakazaki, and Y. Takeda, eds.), pp. 129–137. Saikon, Tokyo.
Benson, H. (1985). "Microbiological Applications: A Laboratory Manual in General Microbiology," 4th Ed. W. C. Brown, Dubuque, Iowa.
Bentjen, S. A., Frederickson, J. K., Van Voris, P., and Li, S. W. (1989). Intact soil-core microcosms for evaluating the fate and ecological impact of the release of genetically engineered microorganisms. *Appl. Environ. Microbiol.* **55,** 198–202.
Bissonnette, G. K., Jezeski, J. J., McFeters, G. A., and Stuart, D. G. (1975). Influence of environmental stress on enumeration of indicator bacteria from natural waters. *Appl. Microbiol.* **29,** 186–194.
Bitton, G. (1980). "Introduction to Environmental Virology." Wiley, New York.
Brigati, D. J., Myerson, D., Leary, J. J., Spaholz, B., Travis, S. Z., Fong, C. K. Y., Hsuing, G. D., and Ward, D. C. (1983). Detection of viral genomes in cultured cells and paraffin-embedded tissue section using biotin-labeled hybridization probes. *Virology* **126,** 32–50.
Bull, A. T. (1980). Biodegradation: some attitudes and strategies of microorganisms and microbiologists. In "Contemporary Microbial Ecology" (D. C. Ellwood, J. N. Hedger, M. J. Latham, J. M. Lynch, and J. H. Slater, eds.), pp. 107–136. Academic Press, London.
Burman, L. G. (1977). Expression of R-plasmid functions during anaerobic growth of an *Escherichia coli* K-12 host. *J. Bacteriol.* **131,** 69–75.

Camper, A. K., and McFeters, G. A. (1979). Chlorine injury and the enumeration of waterborne coliform bacteria. *Appl. Environ. Microbiol.* **37**, 633–641.

Casida, L. E., Jr. (1968). Methods for the isolation and estimation of activity of soil bacteria. In "The Ecology of Soil Bacteria" (T. R. G. Gray, and D. Parkinson, eds.), pp. 97–122. Univ. of Toronto Press, Toronto.

Cassel, D. K., and Nielsen, D. R. (1986). Field capacity and available water capacity. In "Methods of Soil Analysis" (A. Klute, ed.), pp. 901–926. Am. Soc. Agron., Madison, Wisconsin.

Chatterjee, D. K., Kellogg, S. T., Furukawa, K., Kilbane, J. J., and Chakrabarty, A. M. (1981). Genetic approaches to the problems of toxic chemical pollution. In "Recombinant DNA" (A. G. Walton, ed.), pp. 199–212. Elsevier, Amsterdam.

Clewell, D. (1981). Plasmids, drug resistance, and gene transfer in the genus *Streptococcus*. *Microbiol. Rev.* **45**, 409–436.

Colwell, R. R., Brayton, P. R., Grimes, D. J., Roszak, D. B., Huq, S. A., and Palmer, L. M. (1985). Viable but non-culturable *Vibrio cholerae* and related pathogens in the environment: implications for release of genetically engineered microorganisms. *Bio/Technology* **3**, 817–820.

Colwell, R. R., Somerville, C., Knight, I., and Straube, W. (1988). Detection and monitoring of genetically-engineered micro-organisms. In "The Release of Genetically-Engineered Microorganisms" (M. Sussman, C. H. Collins, F. A. Skinner, and D. E. Stewart-Tull, eds.), pp. 47–60. Academic Press, London.

Covarrubias, L., Cervantes, L., Covarrubias, A., Soberon, X., Vichido, I., Blanco, A., Kupersztoch-Portnoy, Y. M., and Bolivar, F. (1981). Construction and characterization of new cloning vehicles. V. Mobilization and coding properties of pBR322 and several deletion derivatives including pBR327 and pBR328. *Gene* **13**, 25–35.

Crosa, J. H., and Falkow, S. (1981). Plasmids. In "Manual of Methods for General Bacteriology" (P. Gerhardt, R. G. E. Murray, R. N. Costilow, E. W. Nester, W. A. Wood, N. R. Krieg, and G. B. Phillips, eds.), pp. 266–282. Am. Soc. Microbiol., Washington, D.C.

Curtiss, R., III (1965). Chromosomal aberrations associated with mutations to bacteriophage resistance in *Escherichia coli*. *J. Bacteriol.* **89**, 23–40.

Curtiss, R., III (1976). Genetic manipulation of microorganisms: potential benefits and biohazards. *Annu. Rev. Microbiol.* **30**, 507–533.

Curtiss, R., III (1981). Gene transfer. In "Manual of Methods for General Bacteriology" (P. Gerhardt, R. G. E. Murray, R. N. Costilow, E. W. Nester, W. A. Wood, N. R. Krieg, and G. B. Philips, eds.), pp. 243–265. Am. Soc. Microbiol., Washington, D.C.

Curtiss, R., III, and Renshaw, J. (1969). F^+ strains of *Escherichia coli* K12 defective in Hfr. *Genetics* **63**, 7–26.

Curtiss, R., III, Charamella, L. J., Stallions, D. R., and Mays, J. A. (1968). Parental functions during conjugation in *Escherichia coli* K12. *Bacteriol. Rev.* **32**, 320–348.

Cuskey, S. M. (1988). Survival, persistence, and colonization. In "The Release of Genetically-Engineered Micro-Organisms" (M. Sussman, C. H. Collins, F. A. Skinner, and D. E. Stewart-Tull, eds.), pp. 231–237. Academic Press, London.

Daughton, C. G., and Hsieh, D. P. (1977). Parathion utilization by bacterial symbionts in a chemostat. *Appl. Environ. Microbiol.* **34**, 174–184.

Devanas, M. A., and Stotzky, G. (1986). Fate in soil of a recombinant plasmid carrying a *Drosophila* gene. *Curr. Microbiol.* **13**, 279–283.

Devanas, M. A., and Stotzky, G. (1988a). Survival of genetically engineered microbes in the environment: effect of host/vector relationship. In "Developments in Industrial Microbiology" (G. Pierce, ed.), Vol. 29 (*J. Ind. Microbiol.*, Suppl. No. 3), pp. 287–296. Elsevier, Amsterdam.

Devanas, M. A., and Stotzky, G. (1988b). Survival of, and gene transfer by, GEMs in soil. *Abstr. Int. Conf. Release Genetically-Eng. Micro-Org., 1st* p. 40.

Devanas, M. A., Rafaeli-Eshkol, D., and Stotzky, G. (1986). Survival of plasmid-containing strains of *Escherichia coli* in soil: effect of plasmid size and nutrients on survival of hosts and maintenance of plasmids. *Curr. Microbiol.* **13,** 269–277.

Doyle, J., Jones, R., Broder, M., and Stotzky, G. (1988). Effects of genetically engineered microbes on microbe-mediated ecological processes in soil. *Abstr. Annu. Meet. Am. Soc. Microbiol.* p. 283.

Drahos, D. J., Hemming, B. C., and McPherson, S. (1986). Tracking recombinant organisms in the environment: β-galactosidase as a selectable non-antibiotic marker for fluorescent pseudomonas. *Bio/Technology* **4,** 439–444.

Drahos, D. J., Barry, G. F., Hemming, B. C., Brandt, E. J., Skipper, H. D., Kline, E. L., Kluepfel, D. A., Jughes, T. A., and Gooden, D. T. (1988). Pre-release testing procedures: US field test of a *lac*ZY-engineered bacterium. *In* "Release of Genetically-Engineered Micro-Organisms" (M. Sussman, C. H. Collins, F. A. Skinner, and D. E. Stewart-Tull, eds.), pp. 181–191. Academic Press, London.

Edlin, G., Lin, L., and Kurdna, R. (1975). Lambda-lysogens of *E. coli* reproduce more rapidly than non-lysogens. *Nature (London)* **225,** 735–737.

Edlin, G., Tait, R. C., and Rodriguez, R. L. (1984). A bacteriophage lambda cohesive ends (*cos*) DNA fragment enhances the fitness of plasmid-containing bacteria growing in energy-limiting chemostats. *Bio/Technology* **2,** 251–254.

Elliot, E. T., Hunt, H. W., Walter, D. E., and Moore, J. C. (1986). Microcosms, mesocosms, and ecosystems: Linking the laboratory to the field. *Proc. Int. Soc. Microbial Ecol., 4th* (F. Megusar and M. Gantar, eds.), pp. 472–480. Slovene Soc. Microbiol., Ljubljana, Yugoslavia.

Forster, A. C., McInnes, J. L., Skingle, D. C., and Symons, R. H. (1985). Non-radioactive hybridization probes prepared by the chemical labelling of DNA and RNA with a novel reagent, photobiotin. *Nucleic Acids Res.* **13,** 745–761.

Frederickson, J. K., Bezdicek, D. F., Brockman, F. J., and Li, S. W. (1988). Enumeration of Tn5 mutant bacteria in soil by using a most-probable-number–DNA hybridization procedure and antibiotic resistance. *Appl. Environ. Microbiol.* **54,** 446–453.

Freifelder, D. (1987). "Microbial Genetics." Jones & Bartlett, Boston, Massachusetts.

Freter, R. (1984). Factors affecting conjugal plasmid transfer in natural bacterial communities. *In* "Current Perspectives in Microbial Ecology" (M. J. Klug and C. A. Reddy, eds.), pp. 105–114. Am. Soc. Microbiol., Washington, D.C.

Gealt, M. A., Chai, M. D., Alpert, K. B., and Boyer, J. C. (1985). Transfer of plasmids pBR322 and pBR325 in wastewater from laboratory strains of *Escherichia coli* to bacteria indigenous to the waste disposal system. *Appl. Environ. Microbiol.* **49,** 836–841.

Germida, J. J., and Khachatourians, G. G. (1988). Transduction of *Escherichia coli* in soil. *Can. J. Microbiol.* **34,** 190–193.

Gile, J. D., Collins, J. C., and Gillett, J. W. (1982). Fate and impact of selected wood preservatives in a terrestrial model ecosystem. *J. Agric. Food Chem.* **30,** 295–301.

Gillett, J. W. (1988). The role of terrestrial microcosms and mesocosms in ecotoxicologic research. *In* "Ecotoxicology—Problems and Approaches" (S. A. Levin, M. Harwell, J. Kelly, and K. D. Kimball, eds.), pp. 367–410. Springer-Verlag, New York.

Godwin, D., and Slater, J. H. (1979). The influence of the growth environment on the stability of a drug resistance plasmid in *Escherichia coli* K12. *J. Gen. Microbiol.* **111,** 201–210.

Grabow, W. O. K., Prozesky, O. W., and Burger, J. S. (1975). Behavior in a river and dam of

coliform bacteria with transferable or non-transferable drug resistance. *Water Res.* **9,** 777–782.
Graham, J. B., and Istock, C. A. (1978). Genetic exchange in *Bacillus subtilis* in soil. *Mol. Gen. Genet.* **166,** 287–290.
Greenberg, E. P., Poole, N. J., Pritchard, H. A. P., Tiedje, J., and Corpet, D. E. (1988). Use of microcosms. *In* "Release of Genetically-Engineered Micro-Organisms" (M. Sussman, C. H. Collins, F. A. Skinner, and D. E. Stewart-Tull, eds.), pp. 265–274. Academic Press, London.
Grunstein, M., and Hogness, D. S. (1975). Colony hybridization: A method for the isolation of cloned DNAs that contain a specific gene. *Proc. Natl. Acad. Sci. U.S.A.* **72,** 3961–3965.
Haas, M. J., and Fleming, D. (1986). Use of biotinylated DNA probes in colony hybridization. *Nucleic Acids Res.* **14,** 3976.
Hakkaart, M. J. J., van Gemen, B., Veltkamp, E., and Nijkamp, H. J. J. (1985). Maintenance of multicopy plasmid Clo DF13. III. Role of plasmid size and copy number in partitioning. *Mol. Gen. Genet.* **198,** 364–366.
Hartl, D. L., Dykhuizen, D., Miller, R. D., Green, L., and DeFramond, J. (1983). Transposable element IS50 improves growth rate of *E. coli* cells without transposition. *Gene* **35,** 503–510.
Holben, W., and Tiedje, J. M. (1988). Application of nucleic acid hybridization in microbial ecology. *Ecology* **69,** 561–568.
Holloway, B. W. (1969). Genetics of *Pseudomonas*. *Bacteriol. Rev.* **33,** 419–443.
Jain, R. K., Burlage, R. S., and Sayler, G. S. (1988). Methods for detecting recombinant DNA in the environment. *CRC Crit. Rev. Biotechnol.* **8,** 33–84.
Jensen, V. (1968). The plate count technique. *In* "The Ecology of Soil Bacteria" (T. R. G. Gray and D. Parkinson, eds.), pp. 158–170, Univ. of Toronto Press, Toronto.
Johnen, B. G., and Drew, E. A. (1977). Ecological effects of pesticides on soil microorganisms. *Soil Sci.* **123,** 319–324.
Johnson, L. F., and Curl, E. A. (1972). "Methods for Research on the Ecology of Soil-Borne Plant Pathogens." Burgess, Minneapolis, Minnesota.
Jones, I. M., Primrose, S. B., Robinson, A., and Ellwood, D. C. (1980). Maintenance of some ColEl-type plasmids in chemostat culture. *Mol. Gen. Genet.* **180,** 579–584.
Jones, R., Broder, M., Doyle, J., and Stotzky, G. (1988). Effects of genetically engineered microorganisms on microbial populations and biochemical processes in soil. *Abstr. Annu. Meet. Soil Sci. Soc. Am.* p. 219.
Jones, S. A., and Melling, J. (1984). Persistence of pBR322-related plasmids in *Escherichia coli* grown in chemostat cultures. *FEMS Microbiol. Lett.* **22,** 239–243.
Kado, C. I., and Liu, S.-T. (1981). Rapid procedure for detection and isolation of large and small plasmids. *J. Bacteriol.* **145,** 1365–1373.
Kelly, W. J., and Reanney, D. C. (1984). Mercury resistance among soil bacteria: ecology and transferability of genes encoding resistance. *Soil Biol. Biochem.* **16,** 1–8.
Khalil, T. A., and Gealt, M. A. (1987). Temperature, pH, and cations affect the ability of *Escherichia coli* to mobilize plasmids in L broth and synthetic wastewater. *Can. J. Microbiol.* **33,** 733–737.
Kleeberger, A., and Klingmüller, W. (1980). Plasmid-free transfer of nitrogen-fixing capability to bacteria from the rhizosphere of grasses. *Mol. Gen. Genet.* **180,** 621–627.
Klein, D. A., and Casida, L. E., Jr. (1967). *Escherichia coli* die-out from normal soil as related to nutrient availability and the indigenous microflora. *Can. J. Microbiol.* **13,** 1461–1470.
Klute, A., ed. (1986). "Methods of Soil Analysis," Part 1. Am. Soc. Agron., Madison, Wisconsin.

Knudsen, G. R., Walter, M. V., Porteous, L. A., Prince, V. J., Armstrong, J. L., and Seidler, R. J. (1988). Predictive model of conjugative plasmid transfer in the rhizosphere and phyllosphere. *Appl. Environ. Microbiol.* **54**, 343–347.
Krasovsky, V. N., and Stotzky, G. (1987). Conjugation and genetic recombination in *Escherichia coli* in sterile and nonsterile soil. *Soil Biol. Biochem.* **19**, 631–638.
Leary, J. J., Brigati, D. J., and Ward, D. C. (1983). Rapid and sensitive colorimetric method for visualizing biotin-labeled DNA probes hybridized to DNA or RNA immobilized on nitrocellulose: Bio-blots. *Proc. Natl. Acad. Sci. U.S.A.* **80**, 4045–4049.
Lee, S. W., and Edlin, G. (1985). Expression of tetracycline resistance in pBR322 derivatives reduces the reproductive fitness of plasmid-containing *Escherichia coli*. *Gene* **39**, 173–180.
Levin, M. (1982). Reviews of environmental risk assessment studies sponsored by EPA. *Recomb. DNA Tech. Bull.* **5**, 177–180.
Levy, S. B., and Marshall, B. M. (1979). Survival of *E. coli* host–vector systems in the human intestinal tract. *Recomb. DNA Tech. Bull.* **49**, 91–97.
Levy, S. B., and Miller, R. V., eds. (1989). "Gene Transfer in the Environment." McGraw-Hill, New York.
Levy, S. B., Marshall, B. M., and Rowse-Eagel, D. (1980). Survival of *Escherichia coli* host–vector systems in the mammalian intestine. *Science* **209**, 391–394.
Lewin, B. (1977). "Gene Expression. Vol. 3: Plasmids and Phages." Wiley, New York.
Liang, L. N., Sinclair, J. L., Mallory, L. M., and Alexander, M. (1982). Fate in model ecosystems of microbial species of potential use in genetic engineering. *Appl. Environ. Microbiol.* **44**, 708–714.
Lin, E. C. C., Goldstein, R., and Syvanen, M. (1984). "Bacteria, Plasmids, and Phages." Harvard Univ. Press, Cambridge, Massachusetts.
Lipson, S. M., and Stotzky, G. (1987). Interactions between viruses and clay minerals. *In* "Human Viruses in Sediments, Sludges, and Soils" (V. C. Rao and J. L. Melnick, eds.), pp. 198–229. CRC Press, Boca Raton, Florida.
Lorenz, M. G., and Wackernagel, W. (1987). Adsorption of DNA to sand and variable degradation rates of adsorbed DNA. *Appl. Environ. Microbiol.* **53**, 2948–2952.
Lorenz, M. G., Aardema, B. W., and Wackernagel, W. (1988). Highly efficient genetic transformation of *Bacillus subtilis* attached to sand grains. *J. Gen. Microbiol.* **134**, 107–112.
McLaren, A. D., Luse, R. A., and Skujins, J. J. (1962). Sterilization of soil by irradiation and some further observations on soil enzyme activity. *Soil Sci. Soc. Am. Proc.* **26**, 371–377.
McPherson, P., and Gealt, M. A. (1986). Isolation of indigenous wastewater bacterial strains capable of mobilizing plasmid pBR325. *Appl. Environ. Microbiol.* **51**, 904–909.
Mancini, P., Fertels, S., Nave, D., and Gealt, M. A. (1987). Mobilization of plasmid pHSV106 from *Escherichia coli* HB101 in a laboratory-scale waste treatment facility. *Appl. Environ. Microbiol.* **53**, 665–671.
Maniatis, T., Fritsch, E. F., and Sambrook, J. (1982). "Molecular Cloning. A Laboratory Manual." Cold Spring Harbor Lab., Cold Spring Harbor, New York.
Marsh, E. B., and Smith, D. H. (1969). R factors improving survival of *Escherichia coli* K-12 after ultraviolet irradiation. *J. Bacteriol.* **100**, 128–139.
Martin, K., Parsons, L. L., Murray, R. E., and Smith, M. S. (1988). Dynamics of soil denitrifier populations: relationships between enzyme activity, most-probable-number counts, and actual N gas loss. *Appl. Environ. Microbiol.* **54**, 2711–2716.
Messing, J. (1979). A multipurpose cloning system based on the single stranded DNA bacteriophage M13. *Recomb. DNA Tech. Bull.* **2**, 43.

Miller, J. H. (1972). "Experiments in Molecular Genetics." Cold Spring Harbor Lab., Cold Spring Harbor, New York.

Miller, R. V. (1988). Potential for transfer and establishment of engineered genetic sequences. Trends Biotechnol. 6/Trends Ecol. Evol. **3**, S23–S27.

Miller, R. V., and Ku, C.-M. C. (1978). Characterization of Pseudomonas aeruginosa mutants deficient in the establishment of lysogeny. J. Bacteriol. **134**, 875–883.

Moodie, H. S., and Woods, D. R. (1973). Anaerobic R factor transfer in Escherichia coli. J. Gen. Microbiol. **76**, 437–440.

Morita, R. Y. (1982). Starvation-survival of heterotrophs in the marine environment. Adv. Microbial Ecol. **5**, 171–198.

Morrison, W. D., Miller, R. V., and Sayler, G. S. (1978). Frequency of F-116 mediated transduction of Pseudomonas aeruginosa in a freshwater environment. Appl. Environ. Microbiol. **36**, 724–730.

Ogram, A., Sayler, G. S., and Barkay, T. (1987). The extraction and purification of microbial DNA from sediments. J. Microbiol. Methods **7**, 57–66.

Omenn, G. S. (1986). Controlled testing and monitoring methods for microorganisms, In "Biotechnology Risk Assessment: Issues and Methods for Environmental Introductions" (J. Fiskel, and V. T. Covello, eds.), pp. 144–163. Pergamon, New York.

O'Morchoe, S. B., Ogunseitan, O., Sayler, G. S., and Miller, R. V. (1988). Conjugal transfer of R68.45 and FP5 between Pseudomonas aeruginosa strains in a freshwater environment. Appl. Environ. Microbiol. **54**, 1923–1929.

Page, A. L., Miller, R. H., and Keeney, D. R., eds. (1982). "Methods of Soil Analysis," Part 2. Am. Soc. Agron., Madison, Wisconsin.

Pertsova, R. N., Kunc, F., and Golovleva, L. A. (1984). Degradation of 3-chlorobenzoate in soil by pseudomonads carrying biodegradative plasmids. Folia Microbiol. **29**, 242–247.

Pettibone, G. W., Sullivan, S. A., and Shiaris, M. P. (1987). Comparative survival of antibiotic-resistant and -sensitive fecal indicator bacteria in estuarine water. Appl. Environ. Microbiol. **53**, 1241–1245.

Polak, J., and Novick, R. P. (1982). Closely related plasmids of Staphylococcus aureus and soil bacilli. Plasmid **7**, 152–162.

Pritchard, H. A. P. (1988). Use of microcosms. In "The Release of Genetically-Engineered Micro-Organisms" (M. Sussman, C. H. Collins, F. A. Skinner, and D. E. Stewart-Tull, eds.), pp. 265–274. Academic Press, London.

Pritchard, H. A. P., and Bourquin, A. W. (1984). The use of microcosms for evaluation of interaction between pollutants and microorganisms. Adv. Microbial Ecol. **7**, 133–215.

Rafii, F., and Crawford, D. L. (1988). Transfer of conjugative plasmids and mobilization of a nonconjugative plasmid between Streptomyces strains on agar and in soil. Appl. Environ. Microbiol. **54**, 1334–1340.

Renz, M., and Kurz, C. (1984). A colorimetric method for DNA hybridization. Nucleic Acids Res. **12**, 3435–3444.

Richter, M. W., Stotzky, G., and Amsterdam, D. (1973). Influence of exogenous agents on R-factor transfer in vivo and in vitro. Proc. Int. Congr. Bacteriol., 1st, Jerusalem p. 275.

Roszak, D. B., and Colwell, R. R. (1987). Survival strategies of bacteria in the natural environment. Microbiol. Rev. **51**, 365–379.

Roszak, D. B., Grimes, D. J., and Colwell, R. R. (1984). Viable but nonrecoverable stage of Salmonella enteritidis in aquatic systems. Can. J. Microbiol. **30**, 334–338.

Saiki, R. K., Gelfand, D. H., Stoffel, S., Scharf, S. J., Higuchi, R., Horn, G. T., Mullis, K. B., and Erlich, H. A. (1988). Primer-directed enzymatic amplification of DNA with a thermostable DNA polymerase. Science **239**, 487–491.

Saye, D. J., Ogunseitan, O., Sayler, G. S., and Miller, R. V. (1987). Potential for transduction of plasmids in a natural freshwater environment: effect of plasmid donor concentration and a natural microbial community on transduction in *Pseudomonas aeruginosa*. *Appl. Environ. Microbiol.* **53**, 987–995.

Sayler, G. S., Shield, M. S., Tedford, E. T., Breen, A., Hooper, S. W., Sirotkin, K. M., and Davies, J. W. (1985). Application of DNA–DNA colony hybridization to the detection of catabolic genotypes in environmental samples. *Appl. Environ. Microbiol.* **49**, 1295–1303.

Sayler, G. S., Jain, R. K., Ogram, A., Pettigrew, C. A., Houston, L., Blackburn, J., and Riggsby, W. S. (1986). Applications for DNA probes in biodegradation research. *Proc. Int. Symp. Microbial Ecol.*, 4th (F. Megusar and M. Gantar, eds.), pp. 499–508. Slovene Soc. Microbiol., Ljubljana, Yugoslavia.

Schilf, W., and Klingmüller, W. (1983). Experiments with *Escherichia coli* on the dispersal of plasmids in environmental samples. *Recomb. DNA Tech. Bull.* **6**, 101–102.

Schofield, P. R., Gibson, A. H., Dudman, W. F., and Watson, J. M. (1987). Evidence for genetic exchange and recombination of *Rhizobium* symbiotic plasmids in a soil population. *Appl. Environ. Microbiol.* **53**, 2942–2947.

Senior, E., Bull, A. T., and Slater, J. H. (1976). Enzyme evolution in a microbial community growing on the herbicide Dalapon. *Nature (London)* **263**, 476–479.

Sherratt, D. J. (1982). The maintenance and propagation of plasmid genes in bacterial populations. *J. Gen. Microbiol.* **128**, 655–661.

Singleton, P. (1983). Colloidal clay inhibits conjugal transfer of R-plasmid Rldrd-19 in *Escherichia coli*. *Appl. Environ. Microbiol.* **46**, 756–757.

Singleton, P., and Anson, A. E. (1981). Conjugal transfer of R-plasmid Rldrd-19 in *Escherichia coli* below 22° C. *Appl. Environ. Microbiol.* **42**, 789–791.

Singleton, P., and Anson, A. E. (1983). Effect of pH on conjugal transfer at low temperatures. *Appl. Environ. Microbiol.* **46**, 291–292.

Sjogren, R. E., and Gibson, M. J. (1981). Bacterial survival in a dilute environment. *Appl. Environ. Microbiol.* **41**, 1331–1336.

Skujins, J. (1984). Microbial ecology of desert soils. *Adv. Microbial Ecol.* **7**, 49–91.

Slater, J. H., and Godwin, D. (1980). Microbial adaptation and selection. In "Contemporary Microbial Ecology" (D. C. Ellwood, J. N. Hedger, M. J. Latham, J. M. Lynch, and J. H. Slater, eds.), pp. 137–160. Academic Press, London.

Slater, J. H., and Somerville, H. J. (1979). Microbial aspects of waste treatment with particular attention to the degradation of organic carbon compounds. In "Microbial Technology: Current Status and Future Prospects" (A. T. Bull, C. R. Ratledge, and D. C. Elwood, eds.), pp. 221–261. Cambridge Univ. Press, London.

Smith, H. W., Parsell, Z., and Green, P. (1978). Thermosensitive antibiotic resistance plasmids in enterobacteria. *J. Gen. Microbiol.* **190**, 37–47.

Smith, P. R., Farrell, E., and Dunican, K. (1974). Survival of R^+ *Escherichia coli* in seawater. *Appl. Microbiol.* **27**, 983–984.

Somerville, C., Knight, I. T., Straube, W. L., and Colwell, R. R. (1988). Probe-directed, polymerization-enhanced detection of specific gene sequences in the environment. *Abstr. Int. Conf. Release Genetically-Eng. Micro-Org.*, 1st, p. 36.

Southern, E. M. (1975). Detection of specific sequences among DNA fragments separated by gel electrophoresis. *J. Mol. Biol.* **98**, 503–517.

Stallions, D. R., and Curtiss, R., III (1972). Bacterial conjugation under anaerobic conditions. *J. Bacteriol.* **111**, 294–295.

Steffan, R. J., and Atlas, R. M. (1988). DNA amplification to enhance detection of genetically engineered bacteria in environmental samples. *Appl. Environ. Microbiol.* **54**, 2185–2191.

Steffan, R. J., Goksoyr, J., Bej, A. K., and Atlas, R. M. (1988). Recovery of DNA from soils and sediments. *Appl. Environ. Microbiol.* **54**, 2908–2915.
Stent, G. S., and Calender, R. (1978). "Molecular Genetics: An Introductory Narrative." Freeman, San Francisco, California.
Stotzky, G. (1960). A simple method for the determination of the respiratory quotient of soils. *Can. J. Microbiol.* **6**, 439–452.
Stotzky, G. (1965a). Microbial respiration. In "Methods of Soil Analysis," Part 2 (C. A. Black, ed.), pp. 1550–1572. Am. Soc. Agron., Madison, Wisconsin.
Stotzky, G. (1965b). Replica plating technique for studying microbial interactions in soil. *Can. J. Microbiol.* **11**, 629–636.
Stotzky, G. (1974). Activity, ecology, and population dynamics of microorganisms in soil. In "Microbial Ecology" (A. I. Laskin and H. Lechevalier, eds.), pp. 57–135. CRC Press, Boca Raton, Florida.
Stotzky, G. (1986). Influence of soil mineral colloids on metabolic processes, growth, adhesion, and ecology of microbes and viruses. In "Interactions of Soil Minerals with Natural Organics and Microbes" (P. M. Huang and M. Schnitzer, eds.), pp. 305–428. Soil Sci. Soc. Am., Madison, Wisconsin.
Stotzky, G. (1989). Gene transfer among bacteria in soil. In "Gene Transfer in the Environment" (S. B. Levy and R. V. Miller, eds.), pp. 165–222. McGraw-Hill, New York.
Stotzky, G., and Babich, H. (1986). Survival of, and genetic transfer by, genetically engineered bacteria in natural environments. *Adv. Appl. Microbiol.* **31**, 93–138.
Stotzky, G., and Krasovsky, V. N. (1981). Ecological factors that affect the survival, establishment, growth, and genetic recombination of microbes in natural habitats, In "Molecular Biology, Pathogenicity, and Ecology of Bacterial Plasmids" (S. B. Levy, R. C. Clowes, and E. L. Koenig, eds.), pp. 31–42. Plenum, New York.
Stotzky, G., and Mortensen, J. L. (1959). Effect of gamma radiation on growth and metabolism of microorganisms in an organic soil. *Soil Sci. Soc. Am. Proc.* **23**, 125–127.
Stotzky, G., Culbreth, W., and Mish, L. B. (1962a). An apparatus for growing plants with aseptic roots for collection of root exudates and CO_2. *Plant Physiol.* **37**, 332–341.
Stotzky, G., Goos, R. D., and Timonin, M. I. (1962b). Microbial changes occurring in soil as a result of storage. *Plant Soil* **16**, 1–19.
Stotzky, G., Schiffenbauer, M., Lipson, S. M., and Yu, B. H. (1981). Surface interactions between viruses and clay minerals and microbes: mechanisms and implications. In "Viruses and Wastewater Treatment" (M. Goddard and M. Butler, eds.), pp. 199–204. Pergamon, Oxford.
Sturtevant, A. B., and Feary, T. W. (1969). Incidence of infectious drug resistance among lactose-fermenting bacteria isolated from raw and treated sewage. *Appl. Environ. Microbiol.* **18**, 918–924.
Syvanen, A. C., Laaksonen, M., and Soderlund, H. (1986). Fast quantification of nucleic acid hybrids by affinity-based hybrid collection. *Nucleic Acids Res.* **14**, 5037–5048.
Talbot, H. W., Yamamoto, D. Y., Smith, M. W., and Seidler, R. J. (1980). Antibiotic resistance and its transfer among clinical and nonclinical *Klebsiella* strains in botanical environments. *Appl. Environ. Microbiol.* **39**, 97–104.
Terawaki, Y., Takayasu, H., and Akiba, T. (1967). Thermosensitive replication of a kanamycin resistance factor. *J. Bacteriol.* **94**, 687–690.
Tremaine, S. C., and Mills, R. L. (1987). Inadequacy of the eucaryotic inhibitor cycloheximide in studies of protozoan grazing on bacteria at the freshwater–sediment interface. *Appl. Environ. Microbiol.* **53**, 1969–1972.
Trevors, J. T. (1987a). Survival of *Escherichia coli* donor, recipient, and transconjugant cells in soil. *Water Air Soil Pollut.* **34**, 409–414.

Trevors, J. T. (1987b). R-plasmid transfer in soil. *Bull. Environ. Contam. Toxicol.* **39**, 74–77.

Trevors, J. T., and Oddie, K. M. (1986). R-plasmid transfer in soil and water. *Can. J. Microbiol.* **32**, 610–613.

Trevors, J. T., Barkay, T., and Bourquin, A. W. (1987). Gene transfer among bacteria in soil and aquatic environments: a review. *Can. J. Microbiol.* **33**, 191–198.

Trieu-Cuot, P., Carlier, C., Martin, P., and Couvalin, P. (1987). Plasmid transfer by conjugation from *Escherichia coli* to Gram-positive bacteria. *FEMS Microbiol. Lett.* **48**, 289–294.

Vandemark, P. J., and Batzing, B. L. (1987). "The Microbes: An Introduction to Their Nature and Importance." Benjamin/Cummings, Menlo Park, California.

van Elsas, J. D., Dijkstra, A. F., Govaert, J. M., and van Veen, J. A. (1986). Survival of *Pseudomonas fluorescens* and *Bacillus subtilis* introduced into two soils of different texture in field microplots. *FEMS Microbiol. Ecol.* **38**, 151–160.

van Elsas, J. D., Govaert, J. M., and van Veen, J. A. (1987). Transfer of plasmid pFT30 between bacilli in soil as influenced by bacterial population dynamics and soil conditions. *Soil Biol. Biochem.* **19**, 639–647.

Vieira, J., and Messing, J. (1982). The pUC plasmids, an M13mp7 derived system for insertion mutagenesis and sequencing with synthetic universal primers. *Gene* **19**, 259–268.

Vieira, J., and Messing, J. (1987). Production of single-stranded plasmid DNA. *Methods Enzymol.* **153**, 3–11.

Walter, M. V., Barbour, K., McDowell, M., and Seidler, R. J. (1987). A method to evaluate survival of genetically engineered bacteria in soil extracts. *Curr. Microbiol.* **15**, 193–197.

Walter, M. V., Olsen, R. H., Prince, V., Seidler, R. J., and Lyon, F. (1989a). Use of catechol dioxygenase for the direct and rapid identification of recombinant microbes taken from environmental samples. *In* "Rapid Methods and Automation in Microbiology and Immunology" (A. Balows, R. C. Tilton, and A. Turano, eds.), pp. 69–77. Brixia Academic Press, Brescia.

Walter, M. V., Porteous, A., and Seidler, R. J. (1989b). Formation of transconjugants on plating medium following *in situ* conjugation experiments. Submitted for publication.

Wang, Z., and Crawford, D. L. (1988). Effects of wildtype and recombinant *Streptomyces* on mineralization of organic carbon in soil. *Abstr. Int. Conf. Release Genet.-Eng. Micro-Org., 1st* p. 13.

Warnes, A., and Stephenson, J. R. (1986). The insertion of large pieces of foreign genetic material reduces the stability of bacterial plasmids. *Plasmid* **16**, 116–123.

Watrud, L. S., Perlak, F. J., Tran, M.-T., Kusano, K., Mayer, E. J., Miller-Wideman, M. A., Obukowicz, M. G., Nelson, D. R., Kreitinger, J. P., and Kaufman, R. J. (1986). Cloning of *Bacillus thuringiensis* subsp. *kurstaki* delta-endotoxin gene in *Pseudomonas fluorescens*: Molecular biology and ecology of an engineered microbial pesticide. *In* "Engineered Organisms in the Environment: Scientific Issues" (H. O. Halvorson, D. Pramer, and M. Rogul, eds.), pp. 40–46. Am. Soc. Microbiol., Washington, D.C.

Watson, B., Currier, T. C., Gordon, M. P., Chilton, M. D., and Nester, E. W. (1975). Plasmids required for virulence of *Agrobacterium tumefaciens*. *J. Bacteriol.* **123**, 255–264.

Weinberg, S. R., and Stotzky, G. (1972). Conjugation and genetic recombination of *Escherichia coli* in soil. *Soil Biol. Biochem.* **4**, 171–180.

Wickham, G. W., and Atlas, R. M. (1988). Plasmid frequency fluctuations in bacterial populations from chemically stressed soil communities. *Appl. Environ. Microbiol.* **54**, 2192–2196.

Wingfield, G. I., Davies, H. A., and Greaves, M. P. (1977). The effect of soil treatment on the response of the soil microflora to the herbicide dalapon. *J. Appl. Bacteriol.* **43**, 39–46.

Wollum, A. G., III (1982). Cultural methods for soil microorganisms. In "Methods of Soil Analysis," Part 2 (A. L. Page, R. H. Miller, and D. R. Keeney, eds.), pp. 781–802. Am. Soc. Agron., Madison, Wisconsin.

Xu, H.-S., Roberts, N., Singleton, F. L., Attwell, R. W., Grimes, D. J., and Colwell, R. R. (1982). Survival and viability of nonculturable *Escherichia coli* and *Vibrio cholerae* in the estuarine and marine environment. *Microbial Ecol.* **8**, 313–323.

Zaske, S. K., Dockins, W. S., and McFeters, G. A. (1980). Cell envelope damage in *Escherichia coli* caused by short-term stress in water. *Appl. Environ. Microbiol.* **40**, 386–390.

Zechman, J. M., and Casida, L. E., Jr. (1982). Death of *Pseudomonas aeruginosa* in soil. *Can. J. Microbiol.* **28**, 788–794.

Zeph, L. R., and Stotzky, G. (1988). Transduction by bacteriophage P1 in soil. *Abstr. Annu. Meet. Am. Soc. Microbiol.* p. 298.

Zeph, L. R., and Stotzky, G. (1989). Use of a biotinylated DNA probe to detect bacteria transduced by bacteriophage P1 in soil. *Appl. Environ. Microbiol.* **55**, 661–665.

Zeph, L. R., Onaga, M. A., and Stotzky, G. (1988). Transduction of *Escherichia coli* by bacteriophage P1 in soil. *Appl. Environ. Microbiol.* **54**, 1731–1737.

Zund, P., and Lebek, G. (1980). Generation time-prolonging R plasmid: correlation between increases in the generation time of *Escherichia coli* caused by R plasmids and their molecular size. *Plasmid* **3**, 65–69.

Zurkowski, W., and Lorkiewicz, Z. (1979). Plasmid-mediated control of nodulation in *Rhizobium trifolii*. *Arch. Microbiol.* **123**, 195–201.

Zwadyk, P., Jr., Cooksey, R. C., and Thornsberry, C. (1986). Commercial detection methods for biotinylated gene probes: comparison with ^{32}P-labeled DNA probes. *Curr. Microbiol.* **14**, 95–100.

Microbial Levan

YOUN W. HAN

United States Department of Agriculture
Agricultural Research Service
Southern Regional Research Center
New Orleans, Louisiana 70179

I. Introduction
II. Occurrence
III. Biosynthesis
    A. Levansucrase
    B. Specificity
    C. Inhibition and Activation
    D. Hydrolysis
IV. Chemical Structure and Properties
    A. Structure
    B. Properties
    C. Composition and Properties of *Bacillus polymyxa* Levan
V. Analytical Methods
    A. Levan
    B. Levansucrase
VI. Production of Levan
VII. Utilization of Levan
    A. Industrial Gums
    B. Blood Plasma Extender
    C. Sweeteners
    D. Other Applications
VIII. Summary
    References

I. Introduction

Industry uses large quantities of natural polysaccharides, and new sources of polysaccharides are being sought. In recent years, attention has been directed toward producing extracellular polysaccharides by microbial fermentation. Examples are dextran and xanthan gum produced by *Leuconostoc mesenteroides* and *Xanthomonas campestris*, respectively. Recently, the sugar industry has faced intense competition from high-fructose corn syrup, which is used as a low-cost alternative sweetener. This chapter investigates the possibility of producing microbial levan from sucrose for industrial applications.

Levans are fructans, natural polymers of fructose. The two main types of fructans are the levans, with mostly $\beta(2 \rightarrow 6)$ linkages, and the inulins, with $\beta(2 \rightarrow 1)$ linkages. Branched fructans with both types of linkages also exist. Levan is a common name for a fructan in which most fructose has $\beta(2 \rightarrow 6)$ linkages. A more descriptive name would be $(2 \rightarrow 6)$-β-D-fructan. Levans and inulins of low molecular weight (M_r; usually <5000) are abundant reserve carbohydrates in many plant tissues (Vandamme and Derycke, 1983). Many microorganisms, when grown on sucrose, produce extracellular levans of high molecular weight.

Microbial levan, like dextran, is often an undesirable by-product of sugar juice processing because it increases the viscosity of the processing liquor (Avigad, 1968; Fuchs, 1959). It was reported by Lippmann as early as 1881, and the name "levulan" was proposed for the compound. Greig-Smith (1901) showed that a strain of *Bacillus*, when grown on sucrose, produced fructans, and the name "levan" was introduced as analogous to "dextran." The term levulan now denotes partially degraded levan fractions. Early reports about levan are confusing because microbial nomenclature was unsystematic and materials were inadequately described. Levans have never had extensive industrial use.

The few existing review articles on levan either are obsolete or are part of reviews of other microbial polysaccharides (Avigad, 1968; Evans and Hibbert, 1946; Hehre, 1955; Pontis and Del Campillo, 1985). In this chapter, we present some of our recent work on microbial production of levan and review the earlier work of others.

II. Occurrence

A variety of microorganisms produce extracellular polysaccharides as a capsule attached to the cell wall or as a slime secreted into the growth medium. These materials are formed as a defense mechanism or as a food reservoir. Some soil microorganisms produce levan, especially *Bacillus* sp. Oral bacteria such as *Rothis dentocariosa*, *Streptococcus salivarius*, and *Odontomyces viscosus* produce and accumulate levan in human dental plaque (Higuchi et al., 1970; Manly and Richardson, 1968; Newbrun, 1969). Several species of yeast (Fuchs et al., 1985; Loewenburg and Reese, 1957) and fungi (Loewenburg and Reese, 1957) also produce levan. Table I lists some of the microorganisms that produce levan.

All species of bacilli belonging to group I in Smith et al. (1952) (*B. subtilis, B. megaterium, B. cereus,* and *B. pumilus*) produce levan from sucrose (Fuchs, 1959). Many strains of *B. polymyxa* that belong to group II also produce polysaccharides—mostly heteropolysaccharides, which consist of D-glucose, D-mannose, D-galactose, D-fructose, glucuronic acid, and pyruvate (McNeely and Kang, 1973; Glukhova et al., 1985; Mitsuda et al., 1981; Madden et al., 1986; Fukui et al., 1985; Ninomiya

TABLE I

Levan-Producing Microorganisms

Microorganisms	References
Acetobacter pasteurianus	Loewnburg and Reese (1957)
Actinomyces viscosus	Pabst (1977); Warner and Miller (1978)
Achromobacter sp.	Lindburg (1957)
Aerobacter aerogenes	Srinivasan and Quastel (1958)
Aerobacter levanicum	Evans and Hibbert (1946); Feingold and Gahatia (1957); Takeshita (1973)
Arthrobacter ureafaciens	Tanaka et al. (1985)
Azotobacter chroococum	Hestrin and Goldblum (1953); Schlubach and Berndt (1964)
Bacillus amyloliquefaciens	Mantsala and Puntala (1982)
Bacillus megaterium	Evans and Hibbert (1946)
Bacillus mesentericus	Hehre (1951)
Bacillus polymyxa	Hestrin el al. (1943); Ninomiya and Kizaki, (1970); Han (1989); Murphy (1952)
Bacillus subtilis	Dedonder (1966); Tanaka et al. (1981); Hestrin et al. (1943); Mantsala and Puntala (1982); Perlot and Monson (1984); Yamamoto et al. (1985)
Corynebacterium levaniformans	Dias and Bhat (1962)
Corynebacterium beticola	Abdou (1969)
Gluconobacter oxydans	Elisashvili (1981)
Leuconostoc mesenteroides	Corrigan and Robyt (1979)
Micobacterium laevaniformans	Fuchs et al. (1985)
Odontomyces viscosus	Krichevsky et al. (1969)
Phytomonas pruni	Evans and Hibbert (1946)
Pseudomonas sp.	Fuchs (1956); Evan and Hibbert (1946); Gross and Rudolph (1987)
Rothis dentocariosa	Lesher (1976)
Streptococcus sp.	Corrigan and Robyt (1979); Shimamura et al. (1987)
Streptococcus salivarius	Kleczkowski and Wierzchowski (1940); Fuchs (1956); Jung et al. (1987); Ribeiro et al. (1988); Dawes et al. (1966); Park et al. (1983); Lyness and Doelle (1983); Long et al. (1975)
Xanthomonas sp.	Dedonder and Peaud-Lenoel (1957); Fuchs (1959)
Zymomonas mobilis	Ribeiro et al. (1988); Dawes et al. (1966); Lyness and Doelle (1983); Viikari (1984)
Yeast	Fuchs et al. (1985); Loewenburg and Reese (1957)
Aspergillus sydawi	Loewenburg and Reese (1957)
Aspergillus versicolor	Loewenburg and Reese (1957)

and Kizaki, 1970). The ratios of various sugars in the polysaccharides depend on the composition of the growth medium and the microbial strain used. Polysaccharide yield increases with sucrose but decreases with glucose and arabinose: mannose has the lowest polysaccharide yield (Glukhova et al., 1985). Bacillus polymyxa grown on sucrose produces homopolysaccharides, consisting of 90–100% fructose, depending on the analytical method used (Murphy, 1952; Han, 1989).

Levans are also found in plants, mainly in monocotyledons, whereas inulins are found in dicotyledons. Levans, a fructan, are closely related to inulins, but have different structure and properties. Plant levans, also called phleins, have a much lower molecular weight than bacterial levans (Pontis and Del Campillo, 1985). Levans are distributed throughout plants: leaves contain small amounts, but large quantities are found in roots, bulbs, tubers, rhizomes, and sometimes in mature fruits, where they serve as storage carbohydrates and may increase resistance to cold and drought (Meier and Reid, 1982, Shiomi et al., 1976). The presence of levans (and fructans in general) in plants has no apparent correlation with the presence or absence of starch. No correlation has been found between the presence of levans (fructans) and the occurrence of C_3 and C_4 photosynthetic pathways (Bender and Smith, 1973). The water-soaked appearance of plants is often caused by levans formed by infected bacteria (Gross and Rudolph, 1988). Levans may help retain moisture and aggregate soil at the plant rhizosphere (Webley et al., 1965).

III. Biosynthesis

A. Levansucrase

The biosynthesis of levan requires the extracellular enzyme levansucrase (sucrose 6-fructosyltransferase, EC 2.4.1.10), which shows specificity for sucrose. Most research on the biosynthesis of levan has been conducted using the enzymes from B. subtilis, Aerobacter levanicum, and S. salivarius. These enzymes have been extensively purified and the mode of action expored (Bauer and Brisk, 1979; Elisashvili, 1984; Feingold et al., 1956; Fouet et al., 1984; LeBrun and Van Rapenbusch, 1980; Mantsala and Puntala, 1982; Pascal and Dedonder, 1972; Reese and Avigad, 1966). Levansucrase of B. subtilis is inducible and exocellular, whereas that of A. levanicum is constitutive and endocellular (Dedonder, 1966). It is uncertain whether levansucrase is one enzyme or a complex of multiple enzymes that synthesize the main chain $\beta(2 \rightarrow 6)$ and the branch $\beta(2 \rightarrow 1)$ linkages.

Chains of levan, like dextran and starch, grow in steps by repeated transfer of a hexosyl group from a donor to a growing acceptor molecule (Hestrin et al., 1956; Hehre, 1955; Sato et al., 1984).

$$n\text{C}_{12}\text{H}_{22}\text{O}_{11} + \text{HOR} \rightleftharpoons \text{H}(\text{C}_6\text{H}_{10}\text{O}_5)n + n(\text{C}_6\text{H}_{12}\text{O}_6)$$
Sucrose   Acceptor   Levan   Glucose

The action of levansucrase is a step-by-step addition of single fructofuranosyl units at the C_6 hydroxyl of the nonreducing fructose terminal unit of a growing levan chain (Fig. 1). The exact nature of this initiation step of the polymer growth from sucrose is not clearly understood. The levansucrase may catalyze the reactions of a readily reversible primary step and a subsequent irreversible step as follows:

$$\text{fru-R} + \text{enz} \rightleftharpoons \text{fru-enz} + \text{R} \qquad (1)$$

$$\text{fru-enz} + \text{acceptor} \rightarrow \text{fru-acceptor} + \text{enz} \qquad (2)$$

where fru is fructose, enz is levansucrase, and R represents a carbonyl of aldose. Possibly, the aldoside part of the substrate molecule is replaced by an enzyme-linked group, and partial decomposition of this levan precursor to aldose and ketose furnishes the energy necessary for levan synthesis. All levan-forming systems, either cell-free or whole-cell, produce fructose and a series of oligosaccharides as products of transfructosylation in addition to glucose and the polymer itself. The yield of

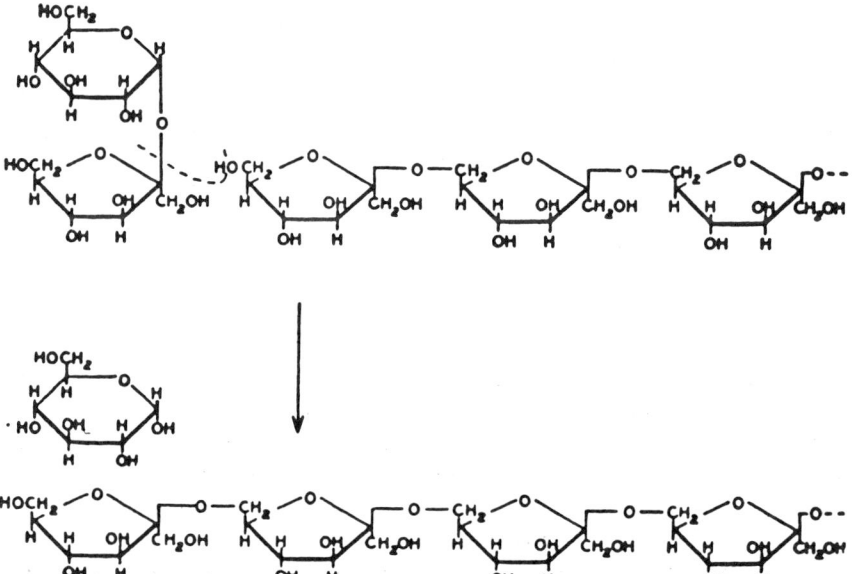

FIG. 1. Unit step in the growth of a levan molecule from sucrose, catalyzed by levansucrase. After Hehre (1951).

levan, therefore, is ~20–30% of sucrose utilized or 40–60% of fructose available (Avigad, 1968).

The enzyme utilizes sucrose as "donor substrate" with the Michaelis constant from 0.02 to 0.06 M. The optimal pH for levansucrase on sucrose is 5.0–6.0, and the activity at pH 4.4 and 7.5 was half the rate at pH 6.0 (Dedonder, 1966; Kiss, 1968). The enzyme activity is stable for several days at pH 5.0. The enzyme gradually loses its activity at 37°C and complete, almost immediate inactivation occurs at 100°C. The addition of metals (e.g., Fe^{3+}, Al^{3+}, Zn^{2+}) increases the stability of the enzyme in relation to temperature. Levansucrase prepared from B. subtilis has a sedimentation coefficient of 2.7, a diffusion coefficient of 6.0×10^{-7}, and a M_r of ~40,000 (Dedonder, 1966). An enzyme freeze-dried or kept at $-20°C$ shows no loss of activity for long periods.

Because hydrolytic breakdown of sucrose and polymerization of fructose occur simultaneously, the exact position of equilibrium in levan synthesis is difficult to define. The ratio of levan to sucrose at equilibrium is generally thought to be >1, because a reversal of the reaction has not been detected. A yield of levan as high as 62% of the theoretical maximum has been reported (Hehre, 1955). The presence of levan is not essential for levansucrase formation, but adding preformed primer (levan) to a levan-synthesizing system accelerates the rate of polymerization, increases final yield, and affects the production of homogeneous, high molecular weight levan (Mattoon et al., 1955; Hestrin, 1956). However, these effects occur only under conditions of low ionic strength (Tanaka et al., 1980). The degree of polymerization of levan is generally regulated by ionic strength, and the enzymatic synthesis of levan could occur in the absence of an "adapter" (preformed primer). Bacillus subtilis levansucrase synthesized levan far more effectively at lower than ambient temperatures (Tanaka et al., 1979).

B. Specificity

The specificity of levansucrase depends not only on the D-fructoside but also on the aldose residue of the substrate (Hestrin et al., 1956). Levansucrase of A. levanicum acts on sucrose, raffinose, and invert sugar (a mixture of glucose and fructose). Terminal fructose is generally belied to be necessary for the enzyme activity. However, some substrates with terminal fructose are not suitable for levan production. The enzyme does not utilize any other common sugars or other substrates having a terminal fructofuranose group (e.g., fructose phosphate, methylfructoside, and inulin). Levan yield from raffinose is about one-third of that from sucrose. The amounts of sucrose consumed are accounted for entirely as levan and reducing sugars. When raffinose (gal-glu-fru) is used as a substrate, the products are levan, melibiose (gal-glu), and fructose, but not sucrose and galactose. A series of glucose-ended, 2,6-linked polysaccharides can be used in synthesis of the levan molecule (Hehre, 1951). However, neither levanbiose, levantriose, nor levantetraose is found to serve as an acceptor substrate. In addition, levansucrase from B. subtilis appeared to have some affinity for glucose but not for fructose.

C. Inhibition and Activation

The formation of levan from sucrose by levansucrase is inhibited by various sugars and sugar alcohols (Table II). Inhibition was caused by both sugars and glycosides, for example, by glucose, methyl-D-glycoside, and others that have a configuration about carbon atom 2 similar to that of D-glucose, such as D-galactose, maltose, and lactose (Hestrin and Avineri-Shapiro, 1944). However, no appreciable inhibition occurred for products like D-mannose, D-fructose, or D-mannitol, for which configuration or structure at carbon atom 2 differs from that of D-glucose. The inhibition of levansucrase activity by glucose is a function of glucose concentration, and complete inhibition of levan production from 1% sucrose occurs at 16% glucose concentration. The inhibitory effect diminished as the glucose concentration fell relative to that of sucrose. Thus, it appears that glucose inhibition is caused

TABLE II

Inhibition of Levansucrase Activity by Sugars and Sugar Derivatives[a]

Sugar	(%) Inhibition
Lactose	76
D-Xylose	76
L-Arabinose	76
D-Glucose	62
L-Sorbose	62
Methyl-D-glucoside	49
D-Galactose	42
Maltose	37
L-Sorbitol	10
D-Fructose	4
D-Mannitol	3
D-Mannose	2

[a] Adapted from Hestrin and Avineri-Shapiro (1944).

by competition with the glucose moiety of sucrose for the enzyme (Tkachenko and Loitsyanskaya, 1976). Polyethylene glycol of M_r 4000 has an activating effect, increasing the maximum velocity by a factor of three (Dedonder, 1966).

D. Hydrolysis

Many levan-forming microorganisms also produce hydrolytic enzymes, levanases, that degrade the levan (Hestrin and Goldblum, 1953; Avigad, 1965). Certain strains of *Pseudomonas, Azotobacter, Aerobacter, Serratia, Bacillus, Clostridium* produce exocellular levanase (Fuchs, 1959). Levansucrase itself is suspected to cause the hydrolysis under certain conditions (Rapoport and Dedonder, 1963). Although indirect evidence of the reversal of enzymatic synthesis of levan can be observed, little is known about the nature of such enzyme degradation. Smith (1976) showed that β-fructofuranosidase present in tall fescue degraded levan by removing one fructose residue at a time until a molecule of sucrose remained. Levansucrase of *B. subtilis* has a hydrolytic effect on small levans (Dedonder, 1966). This hydrolytic action stops at branchpoints. Neither inulin, inulobiose, inulotriose, nor methyl D-fructofuranoside is hydrolyzed, although these substrates are hydrolyzed by inulinase and yeast invertase. This hydrolytic activity may be responsible for the appearance of heterogeneous short-chain polysaccharides, rather than uniform high molecular weight polymers, in the final product of many levan preparations.

IV. Chemical Structure and Properties

A. Structure

Levans are polymers of D-fructose attached by $\beta(2 \rightarrow 6)$ linkages that carry a D-glucosyl residue at the end of the chain. They constitute a series of homologous oligosaccharides and polysaccharides, which can be considered derivatives of sucrose (Hirst, 1957; Tanaka et al., 1981). Figure 2 shows the chemical structures of levan and inulin. They are usually represented by the formula G-F-(F)$_n$, where G-F denotes a sucrosyl group, F the fructose, and n the number of fructose units present in the molecule. Although the structure of levan is represented by a straight chain of $\beta(2 \rightarrow 6)$, linkages of many bacterial levans are branched through $\beta(2 \rightarrow 1)$ bonds (Feingold and Gahatia, 1957; Avigad and Feingold, 1965; Dawes et al., 1966; Khorramian and Stivala, 1982; Lindberg et al., 1973; Marshall and Weigel, 1980). The branch chains are usually short and sometimes consist of one fructose residue.

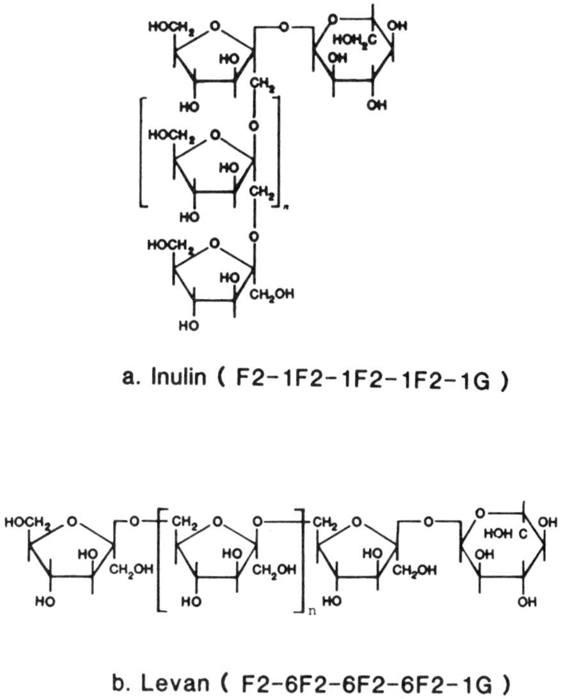

a. Inulin (F2-1F2-1F2-1F2-1G)

b. Levan (F2-6F2-6F2-6F2-1G)

FIG. 2. Chemical structure of (a) inulin and (b) levan.

The structures of levans synthesized in cell-free enzyme systems are similar to those produced by whole-cell systems, but they differ in length. The average chain length of levans produced by the enzyme system was reported to be 10–12 monomeric units, whereas that by whole cell was much longer; sometimes the molecular weight exceeded several million (Evans and Hibbert, 1946; Marshall and Weigel, 1980; Stivala and Zweig, 1981; Tanaka et al., 1983; Pontis and Del Campillo, 1985). In general, levans produced by different organisms have similar structures. The difference may be a varying degree of polymerization and branching of the repeating unit.

Levans are one of the few natural polymers in which the carbohydrate exists in the furanose form. This structural feature plays an important role in the final conformation of the molecules in solution (Marchessault et al., 1980). Moreover, the enhanced flexibility of the furanose ring in comparison with the relatively rigid pyranose ring of the majority of reserve polysaccharide gives additional flexibility to the whole fructan molecule. The linear $2 \rightarrow 6$-linked levan molecules are flexible and tend to be left-hand twist, whereas the twist of inulin is right-handed (French, 1989).

B. Properties

The composition and properties of levan depend greatly on the environmental factors in which the microorganisms are grown (Stivala and Khormanian, 1982). The general properties of levans resemble those of dextrans. Levans are levorotatory, amorphous or microcrystalline, of varying solubility in cold water, very soluble in hot water, and insoluble in absolute ethyl alcohol. Levans are generally more soluble than inulin, which is almost insoluble (<0.5%) in water at room temperature (Phelps, 1965). The high solubility of levan may be a characteristic of $\beta(2 \rightarrow 6)$ linkage compared to $\beta(2 \rightarrow 1)$ linkage. Branching may be only a support factor. Levans are nonreducing, not hydrolyzed by yeast invertase and amylase action, but very susceptible to hydrolysis by acid. They are not colored by iodine, but hydrogen chloride imparts a purple color that distinguishes levan from other polysaccharides not containing fructose (Pontis and Del Campillo, 1985). The molecular weight and viscosity of levans vary depending on the organisms used. Generally, their viscosity in aqueous solution increases sharply when various salts are present, but it decreases drastically with a small increase in temperature (Kang and Cottrell, 1979).

Certain biological properties of levan, such as promotion of infection and necrosis (Feingold and Gehatia, 1957; Shilo, 1959), tumor inhibition and stimulation (Leibovici and Stark, 1984; Yavetz et al., 1985; Keibovici et al., 1986; Stark and Leibovici, 1986), and increase in cell permeability for a cytotoxic agent (Leibovici and Stark, 1985) have attracted attention. These effects are partly caused by suppression of normal inflammatory response. Only levan with $M_r > 10^7$ promotes infection; the effect is lost when the polymer degrades. Levan given intravenously to mice greatly increases the virulence of intraperitoneally injected bacteria. This is partly caused by the intravenous levan sealing the vascular lining, thus affecting its permeability and preventing escape of blood constituents into the peritoneal cavity. The endothelial sealing of levan may have practical importance. Natural levans are serologically active and elicit antibody production, but purified levan preparations are not antigenic.

C. Composition and Properties of *Bacillus polymyxa* Levan

The levan produced by a strain of *B. polymyxa* (NRRL B-18475) consisted of ~98% fructose when analyzed by high-performance liquid chromatography (HPLC) of the acid hydrolysate (Han and Clarke, 1990). The product was readily soluble in water (up to ~5%) at room temperature, but further increase in levan concentration produced a colloid. The product was very susceptible to hydrolysis when heated for 15 minutes in 0.5% oxalic acid. The easy hydrolysis is typical of the fructofuranose structure (Feingold and Gehatia, 1957). Because the initial molecule in levan formation is sucrose, a terminal glucose should be present in levan chains. However, because of the small portion of terminal groups in high molecular weight levan, no significant amount of glucose was observed on hydrolysis.

A 5% aqueous solution of crude levan, after dialysis through a membrane with 12,000 Da cut off, gave one sharp, clean peak just below 2 × 10^6 Da on Sephacryl S-500 (Clarke et al., 1988). This peak was sharper (representing a narrower range of molecular mass) than those of the commercially available dextrans used as gel permeation chromatography (GPC) standards. The uniformity of the product was perhaps caused by long fermentation (≤10 days) and absence of hydrolytic activity in the enzyme system. The compound was stable in aqueous solution at pH 4.5 for ≤36 hours. The B. polymyxa (NRRL B-18475) levan has a specific optical rotation $[a]^{24}$ of $-42.0°$, whereas most other levan preparations were reported to be $-44 \pm 4°$ (Schlubach and Berndt, 1964; Murphy, 1952; Dedonder and Peaud-Lenoel, 1957; Barker et al., 1955). The B. polymyxa levan was nonhygroscopic; lyophilized sheets of levan have been maintained under atmospheric conditions for ≤6 months.

In Fig. 3, carbon-13 nuclear magnetic resonance (^{13}C-NMR) peaks

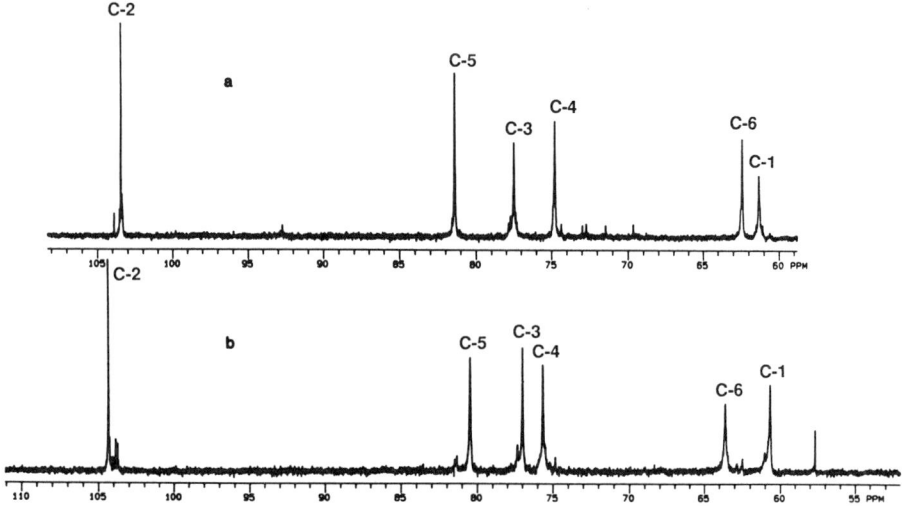

FIG. 3. ^{13}C-NMR spectrum of (a) inulin and (b) levan (spectra by S. Ellzey).

from levan are compared to peaks from inulin. The ^{13}C-NMR spectra showed six main resonances at 104.2, 80.5, 77.0, 75.7, 63.6, and 60.7 ppm, which is almost identical to the peak positions for levan previously identified by Shimamura et al. (1987). The peak positions vary slightly from values in the literature, but the relative spacings between the values obtained from experimental levan and those from the literature are similar. The anomeric peak (C-2) is at ~104 ppm for both. The primary carbons (C-1 and C-6) are more closely grouped in inulin, and the ring carbons (C-3, C-4, and C-5) are more closely grouped in levan; this is characteristic of the differences between inulin and levan (Seymour et al., 1979). Data clearly show the polysaccharide produced by the B. polymyxa (NRRL B-18475) to be levan type with the linkage of $\beta(2 \rightarrow 6)$-fructofuranoside (Table III). Figure 4 shows the infrared spectra of bacterial levan and inulin. The infrared spectra of the bacterial levan were similar to inulin but the bacterial levan showed a characteristically weaker peak at 984 (cm^{-1}) (Barker and Stephens, 1954). Lesher (1976) reported that levan had infrared absorption peaks (cm^{-1}) at 919, 860, and 803, whereas inulin had peaks at 933, 872, and 813.

Methylation analysis revealed that the B. polymyxa levan was made of 71% $\beta(2 \rightarrow 6)$ linkage (1,3,4-trimethyl-D-fructose), 13% terminal group at 1 or 2 position (1,3,4,6- or 2,3,4,6-tetramethyl-D-fructose), 12% branching at 1,2, or 6 position (3:4 dimethyl-D-fructose), and 4% free fructose (Clarke et al., 1988). The branches occurred in the C-1 position with a $\beta(1 \rightarrow 2)$ linkage, with side chains of $\beta(2 \rightarrow 6)$-lined residues. Murphy (1952) also reported a similar composition for B. polymyxa

TABLE III

CHEMICAL SHIFTS FOR ^{13}C-NMR SPECTRA OF INULIN, LEVAN, AND THE POLYSACCHARIDE PRODUCED BY Bacillus polymyxa (NRRL B-18475)

Carbon atom	Chemical shift (ppm)		
	Inulin [a]	Levan [a]	Isolate
C-1	60.9	59.9	60.7
C-2	103.3	104.2	104.4
C-3	77.0	76.3	77.0
C-4	74.3	75.2	75.7
C-5	81.1	80.3	80.5
C-6	62.2	63.4	63.6

[a] Assignment cited from Shimamura et al. (1987).

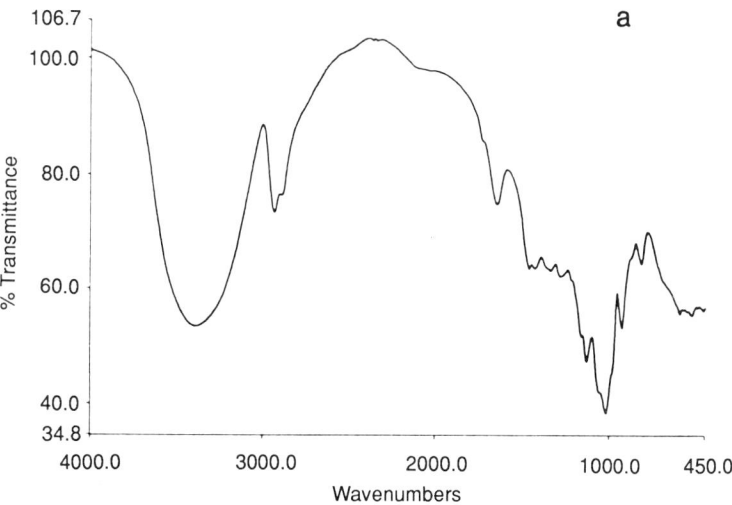

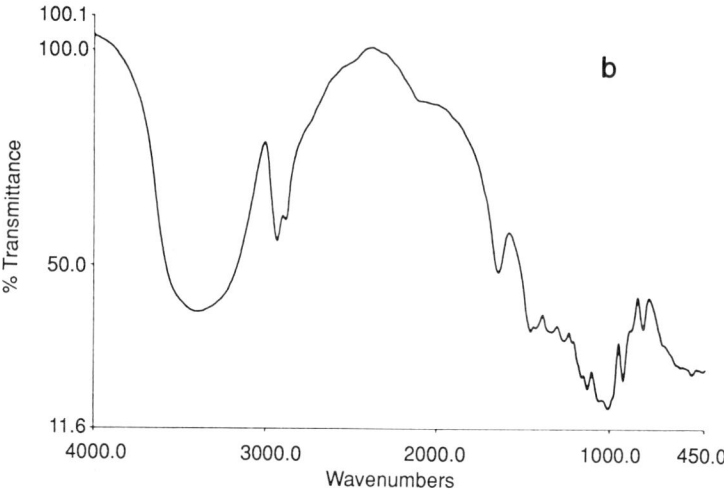

FIG. 4. Infrared spectra of (a) inulin and (b) levan (spectra by N. Morris).

levan. The degree of branching in other microbial levans is reported to be 5–20% (Lindberg et al., 1973).

The *B. polymyxa* polysaccharides exhibit relatively high viscosity and high pseudoplasticity (Kang et al., 1983). The viscosity of the product is highest when grown on sucrose or glucose and lowest on xylose. Dea and Madden (1985) report that *B. polymyxa* (NCIB 11429) polysaccharides contain two components, and the polysaccharides when dissolved in water, exhibit temperature-dependent yield-stress solution and gel-forming properties, depending on the concentration.

V. Analytical Methods

A. LEVAN

Levan may be determined by using any of the colorimetric methods specific for ketose and expressed as their fructose content (Lee et al., 1979). The thiobarbituric acid method (Percheron, 1962) or the acid resorcinol color reagent method (Roe et al., 1949; Yaphe and Arsenault, 1965) may be used. Changes in reducing power brought about by acid hydrolysis (e.g., 0.5% oxalic acid, 100°C, 15 minutes) may also be used. The amount of fructose in the hydrolysate can be determined by the degree of optical rotation. The free sugars and levan in the fermentation mixture can be separated by passing the mixture through a column of cellulose gel filtration medium (e.g., Aminex Matrex Cellufin, GH-25 medium, Amicon Corp, Danvers, Massachusetts). A small amount of a mixture of levan and dextran was analyzed by passing the sample through a polarimeter and using an automated resorcinol–thiourea test (Manly and Cormier, 1970).

A direct method of measuring levan is by HPLC. Levan and free sugars were easily separated on an anion exchange column (e.g., Aminex HPX-87C, Biorad, Richmond, California) (Han and Clarke, 1989). Paper and thin-layer chromatography (TLC) can be used for determination of short-chain components in levan-producing systems or levan hydrolysate. Small quantities of the nonmobile levan can be separated from other keto sugars by TLC. The position of levan can be identified by spraying reagents specific for levan, such as resorcinol HCl and napthoresorcinol HCl (Forsyth, 1950; Yaphe and Arsenault, 1965), urea HCl and urea metaphosphoric acid (Dedonder, 1966), and dimedon [(5,5-dimethyl-1,3-cyclohexanedione)phosphoric acid] (Adachi, 1964). Levan in fermentation broth can be approximated by weighing the precipitate formed by adding ethanol or isopropanol. The precipitate can be purified in advance by redissolving in water, dialyzing, and freeze drying.

The structure of levan can be studied by instrumental analysis. Infrared spectroscopy is widely used to investigate polymer structure and to analyze functional groups. The infrared absorption spectrum of levan has typical bands, which can help in its analysis and differentiate it from inulin and other polysaccharides (Hestrin et al., 1954; Lesher, 1976).

Because D-fructofuranosides are heat-labile and partly decompose during derivatization to forms suitable for analysis, ^{13}C-NMR spectroscopy is a useful method to determine the linkage type of polyfructan (Shimamura et al., 1987; Seymour et al., 1979). The ^{13}C-NMR spectrum of levan and inulin produces six main resonances, and each has its own distinguishing characteristic. These methods are much less involved than classical methylation analysis, in which polysaccharides are exhaustively methylated, hydrolyzed, derivatized with alditol acetate, and the methyl ether obtained is analyzed by gas–liquid chromatography (Hakomori, 1964). These studies provide information on the linkage type and extent of branching, but cannot show the length of side chains that may be of great importance in studying immunochemical properties (Lindberg et al., 1973).

Immunoassay methods utilizing antifructan activity of some myeloma immunoglobulins have been proposed (Glaudemans, 1975). Some are specific for levan (Cisar et al., 1974), whereas others are specific for inulin (Streefkerk and Glaudemans, 1977). The precipitate formed with concanavalin A differentiates levan and inulin (Goldstein and So, 1965). Although no attempts have been made to quantify levan by this method, the method may make it possible to determine levan in a highly specific way.

B. LEVANSUCRASE

Levansucrase is an enzyme responsible for forming levan from sucrose and accumulating glucose in the menstruum. Therefore, levansucrase assay is generally based on the determination of free glucose in the reaction mixture (Chambert et al., 1974; Tanaka et al., 1979). The reaction conditions are set so that the velocity of glucose formation depends only on the amount of enzyme (zero-order reaction). The reaction is stopped by dilution in boiling buffer; the free glucose is determined by the glucose oxidase method or the Somogi–Nelson method (Ceska, 1971). This method can be used for levansucrase in cultural broth, cell extract, or intact cells. However, this method cannot be used to test for activity in cultures grown on glucose as a substrate. The addition of levan primer to the reaction mixture provides more consistent conditions for the assay during purification. Because conditions of the reaction produce significant variations in levan yields, any methods based on levan determination cannot be used to assay the enzyme activity. Such a determination may be useful for the study of the reaction conditions.

VI. Production of Levan

All known production of exopolysaccharides is by aerobic submerged fermentation. Microbial production of polysaccharides requires fermentation and handling of a highly viscous solution, which is not involved in many nonviscous fermentation processes. Unique fermentation parameters of aeration, agitation, and pH and temperature controls should be established for each polysaccharide fermentation. The conditions for producing levan by growing cultures of bacteria vary according to the microorganisms used. Figure 5 shows the flow diagram of levan production by *B. polymyxa*.

Bacillus polymyxa (Strain NRRL B-18475) produces a large quantity of extracellular polysaccharides when grown on 4–16% sucrose solution. Table IV lists the composition of the medium for levan production. Yeast extract enhances the levan yield. The highest level of levan

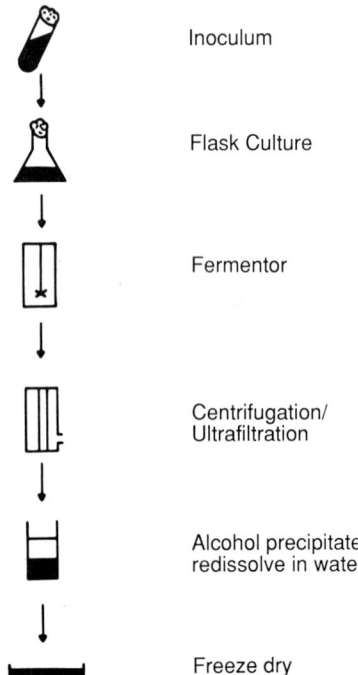

FIG. 5. Flow diagram of levan production by *Bacillus polymyxa* strain NRRL B-18475.

TABLE IV

Composition of Medium for Levan Production

Component	Percentage of total
Sucrose	15.0
Peptone	2.0
Yeast extract	2.0
K_2HPO_4	2.0
$(NH_4)_2SO_4$	2.0
$MgSO_4 \cdot H_2O$	0.2

was obtained in media containing 8–15% sucrose. The organism converted sucrose (S) to levan (L) and accumulated a small amount of glucose (G) in the growth medium (Fig. 6). During fermentation, the sucrose levels dropped and levan started to appear in 2 days; thereafter, sucrose levels gradually decreased as levan increased. Glucose was the major by-product. A small amount of fructose (F) and unidentified fermentation products smaller in molecular weight were also observed. The pH of the growth medium fell from 7.0 to 4.7 because of acid

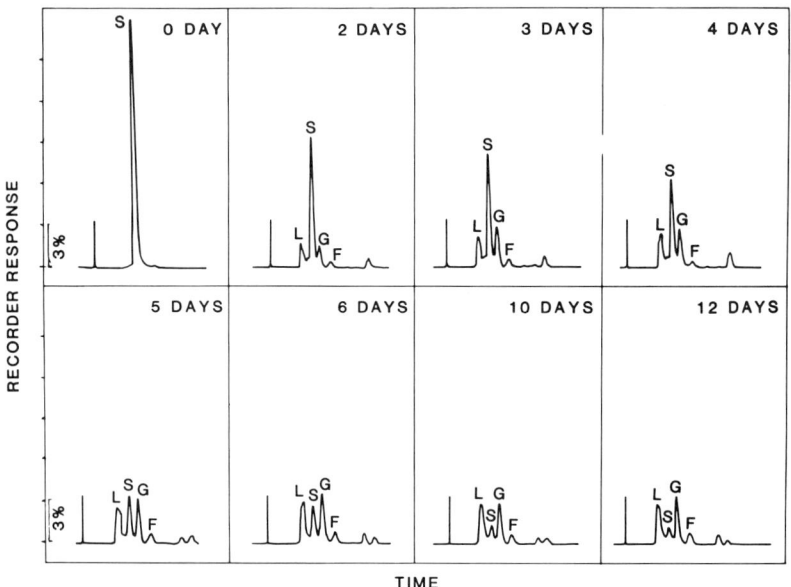

Fig. 6. Levan formation during fermentation of sucrose by *Bacillus polymyxa* strain NRRL B-18475.

production. In other levan production, maintaining pH above 5.5 was important because the optimum pH for levansucrase was 5.5–7.0 and levan may be hydrolyzed at a lower pH (Avigad, 1968). Optimum temperature for growth and levan production was ~30°C.

Aeration is important in biosynthesis of levansucrase (Tkachencho and Loitsyanskaya, 1979). The polysaccharide production was especially pronounced when the culture was gently shaken during cultivation, but vigorous agitation and aeration inhibited levan production (Han, 1989). A small amount of microbial polysaccharide (alcohol precipitate) was also produced when the organism was grown on lactose, maltose, or raffinose, but not on glucose or fructose. The organism produced polysaccharide from sugarcane juice, but the yield was much less than that from sucrose. High sucrose concentration lowered the average molecular weight of the levan synthesized (Dedonder and Peaud-Lenoel, 1957).

Levan was harvested by precipitation from the culture broth by adding ethanol or isopropanol. Acetone and methanol can also be used. The yield and consistency of the product varied depending on the amount of alcohol added. The levan started to precipitate at the medium–alcohol ratio of 1:1.2, and the yield peaked at ~1:1.5. Further increase in the ratio hardened the levan and made the product less fluid. Slightly less isopropanol than ethanol was needed to precipitate levan. Although most of the bacterial cells, unfermented sugars, and other solubles remained in the aqueous alcohol phase, preremoval of microbial cells by centrifuging was needed to obtain a pure form of levan. The product was further purified by repeated precipitation and dissolution in water, followed by dialysis or ultrafiltration. The final product was an off-white, gummy material that could be freeze- or vacuum-dried. Alternate methods of harvesting would be drying by trituration and pulverization with absolute alcohol in a high-speed blender. In a typical fermentation, *B. polymyxa* produced ~3.6 g of levan in 100 ml of 15% sucrose in 10 days (~50% yield on available fructose). Methods of preparing levan by *A. levanicum* and from a cell-free enzyme–sucrose system have been reported (Hestrin *et al.*, 1956; Avigad, 1965).

VII. Utilization of Levan

A. INDUSTRIAL GUMS

Industry consumes natural polymers in tremendous quantities. They contribute important properties to products, even when used in low concentration, and do this at low cost. Useful properties of polysaccharides include providing viscosity; solubility in water and oil; suspending properties; rheological properties; compatibility with salts and surfactants; stability when exposed to heat, acid, and alkali; film formation; holding capacity for water and chemicals; and biological activities. Commercial polysaccharides are water-soluble or water-dispersible hydrocolloids. Their aqueous dispersion usually possesses suspending, dispersing, and stabilizing properties, and can be used to emulsify and stabilize foods.

Natural polysaccharides can be modified to alter their chemical and physical properties. Modification of a low-cost polysaccharide can make it a suitable replacement for a more expensive gum. In general, the properties of neutral polysaccharides can be remarkably altered by introducing a very small amount of substituent groups of either neutral or ionic types.

Several factors are important in selecting a polysaccharide for industrial use. Among these are physical and chemical properties, production cost, consistency in composition and supply, possibility of eventual replacement by another, and acceptability by government agencies if the product is intended for food use (Whistler, 1973). A host of microbial polymers with widely differing properties is available; fermentation technology is expected eventually to develop others to meet many industrial needs. Because it is similar in physical properties and mode of production to dextran, levan could be used in similar ways. Levan should be considered for applications regarding low viscosity, high water solubility, and susceptibility to acid hydrolysis.

B. Blood Plasma Extender

Bacterial dextran has been extensively used as a plasma substitute. The extender is prepared by partial hydrolysis of the high molecular weight polysaccharides and subsequent methanol fractionation of the hydrolyzed product to obtain a material ranging in M_r from 25,000 to 200,000 (Hines et al., 1953). The necessity for hydrolysis, however, is one of the major disadvantages of clinical dextran preparation.

Because levan is similar in physicochemical properties to bacterial dextran, using levan as a blood volume extender has been studied (Mattoon et al., 1955; Schechter and Hestrin, 1963). Native levan was hydrolyzed at pH 3.2 to an extent of 1–2% hydrolysis and M_r between 30,000 and 100,000. After neutralization, fractions of polysaccharide were collected by precipitation with increasing ethanol concentration (Hestrin et al., 1956). A partial acid hydrolysis of levan produced a series of oligosaccharides and levulans (degraded levans) (Feingold and

Gehatia, 1957). Some of the lower members of this series could be separated from a column of charcoal celite by elution with aqueous ethanol and used as a blood plasma volume extender. Because levan injected into the bloodstream is slowly eliminated from the body and is not toxic (Schechter and Hestrin, 1963), it is a potential blood volume extender equivalent to dextran or other currently used materials. However, native levans with high molecular weights are serologically active and elicit antibody formation; hydrolyzed or purified levans are not antigenic (Hehre, 1955; Avigad, 1968).

C. Sweeteners

Because levan is a polymer of fructose, its hydrolytic products may be used as sweetener or precursors of sweetener. Industrial production of ultrahigh-fructose glucose syrups (UHFGS) from inulin has been reported (Vandamme and Derycke, 1983). Jerusalem artichoke was hydrolyzed by acid or enzyme (inulinase), and the resulting fructose was precipitated and purified. The fructose syrup thus obtained was adjusted to the desired fructose content by concentration or blending. Enzymatic hydrolysis was preferred over acidic hydrolysis, because the former produced more fructose and also yielded a product with better flavor and less color (Grootwassink and Fleming, 1980). Acid hydrolysis of a microbial levan produced a series of $\beta(2 \rightarrow 6)$-fructofuranosyl oligosaccharides (di-, tri-, and tetramers) (Kennedy et al., 1989). Although their sweetness and nutritional values have not been studied, their characteristics may resemble those of Neosugar, a fructooligosaccharide, nonnutritive sweetener (Hidaka, 1983; Adachi, 1983). Neosugar is prepared from sucrose by the action of fructosyltransferase of fungi such as *Aspergillus* sp. (Hidaka, 1983; Pazur, 1952), *Fusarium* sp. (Gupta and Bhatia, 1982), and *Aureobasidum pullulans* (Smith et al., 1982; Jung et al., 1987). Neosugar has a pleasant taste with half the sweetness of sucrose but is not utilized by the body (Oku et al., 1984).

D. Other Applications

Levans could possibly be used as an emulsifier, formulation aid, stabilizer and thickener, surface-finishing agent, encapsulating agent, and carrier for flavor and fragrances. Incorporation of levans in photographic emulsion to improve silver granularity has been suggested (Chambers and Overman, 1964). Cross-linked levan preparations may be used for molecular sieves for gel filtration. The low viscosity and high solubility of levan may make it a suitable substitute for gum arabic in a variety of uses in foods, pharmaceuticals, medicine, cosmetics, adhesives, paints, inks, lithography, textiles, and others. The high solubility of gum arabic is responsible for its excellent stabilizing and emulsifying properties when incorporated with large amounts of insoluble materials (Glicksman, 1973).

VIII. Summary

Levans are natural polymers of the sugar fructose found in many plants and microbial products. Like dextrans, they are formed as an undesirable by-product of sugar juice processing. On the other hand, levans, which can only be produced from sucrose, have potential industrial applications as thickeners and encapsulating agents and could provide additional, valuable products from sugarcane juice. A strain of *B. polymyxa* (NRRL B-18475) produced a high yield of polysaccharide when grown on sucrose solution. Hydrolysis and subsequent analyses showed the product to consist entirely of D-fructose. ^{13}C-NMR and methylation analyses indicated the products to be a $\beta(2 \rightarrow 6)$-linked polymer of fructose, with 12% branching. The polysaccharide has a M_r of ~2 million and is readily soluble in water. Levan has not been utilized, but if developed, could be useful in food and other industrial applications.

Aknowledgments

The author acknowledges L. Ban-Koffi, M. Watson, A. French, M. Godshall, C. Tsang, M. Clarke, S. Ellzey, and N. Morris for their assistance in various aspects of preparing this manuscript.

References

Abdou, M. A. F. (1969). *Phytopathol. Z.* **66**, 147–167.
Adachi, S. (1964). *Anal. Biochem.* **9**, 224.
Adachi, T. (1983). *Jpn. Food Sci.* **22**, 71–78.
Avigad, G. (1965). *Methods Carbohydr. Anal.* **5**, 161.
Avigad, G. (1968). *Encyl. Polym. Sci. Technol.* **8**, 711–718
Avigad, G., and Feingold, D. S. (1965). *Biochem. Biophys.* **70**, 178.
Barker, S. A., and Stephens, R. (1954). *J. Chem. Soc.* p. 4550.
Barker, S. A., Pantard, P., Siddigui, I. K., and Stacy, M. (1955). *Chem. Ind. (London)* p. 1450.
Bauer, S., and Brisk, R. (1979). *Isr. J. Med.* **15**, 95–96.
Bender, M. M., and Smith, D. (1973). *J. Br. Grassl. Soc.* **28**, 97–100.
Ceska, M. (1971). *Biochem. J.* **125**, 209–211.
Chambers, V. C., and Overman, J. D. (1964). U.S. Patent 3,137,575 [C.A. **61**, 6568].
Chambert, R., Treboul, G., and Dedonder, R. (1974). *Eur. J. Biochem.* **41**, 285.

Cisar, J., Kafet, E. A., Liad, J., and Potter, M. (1974). *J. Exp. Med.* **139**, 159–179.
Clarke, M. A., Roberts, E. J., Tsang, W. S. C., Godshell, M. A., Han, Y. W., Kenne, L., and Lindburg, B. (1988). *Proc. Sugar Process. Res. Conf., New Orleans, La.*
Corrigan, A. J., and Robyt, J. F. (1979). *Infect. Immun.* **26**, 387–389.
Dawes, E. A., Ribbons, D. W., and Rees, D. A. (1966). *Biochem. J.* **98**, 804.
Dea, I. C. M., and Madden, J. K. (1985). In "New Developments in Industrial Polysaccharides" (V. Crescenji, I. C. M. Dea, and S. S. Stivala, eds.), pp. 27–56. Gordon & Breach, New York.
Dedonder, R. (1966). *Methods Enzymol.* **8**, 500–505.
Dedonder, R., and Peaud-Lenoel, C. (1957). *Bull Soc. Chim. Biol.* **39**, 483.
Dias, F., and Bhat, J. V. (1962). *Antonie van Leeuwenhoeck*, **28**, 63–72.
Elisashvili, V. I. (1981). *Biochemistry* **45**, 14–20.
Elisashvili, V. I. (1984). *Appl. Biochem. Microbiol.* **20**, 82–87.
Evans, T. H., and Hibbert, H. (1946). *Adv. Carbohydr. Chem.* **2**, 253–277.
Feingold, D. S., and Gehatia, M. (1957). *J. Polym. Sci.* **23**, 783–790.
Feingold, D. S., Avigad, G., and Hestrin, S. (1956). *Biochem. J.* **64**, 351–361.
Forsyth, W. G. C. (1950). *Biochem. J.* **46**, 141–145.
Fouet, A., Arnaud, M., Klier, A., and Rapoport, G. (1984). *Biochem. Biophys. Res. Commun.* **119**, 795–800.
French, A. (1989). *J. Plant Physiol.* **134**, 125–136.
Fuchs, A. (1956). *Nature (London)* **178**, 921.
Fuchs, A. (1959). Ph.D. Thesis, Rijksuniv., Leiden.
Fuchs, A., DeBruijin, J. M., and Niedeveld, C. J. (1985). *Antonie van Leeuwenhoek* **51**, 333–351.
Fukui, H., Tanaka, M., and Misaki, A. (1985). *Agric. Biol. Chem.* **49**, 2343–2349.
Glaudemans, C. P. J. (1975). *Adv. Carbohydr. Chem. Biochem.* **31**, 313–346.
Glicksman, M. (1973). In "Industrial Gums" (R. L. Whistler and J. N. Bemiller, eds.), Rev. Ed., Part A. Academic Press, New York.
Glukhova, E. V., Shenderov, B. A., and Yarotskii, S. V. (1985). *Antibiot. Med. Biotechnol.* **30**, 490–495 [C.A. **103**, 84976d].
Goldstein, I. J., and So, L. S. (1965). *Arch. Biochem. Biophys.* **111**, 407.
Greig-Smith, R. (1901). *Proc. Linn. Soc. N.S.* **26**, 589.
Grootwassink, J. W. D., and Fleming, S. (1980). *Enzyme Microb. Technol.* **2**, 45–53.
Gross, M., and Rudolph, K. (1987). *J. Phytopathol.* **119**, 289–297.
Gross M., and Rudolph, K. (1988). *J. Phytopathol.* **120**, 9–19.
Gupta, A. K., and Bhatia, I. S. (1982). *Phytochemistry* **21**, 1249–1253.
Hakomori, S. J. (1964). *J. Biochem. (Tokyo)* **55**, 205.
Han, Y. W. (1989). *J. Ind. Microbiol.* **4**, 447–452.
Han Y. W., and Clarke, M. A. (1990). *J. Agric. Food Chem.* **38(2)**, 393–396.
Hehre, E. J. (1951). *Adv. Enzymol.* **11**, 297–337.
Hehre, E. J. (1955). *Methods Enzymol.* **1**, 178–192.
Histrin, S. (1956). *Ann. N.Y. Acad. Sci.* **66**, 401.
Histrin, S., and Avineri-Shapiro, S. (1944). *Biochem. J.* **38**, 2–10.
Hestrin, S., and Goldblum, J. (1953). *Nature (London)* **172**, 1046–1047.
Hestrin, S., Avineri-Shapiro, D., and Aschner, M. (1943). *Biochem. J.* **37**, 450–456.
Hestrin, S., Shilo, M., and Feingold, D. S. (1954). *Br. J. Exp. Pathol.* **35**, 107.
Hestrin, S., Feingold, D. S., and Avigad, G. (1956). *Biochem. J.* **64**, 340–351.
Hidaka, H. (1983). *Kagaku to Seibutsu* **21**, 291–293.
Higuchi, M., Iwami, Y., Yamada, T., and Araya, S. (1970). *Arch. Oral Biol.* **15**, 565–567.
Hines, G. H., McGhee, R. M., and Shurter, R. A. (1953). *Ind. Eng. Chem.* **45**, 692–705.

Hirst, E. L. (1957). *Proc. Chem. Soc., London* pp. 193–204.
Jung, K. H., Lim, J. J., Yoo, S. J., Lee, J. H., and Yoo, M. Y. (1987). *Biotechnol. Lett.* **9**, 703–708.
Kang, K. S., and Cottrell, I. W. (1979). In "Microbial Technology" (H. J. Peppler and D. Perlman, eds.), 2nd Ed., Vol. 1, pp. 417–481. Academic Press, New York.
Kang, K. S., Veeder, G. T., and Cottrell, I. W. (1983). *Prog. Ind. Microbiol.* **18**, 231–253.
Keibovici, J., Kopel, S., Siegel, A., and Gal-Mor, O. (1986). *Int. J. Immunopharmacol.* **8**, 391–401.
Kennedy, J. F., Stevenson, D. L., and White, C. A. (1989). *Carbohydr. Poly.* **10**, 103–113.
Khorramian, B. A., and Stivala, S. S. (1982). *Carbohydr. Res.* **108**, 1–12.
Kiss, S. (1968). *Rev. Roum. Biol., Ser. Bot.* **13**, 434–438.
Kleczkowski, A., and Wierzchowski, P. (1940). *Soil Sci.* **49**, 193.
Krichevsky, M. I., Howell, A., and Lim, S. (1969). *J. Dent. Res.* **48**, 938–942.
LeBrun, E., and Van Rapenbusch, R. (1980). *J. Biol. Chem.* **255**, 12034–12036.
Lee, K. Y., Nurock, D., and Zlatkis, A. (1979). *J. Chromatogr.* **174**, 187–193.
Leibovici, J., and Stark, Y. (1984). *J. Natl. Cancer Inst.* **72**, 1417–1420.
Leibovici, J., and Stark, Y. (1985). *Cell. Mol. Biol.* **31**, 337–342.
Lesher, R. (1976). Ph.D. Thesis, West Virginia Univ., Morgantown.
Lindberg, B., Lonngren, J., and Thompson, J. L. (1973). *Acta Chem. Scand.* **27**, 1819–1821.
Lindburg, G. (1957). *Nature (London)* **180**, 1141.
Lippmann, E. O. (1881). *Chem. Ber.* **14**, 1509.
Loewenburg, J. R., and Reese, E. T. (1957). *Can. J. Microbiol.* **3**, 643.
Long, L. W., Stivala, S. S., and Ehrlich, J. (1975). *Arch. Oral Biol.* **20**, 504–508.
Lyness, E. W., and Doelle, H. W. (1983). *Biotechnol. Lett.* **5**, 345–350.
McNeeley, W. H., and Kang, K. S. (1973). In "Industrial Gums" (R. L. Whistler and J. N. Bemiller, eds.), Rev. Ed., Part A, pp. 473–497. Academic Press, New York.
Madden, J. K., Dea, I. C. M., and Steer, D. C. (1986). *Carbohydr. Polym.* **6**, 51–73.
Manly, R. S., and Cormier, A. D. (1970). *J. Dent. Res.* **49**, 301–306.
Manly, R. S., and Richardson, D. J. (1968). *J. Dent. Res.* **47**, 1080–1086.
Mantsala, P., and Puntala, M. (1982). *FEMS Microbiol. Lett.* **13**, 395–399.
Marchessault, R. H., Bleha, T., Deslendes, Y., and Revol, J. T. (1980). *Can. J. Chem.* **58**, 2415–2417.
Marshall, K., and Weigel, H. (1980). *Carbohydr. Res.* **80**, 375–377.
Mattoon, J. R., Holmlund, C. E., Schepartz, S. A., Vavra, J. J., and Johnson, M. J. (1955). *Appl. Microbiol.* **3**, 321–333.
Meier, H., and Reid, J. S. G. (1982). *Encycl. Plant Physiol., New Ser.* **13A**, 418–471.
Mitsuda, S., Miyata, N., Hirota, T., and Kikuchi, T. (1981). *Hakko Kagaku Kaishi* **59**, 303–309 [C.A. **95**, 113339].
Murphy, D. (1952). *Can. J. Chem.* **30**, 872.
Newbrun, E. (1969). *J. Dent. Child.* **14**, 239–248.
Ninomiya, E., and Kizaki, T. (1970). *Nippon Nogei Kagaku Kaishi* **44**, 270–274.
Oku, T., Tokunaga, T., and Tosoya, N. (1984). *J. Nutr.* **114**, 1574–1581.
Pabst, M. J. (1977). *Infect. Immun.* **15**, 518–526.
Park, Y. K., Mortatti, M. P. L., and Sato, H. H. (1983). *Biotechnol. Lett.* **5**, 515–518.
Pascal, M., and Dedonder, R. (1972). *Carbohydr. Res.* **24**, 365–377.
Pezur, J. H. (1952). *J. Biol. Chem.* **199**, 217–225.
Percheron, F. (1962). *C. R. Hebd. Seances Acad. Sci.* **255**, 2521–2522.
Perlot, P., and Monson, P. (1984). *N.Y. Acad. Sci.* **434**, 468–471.
Phelps, C. F. (1965). *Biochem. J.* **95**, 41–47.
Pontis, H. G., and Del Campillo, E. (1985). In "Biochemistry of Storage Carbohydrates in

Green Plants" (P. M. Dey and R. A. Dixon, eds.), pp. 205–227. Academic Press, New York.
Rapoport, G., and Dedonder, R. (1963). *Bull. Soc. Chim. Biol.* **45**, 493–515.
Reese, E. T., and Avigad, G. (1966). *Biochim. Biophys. Acta* **113**, 79–83.
Ribeiro, J. C. C., Guimarues, W. V., Borges, A. C., Silv, D. O., and Crug, C. D. (1988). *Rev. Microbiol.* **19**, 196–201.
Roe, S. H., Epstein, J. H., and Goldstein, N. P. (1949). *J. Biol. Chem.* **178**, 839.
Sato, S., Koga, T., and Inoue, M. (1984). *Carbohydr. Res.* **134**, 293–304.
Schechter, I., and Hestrin, S. (1963). *J. Lab. Clin. Med.* **61**, 962.
Schlubach, H. H., and Berndt, J. (1964). *Ann. Chem.* **677**, 172.
Seymour, F. R., Knapp, R. D., and Jeanes, A. (1979). *Carbohydr. Res.* **72**, 222–228.
Shilo, M. (1959). *Annu. Rev. Microbiol.* **13**, 255.
Shimamura, A., Tsuboi, K., Nagase, T., Ito, M., Tsumori, H., and Mukasa, H. (1987). *Carbohydr. Res.* **165**, 150–154.
Shiomi, N., Yamada, J., and Izawa, M. (1976). *Agric. Biol. Chem.* **40**, 567–575.
Smith, A. E. (1976). *J. Agric. Food Chem.* **24**, 476–478.
Smith, J. A., Grove, D., Luenser, S. J., and Park, L. G. (1982). U.S. Patent 4,309,505.
Smith, N. R., Gordon, R. E., and Clark, F. E. (1952). *U.S. Dep. Agric., Agric. Monogr.* No. 16.
Srinivasan, S., and Quastel, J. H. (1958). *Science* **127**, 143.
Stark, Y., and Leibovici, J. (1986). *Br. J. Exp. Pathol.* **67**, 141–148.
Stivala, S. S., and Khormanian, B. A. (1982). *Abstr. Pap. Am. Chem. Soc.* No. 183 (O).
Stivala, S. S., and Zweig, J. E. (1981). *Biopolymers* **20**, 606–620.
Streefkerk, D. G., and Glaudemans, C. P. J. (1977). *Biochemistry* **16**, 3760–3765.
Takeshita, M. (1973). *J. Bacteriol.* **116**, 503–506.
Tanaka, T., Susumu, O., and Yamamoto, T. (1979). *J. Biochem. (Tokoyo)* **85**, 287–293.
Tanaka, T., Oi, S., and Yamamoto, T. (1980). *J. Biochem. (Tokoyo)* **87**, 297–303.
Tanaka, T., Yamamoto, S., Oi, S., and Yamamoto, T. (1981). *J. Biochem. (Tokoyo)* **90**, 521–526.
Tanaka, K., Karigane, T., and Yamaguchi, F. (1983). *J. Chromatogr.* **265**, 374–377.
Takaka, K., Karigane, T., Fujii, S., Chinzaka, T., and Nagamura, S. (1985). *J. Biochem. (Tokoyo)* **97**, 1679–1688.
Tkachenko, A. A., and Loitsyanskaya, M. S. (1976). *Microbiology (Engl. Transl.)* **45**, 387–391.
Tkachenko, A. A., and Loitsyanskaya, M. S. (1979). *Appl. Biochem. Microbiol.* **14**, 502–505.
Vandamme, E. J., and Derycke, D. G. (1983). *Adv. Appl. Microbiol.* **29**, 139–176.
Viikari, L. (1984). *Appl. Microbiol. Biotechnol.* **19**, 252–255.
Warner, T. N., and Miller, C. H. (1978). *Infect. Immun.* **19**, 711–719.
Webley, D. M., Duff, R. B., Bacon, J. S. D., and Farmer, V. C. (1965). *J. Soil. Sci.* **16**, 149.
Whistler, R. L. (1973). In "Polysaccharides and Their Derivatives" (R. L. Whistler, ed.), pp. 5–25. Academic Press, New York.
Yamamoto, S., Iizuka, M., Tanaka, T., and Yamamoto, T. (1985). *Agric. Biol. Chem.* **49**, 343–350.
Yaphe, W., and Arsenault, G. P. (1965). *Anal. Biochem.* **13**, 143–148.
Yavetz, H., Daridal, G., Leibovici, J., and Wolman, M. (1985). *Cell. Mol. Biol.* **31**, 343–348.

Review and Evaluation of the Effects of Xenobiotic Chemicals on Microorganisms in Soil[1]

R. J. Hicks,* G. Stotzky,† and P. Van Voris‡

*Groundwater Technology, Inc.
Concord, California 94520

†New York University
New York, New York 10003

‡Pacific Northwest Laboratory
Richland, Washington 99352

I. Introduction
II. Soil as a Microbial Habitat
    A. Solid Phase
    B. Water Phase
    C. Gaseous Phase
III. Microorganisms and Their Activities in Soil
    A. Abundance and Types
    B. Activities in Soil
IV. Xenobiotics
    A. Classification and Sources
    B. Interactions of Xenobiotics with Soils
V. Environmental Factors That Influence Interactions between Microorganisms and Xenobiotics
    Physicochemical Factors
VI. Effects of Xenobiotics on Microorganisms in Soil
    A. Microbiological Processes
    B. Mode of Action
VII. Methods of Assessing the Effects of Xenobiotics on Microorganisms
    A. Laboratory Test Systems
    B. Methods for Assessing Xenobiotic Impacts on Relevant Microbial Processes and Properties
    C. Assessing the Efficacy of Test Methods
VIII. Conclusions and Recommendations
    References

The primary objective of this document was to review and evaluate the relevance and quality of existing xenobiotic data bases and test

[1] Although the information in this document has been funded wholly or in part by the United States Environmental Protection Agency under Interagency Agreement DW89931928894-01-0 to Pacific Northwest Laboratory, operated for the U.S. Department of Energy by Battelle Memorial Institute, it does not necessarily reflect the views of the Agency and no official endorsement should be inferred.

methods for evaluating (1) direct and indirect effects (both adverse and beneficial) of xenobiotics on the soil microbial community, (2) direct and indirect effects of the soil microbial community on xenobiotics, and (3) adequacy of test methods used to evaluate these effects and interactions. Xenobiotic chemicals are defined here as those compounds, both organic and inorganic, produced by human beings and introduced into the environment at concentrations that cause undesirable effects. Because soil serves as the main repository for many of these chemicals, it has a major role in determining their ultimate fate. Once released, the distribution of xenobiotics between environmental compartments depends on the chemodynamic properties of these compounds, the physicochemical properties of the soils, and the transfer between soil–water and soil–air interfaces and across biological membranes. Abiotic and biotic processes can transform the chemical compound, thus altering its chemical state and, subsequently, its toxicity and reactivity. Ideally, the conversion is to carbon dioxide, water, and mineral elements, or, at least, to some harmless substance. However, intermediate transformation products, which can become toxic pollutants in their own right, can sometimes be formed.

When exposed to xenobiotic compounds, various segments of the soil microbial community are affected to different extents. The degree to which a xenobiotic affects microbial activities is largely dependent on the chemical, its dosage, and the particular physicochemical parameters of the environment, such as soil type, temperature, water content, pH, method of application, and other factors. Soil physicochemical factors are particularly important and probably account for the variations in toxic effects often seen with the same compound. A strong correlation between compound class and its effects on soil microorganisms cannot be established because of a paucity of data on the effects of organic compounds other than pesticides. However, a few generalizations have emerged. Broad-range biocidal compounds, such as soil fumigants, appear to affect detrimentally all microbial processes, at least temporarily. This effect is also observed with most heavy metals. Compounds such as herbicides and insecticides are less detrimental to microbial activity and, under certain conditions, may stimulate activity.

Some soil microbial processes and properties appear to be more sensitive to xenobiotics than others. For example, nitrification appears to be highly sensitive, whereas nitrogen mineralization is relatively resistant to xenobiotics. These results reflect the differences in the diversity of microorganisms mediating these processes.

The present state of knowledge of the cytological and biochemical effects of xenobiotics does not provide any definitive evidence regarding their mode of action. However, available data suggest that xenobiotics may interfere with photosynthesis, oxidative metabolism, and the synthesis of cellular constituents. In addition, certain compounds, such as the chlorinated aromatics, do alter the composition of the cellular membrane, thereby changing cell membrane permeability and altering cellular physiology.

Numerous methods exist for measuring different microbial processes, microbial populations, and soil enzymes, and most of these methods can be applied to assess the effects of xenobiotics. However, one of the major research needs is to establish which of the available methods are the most valid for investigating particular classes of xenobiotic compounds. Additional studies are also needed to identify and select the microbially mediated ecological processes that can best be used to generate meaningful data of the short- and long-term effects of xenobiotics before selection and standardization of techniques can be accomplished. Finally, a systemic examination of those classes of xenobiotics that have not been evaluated for their effects on microorganisms needs to be performed to allow development of a predictive model for environmental risk assessment.

I. Introduction

Xenobiotics are defined here as both organic and inorganic anthropogenic compounds that are introduced into the environment at concentrations that cause undesirable effects. Some of these compounds may be purposefully released and are designed to be beneficial to humanity (e.g., pesticides). Alternatively, they may be accidentally released into the environment as wastes or residues from industrial manufacturing and processing of fossil fuels. These compounds may interact at a number of differing trophic levels within an ecosystem. To estimate the environmental risks to the soil ecosystem associated with release, either purposeful or accidental, of xenobiotics, two questions were posed: first, are there sufficient data in the literature to assess accurately the significance of the impacts of xenobiotics on the microbiological portion of the soil system; and second, are the experimental approaches used in assessing those effects adequate, or are new or improved methods needed? Because of limitations in scope and budget, effort was focused only on the microbiological component of the soil system. The principal objective of this assessment was to review and evaluate the relevance and quality of existing data bases on the effects of xenobiotics on soil microorganisms and determine the adequacy of test methods relating to the following:

1. Direct and indirect effects (both beneficial and adverse) of xenobiotics on the soil microbial community
2. Direct and indirect effects of the soil microbial community on xenobiotics
3. Adequacy of test methods used to evaluate these effects and interactions

This effort has produced a series of conclusions regarding the quality and breadth of published data on xenobiotics in soil systems and the suitability of current methods used to measure any potential impacts. On the basis of these conclusions, areas that need additional research have been identified.

II. Soil as a Microbial Habitat

Soil is one of the most dynamic sites of biological interactions in nature, serving as a growth medium for vegetation and as a habitat for fauna. In addition, soil serves as a recepticle for the multitude of organic and inorganic chemicals released by human beings— either intentionally, as in the case of agricultural chemicals, or accidentally—and thus has a major role in determining the overall quality of our environment. The soil environment is dominated by a solid phase composed of (1) inorganic minerals; (2) plant, animal, and microbial residues in various stages of decay; and (3) a living and metabolizing microbiota (Stotzky, 1986). The solid phase is surrounded by fluctuating aqueous and gaseous phases. The proportion and physicochemical properties of these three phases strongly influence the growth, activity, and population dynamics of microorganisms in soil and also modify the effects of xenobiotics on the microbial community. Therefore, it is important to review briefly the nature and properties of soils before describing the effects of xenobiotics on microorganisms in soil.

A. Solid Phase

Soil solids are composed of mineral material, such as sand, silt, and clay particles, and of organic matter, consisting of living and dead biomass. The proportion of each component in a particular soil controls its physical and chemical properties, such as porosity, water-holding capacity, cation exchange capacity, cation–anion exchange ratio, and aggregate stability. In terms of the behavior of xenobiotics in soil and their effects on indigenous microbial populations, the most influential soil components appear to be the colloid-sized clay minerals and organic particles (Stotzky and Burns, 1982). This influence results, in part, from their high surface–volume ratios, their ionic properties, and their high affinity for water molecules (Burns, 1983; Stotzky, 1986). Consequently, xenobiotics, as well as microorganisms, tend to concentrate at colloid–water interfaces. In addition to clays and organic matter, microorganisms are also a component of soil solids that can influence the behavior of xenobiotics.

1. Clays

The building blocks of clay minerals are two basic units composed of oxygen or hydroxide and silicon or aluminum. One unit is the silicon tetrahedron in which oxygen atoms form the corners of a tetrahedron held together by a silicon ion in the center. It is possible for isomorphic substitutions (i.e., substitution by an ion of similar size but of a lower valence with essentially no modification in crystalline structure) to occur between the tetravalent silicon ion, and trivalent ions, such as aluminum, resulting in a net negative charge. The other unit is the aluminum octahedron in which six hydroxyl groups or oxygen atoms form the corners of an octahedron held together by an aluminum ion in the center. As in the silicon tetrahedron, isomorphic substitutions between the trivalent aluminum atom and divalent ions, such as magnesium or ferrous iron atoms, can occur, creating a negative charge within the unit. Each of these basic units (e.g., silica tetrahedron or aluminum octahedron) can link together horizontally to form sheets commonly referred to as tetrahedral or octahedral sheets, respectively.

Clay minerals consist of silicon tetrahedral and aluminum octahedral sheets held together by shared oxygen atoms. The minerals consist of two main types of unit layers, depending on the ratio of tetrahedral to octahedral sheets, whether 1 : 1 (Si–A) or 2 : 1 (Si–Al–Si). The physical and chemical properties of a particular clay mineral depend on the ratio of tetrahedral to octahedral sheets and on the nature and location of the isomorphic substitutions that occur within the crystalline structure.

Isomorphic substitutions within the basic unit impart a net negative charge to most clay minerals. In 2 : 1-type clays, isomorphic substitution accounts for the majority of the negative charges. In 1 : 1 clays, there is little isomorphic substitution, and unsatisfied charges resulting from broken edges on the clay are primarily responsible for the negative charges. The negative charges are compensated by exchangeable cations (e.g., Al^{3+}, Fe^{3+}, CA^{2+}, Mg^{2+}, K^+, Na^+, H^+, NH_4^+) in the ambient soil solution. The amount of charge-neutralizing cations that can be retained by clays is termed the cation exchange capacity (CEC) and is expressed in units of milliequivalents of cations a clay can adsorb per unit weight.

Clay minerals differ in the position (whether mainly in the tetrahedral or octahedral sheet) and amount of isomorphic substitution (e.g., density of charge) occurring within the lattice structure, as well as in the type and amount of hydration of charge-compensating cations. These variables influence the amount and degree of interlayer associations that occur, which, in turn, influence the degree of expansion a clay undergoes on wetting. Some 2:1-type clays such as montmorillinite expand considerably on wetting, resulting in a large surface area that can serve as a site of adsorption for inorganic and organic molecules. Other 2:1 clays, such as illite, as well as 1:1 clays, do not normally expand on wetting and consequently have a lower surface–volume ratio. Some 2:1 clays, such as vermiculite, exhibit limited expansion. The increased surface area of expanding clays coupled with the high amount of isomorphic substitution that often occurs in these types of clays usually results in higher CEC and greater adsorption of water and gases on these clays than in the nonexpanding types. Certain clays that often exhibit positive charge sites at lower pH values can participate in the adsorption of anions from solution. These positive charge sites primarily originate from broken bonds in the octahedral sheet that expose Al^{3+} groups on the edges of the clay mineral.

The preceding section is not intended to be an exhaustive review on clays, but rather a brief summary of some of the physical and chemical properties of clay minerals that may have a role in determining the effect of xenobiotics on microorganisms in soil. For more detailed reviews of the aforementioned material the reader is referred to the excellent works of Baver et al. (1972), Dixon and Weed (1977), or Russell (1973).

2. *Organic Matter*

Soil organic matter includes a broad spectrum of organic constituents that can be divided into two major types: (1) nonhumic substances, consisting of essentially unaltered plant, animal, and microbial debris and of compounds belonging to the well-known classes of organic chemistry (e.g., aliphatic and aromatic acids, amino acids, carbohydrates, fats, and waxes); and (2) humic substances, a series of high molecular weight, dark-colored substances formed by secondary synthesis reactions, many of which are mediated by microbes (Stevenson, 1985). Humic substances rank with colloidal clays in terms of importance to microbial activity and xenobiotic behavior.

A variety of functional groups, including COOH, phenolic OH, enolic OH, alcoholic OH, quinone, amine, hydroxyquinone, and lactone, are found in humic substances (Stevenson, 1985). The pK_a values of most of the functional groups are such that soil organic colloids are predominantly negatively charged in most soils and, therefore, contribute to the CEC of the soil. However, positively charged sites are often found on soil organic matter, particularly at low pH. Thus, as with clays, soil organic matter can also have an associated anion exchange capacity (AEC). There is little information on the importance of the AEC of organic colloids or clays to the behavior of xenobiotics or microorganisms in soils. The AEC of organic colloids and clays is often overshadowed by their CEC; however, as most xenobiotics and microorganisms are net negatively charged, the AEC of organic and inorganic colloids may be important in the interactions of these colloids with xenobiotics and microbes.

In addition to their ionic nature, humic substances have a number of other properties that are important in the behavior of xenobiotics. Humic substances are polydisperse materials that expand on wetting and therefore have extensive internal surface area (Stotzky and Burns, 1982). The large surface area coupled with the large number of functional groups make humic colloids important in the adsorption of xenobiotics in soils through such mechanisms as H-bonding and van der Waals forces. Organic colloids also contain hydrophobic sites that are important in the adsorption of nonpolar organic compounds.

A more detailed description of the types and activities of microorganisms in soil is deferred to a later section; however, it is important to include microbes in this section because they do contribute to the solid phase of the soil. The surfaces of microorganisms are predominantly negatively charged (Burns, 1979); however, as with organic colloids, positively charged sites can occur. Therefore, the possibility exists for the adsorption of both anionic and cationic xenobiotics to living organisms. In addition, portions of the outer surfaces of many microorganisms are hydrophobic and, thus, may contribute to the binding of nonpolar xenobiotics.

B. Water Phase

The variable amount and the energy state of water in soil are important factors that affect the growth of microorganisms in soil and also, because of the solvent properties of water, the behavior of xenobiotics. Soil water governs the air content and gas exchange in soil, thus affecting the activity of microorganisms and the chemical state of the soil (e.g., redox potential). In addition, soil water content affects the swelling of clays and, thus, the specific surface area of soils available for interaction with xenobiotics.

The amount of water contained in a unit mass or volume of soil can be characterized in terms of water content. The physicochemical condition or state of soil water is characterized in terms of its free energy or potential. Water potential is the free energy of water in a system, relative to the free energy of a reference pool of pure, free water (Papendick and Campbell, 1980). The availability of water (e.g., activity of water) for physiological processes decreases as the water potential decreases.

In soils, sand- and silt-sized particles do not retain water against gravitational pull; therefore, it is primarily the clay fraction that retains enough water to sustain microbial growth (Stotzky, 1986). Stotzky (1986) has speculated that this is the reason for the apparent correlation of microbial activity with the clay fraction. Soil organic matter also retains water; however, little is known about the importance of organic matter-associated water to microbial events in soils.

Water adjacent to the surface of clays is presumed to be highly ordered because of charge interactions between the water, the clay surface, and cations associated with the surface (Russell, 1973; Low, 1961, 1979; Farmer, 1978). This ordering of water lowers the activity of water so that it is probably unavailable to microorganisms. Therefore, it is likely that microorganisms are growing some distance from the clay surface in the region where water is still under the attraction of the clays but where its activity is high enough to support microbial growth (Stotzky, 1986). It is probable that xenobiotics would also be concentrated in this region (Stotzky, 1986). Additional information is needed about the effects of the physicochemical characteristics of soil water on the interactions among water, microorganisms, and xenobiotics in soils.

C. Gaseous Phase

The soil pores that are not filled with water contain gases that constitute the soil atmosphere. Plant roots and organisms living in the soil remove O_2 from the soil atmosphere and respire CO_2 into it; therefore, the composition of the atmosphere within the soil pores usually differs from that of the atmosphere above the soil in being richer in CO_2 and poorer in O_2. The magnitude of this difference depends on the rate of removal of O_2 and the rate of gas exchange between the atmosphere and the soil pores.

In well-aerated soils, the rate of O_2 transfer from the atmosphere is nearly equal to the rate of its removal by organisms. Under these conditions, the dominant metabolic activities occurring in soil are those in which O_2 is used as the terminal electron acceptor and as a substrate for oxygenase enzymes. However, even in well-aerated soils, and particularly in poorly aerated soils, microsites can occur within the soil wherein O_2 consumption exceeds O_2 replacement. The O_2 concentration can then fall nearly to zero, and prolonged anaerobic conditions can result in metabolic activities dominated by reduction reactions such as fermentation, denitrification, sulfate reduction, and methane formation. In O_2-deficient soils, other gases, including aldehydes, alcohols, and ethylene, N_2O and N_2 from denitrification, CH_4 from methanogenesis, and H_2S from anaerobic sulfate reduction, can occur in high concentrations in the soil atmosphere (Stotzky and Schenck, 1976).

III. Microorganisms and Their Activities in Soil

The soil biota consists of microscopic and macroscopic inhabitants that interact as a distinct biological community. Microscopic inhabitants include bacteria, fungi, algae, and protozoa, and macroscopic communities include nematodes, oligochaetes (earthworms), arthropods (micro and macro), and gastropods (snails). Although all these organisms have an important role in the global cycling of nutrients, the microscopic inhabitants (particularly the bacteria and fungi) of soil have a unique role because of their metabolic diversity (Alexander, 1977). Therefore, this section focuses on these populations.

A. ABUNDANCE AND TYPES

A diverse range of microorganisms exists in soils. Numbers of microorganisms in soil habitats are normally higher than in other habitats, such as freshwater or marine environments.

1. Bacteria

The numbers of bacteria occurring in soils are usually higher than those of the other groups; however, because of their small size in relation to the large cell size and extensive filaments of the other groups, bacteria account for less than half of the total microbial biomass in soil (Alexander, 1977). Typically, there are between 10^6 and 10^9 bacteria per gram of soil. In well-aerated soils, both bacteria and fungi are present; however, if O_2-limited conditions are present, bacteria account for most of the microbial community biomass.

Common bacterial genera found in soils include *Acinetobacter, Agrobacterium, Alcaligenes, Arthrobacter, Bacillus, Brevibacterium, Caulobacter, Cellulomonas, Clostridium, Corynebacterium, Flavobacterium, Micrococcus, Mycobacterium, Pseudomonas, Staphylococcus, Streptococcus,* and *Xanthomonas* (Atlas and Bartha, 1987). The relative proportion of individual bacteria varies widely in different soils (Table I).

TABLE I

Relative Proportion of Bacterial Genera
Commonly Found in Soils [a]

Genus	Percentage
Arthrobacter	5–60
Bacillus	7–67
Pseudomonas	3–15
Agrobacterium	1–20
Alcaligenes	1–20
Flavobacterium	2–12
Corynebacterium	2–10
Micrococcus	<5
Staphylococcus	<5
Xanthomonas	<5
Mycobacterium	<5

[a] Adapted from Atlas and Bartha (1987).

Actinomycetes can compose between 10% and 33% of the bacterial population in soil (Alexander, 1977). The most common genera include *Streptomycetes, Nocardia, Micromonospora*, and *Actinomyces*. Actinomycetes are relatively resistant to adverse conditions, such as desiccation, extremes in pH, and the lack of easily metabolizable carbon sources.

A number of bacteria in soil are plant or animal pathogens. Some pathogenic bacteria are allochthonous and enter the soil in association with diseased plant or animal tissues. Examples of plant pathogens found in soil are *Agrobacterium, Corynebacterium, Erwinia*, some *Pseudomonas*, and *Xanthomonas;* animal pathogens include *Klebsiella, Clostridium, Streptococcus,* and *Salmonella* (Atlas and Bartha, 1987). Most of these allochthonous organisms are normally unable to compete with saprophytic bacteria and are eliminated through competitive exclusion processes; however, many are nonobligate pathogens or have evolved a permenent soil phase and are able to reproduce and grow.

Soil bacteria are particularly significant to, and occupy a key position in, the global cycling of carbon and other elements. Their diverse metabolic capabilities enable them to expoit many sources of energy and cell carbon. Unique metabolic features of bacteria include anaerobic respiration, chemolithotrophic growth, fixation of molecular nitrogen, and utilization of methane (Schlegel and Jannasch, 1981). These unique metabolic features render soil bacteria the principal agents for the global cycling of many inorganic compounds, especially nitrogen, sulfur, and phosphorus, but also metals and metalloids such as arsenic, iron, mercury, manganese, and selenium.

2. *Fungi*

Although numerically much less abundant (between 10^4 and 10^6 fungal propagules per gram soil) than bacteria, fungi are the major contributors to soil biomass and can account for as much as 70% by weight of the biomass (Lynch, 1983). Soil fungi can occur free-living or in mycorrhizal association with plant roots. Members of the class of Fungi Imperfecti such as *Aspergillus, Geotrichum, Penicillium,* and *Trichoderma* are the fungi most frequently isolated; however, numerous ascomycetes and basidiomycetes also occur in soil (Atlas and Bartha, 1987). The presence of yeast can be demonstrated in most soils, and some species have been isolated exclusively from soils (Alexander, 1977; Atlas and Bartha, 1987).

Most fungi in soil are opportunistic. They grow and conduct their metabolic activities when environmental conditions (e.g., nutrients, moisture, temperature, aeration) are favorable. Soil fungi are active in the transformation of cellulose and are the principal agents for the transformation of lignins produced by plants. The breakdown of these polymers releases single molecules that are subsequently used by other soil organisms, particularly bacteria.

3. *Algae and Protozoa*

A number of genera of algae and protozoa live in soil or on the soil surface. Population densities have been estimated to be between 10^1 and 10^6 per gram soil for algae and between 10^4 and 10^5 per gram soil for protozoa (Alexander, 1977; Atlas and Bartha, 1987). The abiotic environmental parameter most influential in regulating the growth of algae and protozoa in soils include sunlight and CO_2 for algae and O_2 for protozoa. Protozoa are important predators in soil and help to regulate the size of bacterial populations (Alexander, 1977). Algae contribute to the organic carbon input of soil and also contribute to soil structure and erosion control (Alexander, 1977).

4. *Root-Associated Microorganisms*

Many microorganisms in soil interact with plant roots on the root surface (rhizoplane) or within the region directly influenced by the root (rhizosphere). Microbial populations within the rhizosphere are usually higher in number per unit weight or volume of soil and physiologically different than free-living microorganisms (Atlas and Bartha, 1987). These differences have been attributed to the release of substances from roots (root exudates) that modify the soil environment.

The interactions between microorganisms and plants can be mutually beneficial to the plant and its associated microorganisms. For example, the roots of many plants establish a mutualistic relationship with fungi, called mycorrhizal fungi, in which the fungus becomes an integral part of the plant root. The mycorrhizae enhance the uptake of mineral nutrients and enable plants to grow in habitats in which they otherwise would not grow. In return, the fungus obtains organic carbon and possibly other nutrients from the plant host. Other beneficial associations include the symbiosis between nitrogen-fixing bacteria (e.g., *Rhizobium*) and leguminous plants.

The microbial community of the rhizosphere is composed mainly of nonpathogenic microorganisms; however, plant pathogens do exist in the soil and under certain conditions can invade and form harmful relationships (from the plant's viewpoint) with the host plant. Variables that control the relationship between plant pathogenic microorganisms and the host plant include the activity of other soil microorganisms as antagonists to the pathogen, the physiological status of the plant, the presence of protective surfaces on the root, and root exudates (Alexander, 1977).

5. *Cell-Free Enzymes in Soil*

The overall biochemical activity of soil results from a series of reactions catalyzed by enzymes, either as intracellular components of the microbial community or as extracellular (i.e., cell-free) enzymes. Cell-free enzymes exist in soil as a result of their excretion into the soil by living cells or after the lysis of dead plant or microbial cells. The activity of >50 enzymes, some of which are listed in Table II, has been demonstrated in soil (Lynch, 1983).

B. ACTIVITIES IN SOIL

Soil microorganisms have a primary catabolic role in the environment through degradation of plant and animal residues, which contributes to the cycling of nutrients. The activities of microorganisms in soil are essential to the global cycling of carbon, nitrogen, sulfur, phosphorus, and other elements, because many substances cannot be degraded by organisms other than microbes (Doetsch and Cook, 1974).

TABLE II

Enzymes in Soil [a]

Enzyme category	Enzyme names
Oxidoreductases	Catalase
	Catechol oxidase
	Dehydrogenase
	Diphenol oxidase
	Glucose oxidase
	Peroxidase
	Urate oxidase
Transferases	Transaminase
	Transglycosylase
Hydrolases	Acetylesterase
	Amylase
	Asparaginase
	Cellulase
	Deamidase
	Invertase
	Galactosidase
	Urease
	Lipase
	Protease
	Pyrophosphatase
	Nucleotidase

[a] From Burns (1986).

1. *Carbon Transformations*

The most important element in the biosphere, and the foundation of the structure of all cells, is carbon. Inorganic CO_2 is converted into organic forms by photosynthesis by plants and some microorganisms. These organic forms of carbon are subsequently used by animals in the generation of new cell material. After the death of plants, animals, and microbes, the metabolic activities of soil microorganisms transform this organic carbon into CO_2, microbial biomass, and soil organic matter. The cycling of carbon is shown in Fig. 1.

2. *Nitrogen Transformations*

Nitrogen has considerable biological and economic importance. As a key building block of proteins, it is an indispensable and often limiting component of plants, animals, and microorganisms, and vast quantities of nitrogen are used as agricultural fertilizers. As does carbon, nitrogen

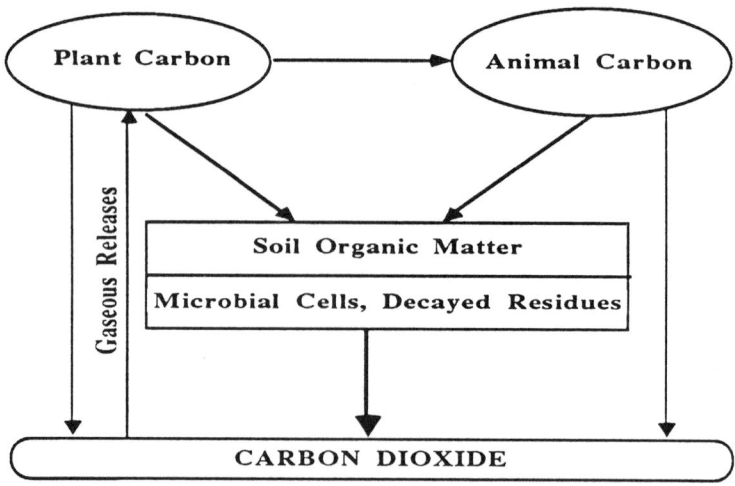

FIG. 1. The carbon cycle. From Alexander (1977).

undergoes a variety of transformations in soil in which the element is shuttled between organic and inorganic forms (Alexander, 1977). The microbial population of soil has an essential role in this cycle. Atmospheric N_2 is converted to organic forms by the action of bacteria that are free-living or live in symbiosis with plant roots or leaves. During the decay of plant, animal, and microbial biomass, the organic nitrogen is mineralized to ammonium, which can be immobilized by plants and microorganisms or oxidized to nitrate by autotrophic bacteria and some fungi. Nitrate can be lost from soil by leaching, which may cause groundwater pollution, can be immobilized by plants or microorganisms, or, under anaerobic conditions, can be utilized as the terminal electron acceptor by facultative anaerobic bacteria and thus be reduced to gaseous nitrogen compounds. Some gaseous nitrogen compounds are important atmospheric pollutants. The cycling of nitrogen is shown in Fig. 2.

The mineralization, immobilization, and, to a limited extent, denitrification reactions of the nitrogen cycle are conducted by numerous microbial species. Therefore, as in carbon cycling, these reactions are not species-specific, and monitoring these processes after the addition of a xenobiotic to soil does not indicate which microbial species are affected by the xenobiotic. In contrast, nitrification and nitrogen fixation result from the metabolic activities of highly specialized microbial groups. Therefore, monitoring the changes in the metabolic activities of these microorganisms enables the assessment of the effects on specific species by xenobiotics.

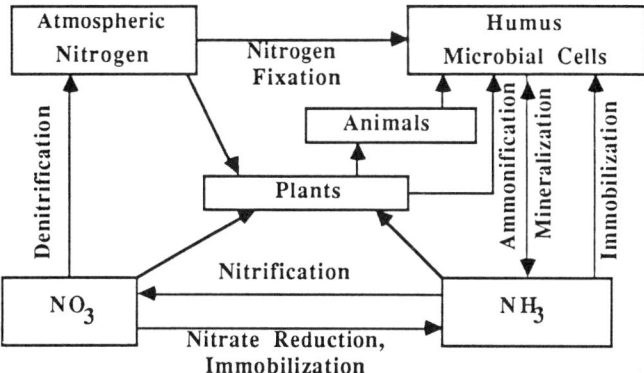

FIG. 2. The nitrogen cycle. Adapted from Alexander (1977).

3. Transformations of Other Elements

Soil microorganisms also have an important role in the mineralization and transformation of other elements, such as phosphorus, sulfur, iron, and manganese, in addition to carbon and nitrogen. Phosphorus is often limiting for plant growth because the concentration of the major form of phosphorus available to plants (PO_4^{3-}) is often very low. Soil microorganisms, particularly mycorrhizal fungi, can be important to the phosphorus nutrition of plants. Sulfur is cycled in the environment primarily through the activities of microorganisms that assimilate sulfate, mineralize organic sulfur compounds and liberate H_2S, or oxidize H_2S to elemental sulfur or sulfate. Sulfate is also used by some strict anaerobic bacteria as a terminal electron acceptor. Iron and manganese, as well as some other metals, are cycled in the environment between their oxidized and reduced forms by the action of microorganisms that can use these elements as a source of electrons or as terminal electron acceptors (Atlas and Bartha, 1987; Ghiorse, 1984).

IV. Xenobiotics

Literally defined, "xenobiotic" compounds are chemicals that are foreign to the biosphere. Under this definition, many compounds, such as metals, some pesticides, and many nonpesticidal organic chemicals, cannot be considered xenobiotic because they occur naturally in the environment. However, this restricted definition fails to take into account human activities, which can increase the concentration of "natural" compounds in an environmental compartment to levels that cause undesirable effects. For example, the essential element phosphorus is not normally a xenobiotic. However, if it is released into aquatic environments in high concentrations, undesirable eutrophication results. Therefore, for the purposes of this review, the definition proposed by Hutzinger and Veerkamp (1981) will be used: a xenobiotic is any compound released anthropogenically into an environmental compartment at a concentration that causes an undesirable effect.

A. Classification and Sources

1. Organic Xenobiotics

In terms of quantity produced, diversity, and potential adverse effects on the environment, the most important types of xenobiotic compounds are those of which the molecular structure is carbon-based (i.e., organic). Of the 65 classes of organic compounds that are considered hazardous, 114 organic componds have been designated by the United States Environmental Protection Agency (EPA) as priority pollutants (Kobayashi and Rittmann, 1982). Table III lists some examples of environmentally relevant organic compounds that have been detected in soil systems and which represent a potential hazard to this ecosystem.

A number of the compounds listed in Table III represent classes of compounds that are known or suspected carcinogens, teratogens, and/ or mutagens. The halogenated organic compounds, especially the chlorinated aromatic compounds, are of particular concern because of the deleterious effects many of these compounds have on biota and their recalcitrant behavior in the environment (Alexander, 1977). Most chlorinated compounds were absent from the biosphere before their anthropogenic synthesis and, thus, can be considered to be true xenobiotics. Consequently, they persist in the environment because they are not susceptible to the normal rate of biological transformation (Reineke, 1984).

The agricultural industry is an important source of organic xenobiotic contaminants in soil systems. Other sources of organic xenobiotic compounds include sewage effluent disposal, sludge disposal, the petroleum industry (e.g., leakage from home fuel and service station storage tanks), the mining industry, and other nonagricultural industrial wastes (Keswick, 1984). Figure 3 shows the distribution of hazardous-waste generation by standard industrial classification. A classification system for herbicides, the most common of the agricultural chemicals, is provided in Table IV.

2. Fossil Fuel-Related Xenobiotics

Fossil fuels represent the major source of feedstocks for the production of organic chemicals. The large number and types of organic chemi-

TABLE III

EXAMPLES OF IMPORTANT XENOBIOTIC COMPOUNDS

Compound type	Examples
Aliphatic (halogenated)	Trichloroethane
	Trichloromethane
	Methylene chloride
	Tetrachloroethane
Aliphatic (nonhalogenated)	Acylonitrile
Aromatic (nonhalogenated)	Toluene
	Benzene
	Nitrobenzene
	Phenol
	Cresol
Aromatic (halogenated)	Pentachlorophenol
	Chlorobenzoate
	Hexachlorophenol
	Dichlorobenzoate
Polycyclic (nonhalogenated)	Naphthalene
	Benzo[a]pyrene
	Anthracene
	Biphenyl
	Phenanthracene
	Benzo[a]anthracene
Polycyclic (halogenated)	Polychlorinated biphenyls (PCBs)
Pesticides	Toxaphene
	Lindane
	DDT
	Heptachloroborane
	Dieldrin
	2,4-D

cals composing this group make prediction of their behavior in soils extremely difficult. However, a classification scheme was developed to provide an approach for systematic study of a wide range of organic constituents using representative compound classes (Zachara et al., 1984). This classification scheme, similar to that for herbicides, is based on the physicochemical properties of organic residues (Table V). Eight classes of compounds were selected, based on specific criteria. These criteria included:

1. The chemical composition of a wide variety of liquid wastes
2. Potential environmental concentrations for individual classes
3. Water solubility
4. Chemical complexity

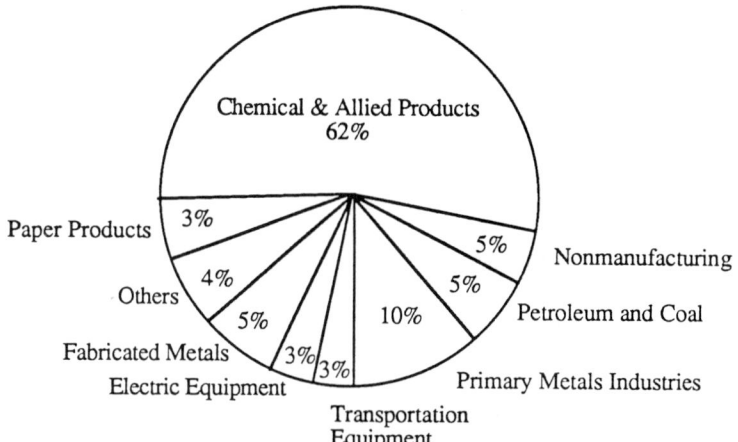

FIG. 3. Percentage of hazardous substances generated by standard industrial classification in 1980. Based on Keswick (1984).

TABLE IV

CLASSIFICATION OF HERBICIDES BASED ON CHEMICAL GROUP AND CLASS [a]

Chemical group	Chemical type	Representative compounds
Aliphatics	Chlorinated	TCA, Dalapon
	Arsenicals	DSMA, MAA, MAMA, MSMA
	Other	Acrolein, Glyphosate
Amides	Chloroacetamides	Alachlor, Metolachlor, Terbuchlor
	Other	Diphenamid, Propanil
Benzoics		Dicamba, TBA
Bipyridiliums		Diquat, Paraquat
Carbamates		Barban, Propham
Dinitroanilines		Benifin, Oryzalin, Trifluralin
Diphenyl ethers		Fluorodifen, Nitrofluorofen
Hydrocarbon/oils	Unsaturated ring	Benzene, Naphthalene
	Saturated ring	Cyclopentane, Cyclohexane
	Unsaturated, polar	Trimethyl Benzene
Nitriles		Bromoxynil, Dichlorbenil
Phenoxys		2,4-D, 2,4,5-T, Silvex
Thiocarbamates		Cycloate, Vernolate, Metham
Triazines		Atrazine, Desmetryn, Simazine
Uracils		Bromocil, Terbacil
Ureas		Diuron, Fenuron, Tebuthiuron

[a] Based on Hartley and Kidd (1983) and Beste (1983).

TABLE V

CHEMICAL CLASSIFICATION OF FOSSIL FUEL-RELATED ORGANIC RESIDUES [a,b]

Compound class	Representative compounds
Amine	Aniline, 1-Aminonaphthalene, 1-aminoanthracene, 2-aminobenzo[a]anthracene
Basic aromatic N-heterocycles	Pyridine, quinoline, acridine, benzo[a]acridine
Phenols	Phenol, 1-naphthol, 1-hydroxyanthracene, 2-hydroxybenzo[a]anthracene
Neutral aromatic N-heterocycles	Indole, carbazole, benzo[c]carbazole
Nitroaromatics	Nitrobenzene, 1-nitronaphthalene, 1-nitroanthracene, 2-nitrobenzo[a]anthracene
Thiophenes	Benzo[b]thiophene, dibenzo[b,d]thiophene, Benzo[b]naphtho(1,2-d)thiophene
Neutral aromatic hydrocarbons	Naphthalene, anthracene, benzo[a]anthracene
Furans	Benzo[b]furan, dibenzo[b,d]furan, benzo[b]naphtho(1,2)furan

[a] From Zachara et al.(1984).
[b] Aromatic classes and suitable representative compounds are listed in order of increasing complexity.

3. Inorganic Xenobiotics

Many inorganic compounds formed as waste products of modern industry, such as the oxides of sulfur and nitrogen, can be considered to be xenobiotics because of the elevated concentrations released and the deleterious effects that they have on the biosphere. However, the principal inorganic xenobiotic chemicals are the heavy metals and metalloids (Keswick, 1984). Although many metals are reasonably abundant in the earth's crust, anthropogenic activities (including fuel combustion, mining, smelting, and agricultural practices) often increase their concentration to toxic levels. Metals that are toxic at sufficiently high concentrations and therefore are potentially hazardous include arsenic, boron, cadmium, chromium, copper, lead, mercury, molybdenum, nickel, selenium, tin, vanadium, and zinc (Babich and Stotzky, 1982, 1985a; Chang and Broadbent, 1982).

B. INTERACTIONS OF XENOBIOTICS WITH SOILS

After xenobiotics are released into the environment, an irreversible chain of dynamic events is set into motion. The xenobiotic can modify the abiotic and biotic processes that occur in soil and can, in turn, be acted on by these processes. Many different interactions of xenobiotics with soils have been recognized (Fig. 4). The chemical properties of the xenobiotic and its environmental concentration determine, to a large extent, the interactions that occur.

When a xenobiotic is released into soil, it may be transported or chemically modified by biotic or abiotic processes. The behavior of the xenobiotic will depend on its chemodynamic properties and on the physical, chemical, and biological properties of soil. Table VI lists some of the processes that affect the behavior of a xenobiotic in soil.

Adsorption, leaching, bioconcentration, and volatilization are processes that affect the transport of the original compound within soil, whereas hydrolysis, oxidation–reduction, and microbial transformation affect its modification. However, these processes do not occur independently of one another, and the rate at and extent to which one process occurs will govern the rate and extent of the other processes. For example, the adsorption of 2,4-D on clay minerals and organic matter has been shown to decrease its biodegradability (Ogram et al., 1985).

The most versatile and active of systems that affect the modification of xenobiotics in soil are biotic (Tinsley, 1979). Many xenobiotic compounds have little structural resemblance to natural compounds, and degradation of these xenobiotic compounds will be dependent on (1) the ability of existing microbial enzyme systems to act on those xenobiotics that are similar, but not identical, to chemicals found in nature; or (2) the ability of the xenobiotic to induce the synthesis of necessary degradative enzymes. Biodegradation is less likely for a molecule with structural features seldom or never encountered in natural products. Some of the relationships between chemical structure and biodegradability are outlined in Table VII.

In addition to the structural features of the xenobiotic, environmental conditions (e.g., the presence or absence of oxygen, the content of usable water, pH, and temperature) must be conducive to the activity of those

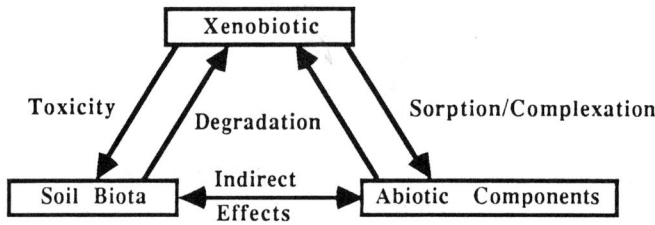

FIG. 4. Interactions of xenobiotic chemicals with soil.

TABLE VI

Processes in Soil That Affect the Behavior of Xenobiotics [a]

Process	Governing factors
Hydrolysis	pH
Microbial transformation	Presence of degradative enzymes; appropriate environmental conditions
Volatilization	Equilibrium vapor pressure
Oxidation–reduction	E_h
Leaching	Solubility
Adsorption	Partition coefficient pK_a of adsorbate; types of adsorbents available; solubility
Bioconcentration	Partition coefficient; pK_a of adsorbate

[a] Modified from Hutzinger and Veerkamp (1981).

microorganisms containing the appropriate enzymes. A discussion of all possible transformations that could occur in soil is beyond the scope of this summary. However, some of the more common types of reactions catalyzed by microorganisms include dehalogenation, deamination, decarboxylation, methyl oxidation, hydroxylation, β-oxidation, reduction of triple and double bonds, sulfur oxidation, hydration of double bonds, polymerization, and nitro metabolism (Alexander, 1981).

The effects of xenobiotics on the abiotic properties of soil, both physical and chemical, have not received much attention. Adsorption of

TABLE VII

Influence of Structure on Biodegradability of Xenobiotics [a]

Type of compound or substituents	More degradable	Less degradable
Hydrocarbons	Higher alkanes (~ 12 C)	Lower alkanes
	Alkanes	High molecular weight alkanes
	Straight-chain paraffinic	Branched paraffinic
	Paraffinic mono- and bicyclic aromatic	Aromatic
		Polycyclic aromatic
Aromatic substitutes	-OH	-F
	-COOH	-Cl
	-NH$_2$	-Br
	-OCH$_3$	-NO$_2$
		-CF$_3$, sO$_3$H
Aliphatic chlorine	-Cl > 6 carbon atoms from terminal C	-Cl < 6 carbon atoms from terminal C

[a] From Hutzinger and Veerkamp (1981).

a xenobiotic on mineral and organic surfaces can change the chemical properties of the soil by increasing or decreasing the CEC, AEC, pH, and percentage base saturation, and by altering the hydrophilic/hydrophobic characteristics of the adsorbent (i.e., mineral or organic surface). Effects of xenobiotics on the physical properties of soil include (1) changes in porosity by occupation of pore space, thereby reducing the ability of the soil to transmit water; (2) reduction in water-holding capacity caused by the exclusion of water; and (3) reduction in stability of soil aggregates, as a result of the disruption of mineral–organic matter interactions. In addition to the direct effects that xenobiotics can have on abiotic soil properties, they may also affect these properties indirectly. For example, bacteria using a xenobiotic as a nutrient source can cause a decrease in the hydraulic conductivity of soil by excreting metabolic products that clog soil pores (Frankenberger et al., 1979).

V. Environmental Factors That Influence Interactions between Microorganisms and Xenobiotics

Interactions between xenobiotics and microorganisms in soil are influenced by a variety of environmental factors, such as pH, E_h, temperature, water content, soil type, CEC, and AEC. The influence of environmental factors on the growth and activities of microorganisms in soil has been studied and reviewed extensively (Atlas and Bartha, 1987; Domasch et al., 1983; Doetsch and Cook, 1974; Stotzky, 1974). In addition, the abiotic factors that affect interactions beween microorganisms and inorganic xenobiotics (e.g., heavy metals) have been reviewed by Babich and Stotzky (1982, 1985a). However, the influence of environmental factors on microbe–organic xenobiotic interactions has not received sufficient attention.

Physicochemical Factors

1. *Clay Minerals*

The type and concentration of clay minerals present in a particular soil have a great effect on the toxicity of various inorganic and organic zenobiotics to microorganisms. Heavy metals can be adsorbed on clay minerals, thereby reducing their concentration in the soil solution and attenuating their adverse effect on microbial populations in soil (Stotzky, 1986). The toxic effects of cadmium, lead, nickel, mercury, and zinc have been shown to be reduced in the presence of clays (Babich and Stotzky, 1977, 1978, 1979, 1982, 1985a; Babich et al., 1983; Debosz et al., 1985; Stotzky, 1986), primarily as a result of the CEC of the clays.

Few studies have investigated the effect of clay minerals on the toxicity of organic xenobiotics. Organic compounds, especially at pH values below their pI, may be adsorbed on clay minerals, thereby reducing their inhibitory effects on microorganisms. Clay minerals have been shown to decrease the toxicity of some antibiotics and pesticides in soil (Stotzky, 1986). On the other hand, adsorption of organic compounds on clay surfaces may increase toxicity by concentrating the compound in the region of high microbial density; however, this hypothesis has not been proven.

In addition to a CEC, clays also have an AEC, the magnitude of which is dependent on the pH of the soil. The AEC of clay minerals can be important in the adsorption of anionic xenobiotics.

2. *Organic Matter*

Organic matter contributes to both the CEC and the AEC of soil, and therefore the effect of this fraction on the toxicity of xenobiotics would be similar to that of clays. For example, organic matter has been shown to decrease the toxic effect of some heavy metals (Babich and Stotzky, 1982), presumably, in part, by binding the heavy metals by an ion exchange mechanism. In addition, soluble organic matter in the soil solution may reduce the toxicity of heavy metals by chelating the metal and thus reducing its availability to microorganisms (Babich and Stotzky, 1985a; Stevenson and Finch, 1986).

The behavior of organic xenobiotics may also be altered by the presence of organic matter. Reduction in solution phase concentration can occur through a variety of mechanisms, including ion exchange, protonation, covalent bonding, H-bonding, van der Waals forces, and coordination through an attached metal ion (ligand exchange) (Stevenson, 1985). In addition, nonpolar xenobiotics can be partitioned onto organic surfaces by hydrophobic mechanisms.

3. *pH*

The pH of soil affects interactions between microorganisms and xenobiotics in various ways. First, pH influences the sorptive behavior of both inorganic and organic compounds. For example, the adsorption of some heavy metals (e.g., copper, lead, zinc) increased with increasing pH, whereas the adsorption of other metals (e.g., mercury) decreased with increasing pH (Farrah and Pickering, 1978a,b). Adsorption of some organics is a function of their pI; therefore, pH levels that increase the cationic nature of organic xenobiotics should increase their adsorption on clay and organic surfaces. Second, pH influences the chemical speciation, mobility, and toxicity of metals (Babich and Stotzky, 1982) and, possibly, of organics. Third, the pH of soil can alter the CEC : AEC ratio of organic matter and thus influence the binding of xenobiotics on this fraction. Fourth, pH affects the extent of complexation of metals with soluble organics. Fifth, the physiological state and the metabolic activities of microorganisms are affected by soil pH (Atlas and Bartha, 1987; Babich and Stotzky, 1982), and this can affect their sensitivity to xenobiotics.

4. Temperature

Temperature has been shown to modify the toxicity of some inorganic xenobiotics (Babich and Stotzky, 1982). However, the particular effect depends on the metal and the organism. For example, the toxicity of chromium(III) to *Navicula seminulum* decreased as the temperature increased from 22° to 30°C; however, the toxicity of chromium(III) to *Cyclotella meneghiniana* increased as the temperature increased from 5° to 25°C (Cairns et al., 1978). Although these studies were performed in water, it is probable that similar relations exist in soil. Temperature probably also indirectly modifies the toxicity of an organic xenobiotic by altering the physiological state of the microorganisms and, thus, their sensitivity or resistance to the xenobiotic (Babich and Stotzky, 1985b).

Few studies have been conducted to assess the effect of temperature on microbe–organic interactions. The solubility of most organics and thus availability decreases with decreasing temperatures (Tinsley, 1979). Therefore, the toxicity of some organic xenobiotics may increase with increasing temperature, but increasing temperatures increases the volatility of organics. This would have the effect of transferring the organic from the soil solution into the soil atmosphere and, possibly, away from susceptible microorganisms.

5. Redox Potential

Most inorganic and organic compounds can either accept electrons and be reduced, or donate electrons and be oxidized. This alteration in oxidation state is important in microbe–xenobiotic interactions because the oxidized or reduced form of the xenobiotic often affects its toxicity and environmental behavior (e.g., adsorption, volatility, solubility). In addition, the redox potential can affect the types of microorganisms active in soil and thus the susceptibility of populations at given times.

Numerous studies have examined the effects of microbial activity on the degradation and fate of organic compounds under different redox conditions (e.g., Gibson, 1984); however, few studies have examined the effects of redox potential on the toxicity of xenobiotics to microorganisms. Nevertheless, redox potential may be an important factor in influencing the toxicity of organic xenobiotics to microorganisms in soil.

6. Interactions among Xenobiotics

Xenobiotics are rarely, if ever, present in soil as individual constituents. Xenobiotic interactions, whether synergistic, additive, or antagonistic, are likely to affect microorganisms differently than a particular xenobiotic individually. Examples of synergistic, additive, and antagonistic interactions between heavy metals have been reviewed by Babich and Stotzky (1982). Antagonistic interactions can result from competition between cations for common sites on the surface of a susceptible microorganism, whereas synergistic or additive effects may result from the increased adsorption of one metal onto the cell caused by the presence of another metal (Babich and Stotzky, 1982). Possible synergistic, additive, or antagonistic interactions between organic xenobiotics have received little attention (Babich and Stotzky, 1985b). Nevertheless, organic–organic interactions probably have a major influence on the susceptibility of microorganisms to particular xenobiotics.

In closing this section of the review, it should be emphasized that more research is needed to determine the influence of physicochemical factors on interactions between microorganisms and xenobiotics in soil, especially organic xenobiotics. Better understanding of the physicochemical characteristics of a particular soil that affect microbe–xenobiotic interactions is necessary to develop models for predicting the effects of xenobiotics (existing and new) in different soils, which can then be used to identify "high-risk" and "low-risk" soils, i.e., those soils that accentuate or attenuate the effects of a particular xenobiotic (Babich and Stotzky, 1982). One benefit gained by this understanding would be help in designing and siting new industries to match deleterious manufacturing outputs with soil types capable of reducing the adverse effects of the outputs. For example, because the toxic effects of a number of heavy metals are reduced in soils having a high CEC (Stotzky, 1986), it would be wiser to place new refineries at locations where the surrounding soils contain clays that have a high CEC rather than where the soils are dominated by sand or clays having a low CEC. In addition, an understanding of the influence of physicochemical factors on microbe–xenobiotic interactions would help in modifying existing facilities to handle their pollutant output in a manner more environmentally sound and in detoxifying soils already polluted by xenobiotics (Babich and Stotzky, 1982, 1983, 1985a).

VI. Effects of Xenobiotics on Microorganisms in Soil

Xenobiotics, particularly pesticides, are often applied directly to soil to control organisms considered deleterious to agricultural crops, livestock, and humans. In addition, large quantities of organic and inorganic compounds enter the soil from the disposal of waste products from industrial, energy, agricultural, domestic, and national defense programs. There is concern among soil microbiologists and other scientists that, once in the soil, these compounds may adversely influence the growth and activity of beneficial indigenous microorganisms. The majority of biochemical transformations in soil result from microbial activity (Alexander, 1977), and any compound that alters the number or activity of microbes could affect soil biochemical processes and, ultimately, influence soil fertility and plant growth.

A. Microbiological Processes

1. Microbial Respiration

Carbon dioxide (CO_2) evolution and oxygen (O_2) uptake are frequently used to assess the effect of xenobiotics on the overall metabolic activity of the soil microbial population. Respiration is nonspecific, and estimations of this process give no indication of the selective suppression of sensitive species. Therefore, cautious interpretation of experimental results is necessary.

There is little doubt that xenobiotic compounds can influence soil respiration. However, the specific influence (whether stimulatory or inhibitory) depends on the type of compound, its concentration, and the particular physicochemical factors of the soil to which the compound is added.

a. Type of Compound. It is difficult to determine from the literature a clear relation between compound structure and its effect on microbial processes, such as respiration. In general, low concentrations of recalcitrant compounds, such as the chlorinated aromatic hydrocarbons, exert little influence on soil respiration (Parr, 1974), indicating that they are not toxic and do not serve as a carbon source and, thus, do not influence CO_2 production or O_2 consumption. However, at higher concentrations, chlorinated aromatics are toxic to microorganisms (Boyd and Shelton, 1984), and thus inhibition of soil respiration is to be expected. For example, pentachlorophenol at 200 ppm has been reported to inhibit completely O_2 uptake by soil (Grossbard, 1976). Less persistent organic compounds, such as the carbamate and phenylurea pesticides, appear to depress respiration; however, their effect is also concentration-dependent (Bartha et al., 1967). Soil respiration is depressed to the greatest degree by the nonselective eradicant-type organic compounds, such as fungicides (Parr, 1974).

At low concentrations, other organic xenobiotic compounds have been shown to stimulate O_2 consumption (Grossbard, 1976), possibly because these compounds are utilized as a carbon and/or energy source or because of the solubilization of soil organic compounds by the added organic xenobiotic and their subsequent utilization by the microbial biomass. The increase in O_2 consumption may also be caused by the uncoupling of oxidative phosphorylation from the electron transport chain, thereby increasing microbial O_2 consumption; this was suggested because the initial stimulation of O_2 uptake by some compounds is followed by a marked inhibition (Bartha et al., 1967).

The inhibitory effect of xenobiotics is often temporary, and after inhibition, respiration often tends to increase above the controls (Wainwright, 1978), probably as the result of an initial kill of some soil microorganisms by the xenobiotic followed by utilization of the dead microbial tissue as substrate by surviving microbial populations. This phenomenon has been used as the basis of the soil fumigation method for the estimation of total microbial biomass in soil (Jenkinson and Powlson, 1976). A temporary depression in soil microbial activity may be beneficial from an economic standpoint, because soil microorganisms often compete with agronomic plants for vital nutrients. If the reduction in soil microbial activity coincides with a critical stage in the growth of a crop, the plant may be better able to compete for these nutrients.

b. *Concentration.* The concentration in soil of a xenobiotic determines whether it will affect soil microbial respiration. At sufficiently high concentrations, few xenobiotics are without effect (Grossbard and Davies, 1976), whereas at low concentrations, most xenobiotics have few acute inhibitory effects on soil respiration (Gaur and Misra, 1977). However, little is known about the chronic effects of repeated applications of low concentrations of xenobiotics on soil microorganisms. Contradictory results have been obtained with the same compounds (Table VIII).

c. *Environmental Parameters.* Some of these discrepancies can be explained by differences in the soil type used in the research. The "effective concentration" (i.e., solution-phase concentration) of metals, such as cadmium or zinc, is a function of the pH of the soil solution and the availability of cation or anion exchange sites. Soil pH influences the speciation form of the metal, which, in turn, influences the charge and, therefore, whether the metal adsorbs to cation and anion exchange sites on clay minerals or organic colloids. The CEC or AEC of soil will determine the extent of adsorption. Soils with a high exchange capacity would bind these metals to greater extent, thus lowering their availability susceptible microbial populations. Organic compounds, such as

TABLE VIII

Variations in the Effects of Selected Xenobiotics on Soil Respiration

Compound	Concentration (ppm)	Effect	Reference
Simazine	8000.0	None	Grossbard (1976)
	10.0	Inhibition	Wainwright (1978)
	5.0	Stimulation	Smith and Weeraratna (1974)
Cadmium	1000.0	None	Bewley and Stotzky (1983)
	1000.0	Inhibition	Doelman and Haanstra (1984)
Zinc	1000.0	None	Bewley and Stotzky (1983)
	1000.0	Inhibition	Doelman and Haanstra (1984)

simazine, usually bind on soil organic matter, thus soils with high content of organic matter would be more effective in lowering the toxicity of added organic xenobiotics. Other environmental factors that influence the effect of xenobiotics on soil respiration include pH, E_h, ionic composition, temperature, and water content (Babich and Stotzky, 1982; Grossbard, 1976).

Time is a factor often overlooked. The effect of most xenobiotic compounds on soil microbial processes, such as respiration, is time-dependent (Babich et al., 1983). For example, Doelman and Haanstra (1984) examined the long- and short-term effects of cadmium, chromium, copper, nickel, lead, and zinc on soil microbial respiration and found that the reduction in respiration caused by these metals remained significant even after 18 months. In contrast, Babich et al. (1983) and Debosz et al. (1985) found that the inhibitory effects of some heavy metals were transitory. For this reason, Domsch et al. (1983) suggested that specific monitoring periods be established in assessing the effect of xenobiotics on microbial populations: effects lasting <30 days can be considered negligible; effects lasting ≤60 days, tolerable; and those effects extending over >60 days, critical. These guidelines were based on the time required (estimated at 30 days) for microbial communities to recover from natural stress events, such as fluctuations in temperature, pH, and water content.

2. *Carbon Transformations*

Soil respiration provides an overall indication of the effects of xenobiotic compounds on soil microbial activities. However, it is also important to determine their effects on the utilization of specific carbon compounds. The breakdown of cellulose is particularly important, because it is the principal component of plant litter. The assessment of the effects of xenobiotic compounds on cellulose degradation would, therefore, indicate the extent to which these compounds influence the ability of indigenous microorganisms to decompose organic matter.

Inasmuch as the initial decomposition of cellulose is usually attributed to soil fungal populations, it is not surprising that fungicidal compounds have the greatest impact on cellulose degradation (Grossbard, 1976). Nonfungicidal compounds, such as herbicides, have also been shown to inhibit cellulose degradation (Wainwright, 1978). Grossbard (1973a,b) used buried calico strips and cellulose powder to examine the effects of 50 and 500 ppm aminotriazole, Linuron, and Metoxuron. The assessment of cellulolytic activity was made by measuring the tensile strength of the calico strips after incubation. In general, these organic compounds inhibited cellulose degradation at the high concentration but had no effect at concentrations closer to field rates (50 ppm).

Certain heavy metals also inhibit the cellulolytic activity of soil-derived organisms. In experiments in which the activity of cellulose-decomposing fungi was assessed by measuring the release of dye from remazol blue-dyed cellophane films overlaid on an agar medium, Khan and Frankland (1984) observed that zinc, copper, nickel, lead, and cadmium inhibited cellulolytic activity at all concentrations tested (10–1000 ppm). The degree of inhibition varied, depending on the concentration and type of metal applied. For example, cadmium appeared to be much more toxic than lead. The inhibition of cellulolytic activity in soils by heavy metals has also been observed by Martin et al. (1982) and Tyler (1975).

3. Nitrogen Transformations

The transformation of organic nitrogen to inorganic forms is an important microbial function contributing to the fertility of soil and a transformation that has become a major indicator in assessing the effects of xenobiotics. The major nitrogen transformations mediated by soil microorganisms include ammonification, nitrification, denitrification, and nitrogen fixation (see Fig. 2). Ammonification, nitrification, and nitrogen fixation represent input processes in which nitrogen is converted to forms available for uptake by plants. Denitrification is an output process in which inorganic nitrogen is reduced to gaseous products by facultative bacteria that use it as a terminal electron acceptor, thereby causing it to be lost from soil.

Ammonification, the production of ammonium-N from organic forms such as proteins, amides, and amino acids, does not generally appear to be inhibited by organic xenobiotic compounds and, in many cases, appears to be stimulated. For example, herbicides and insecticides have been shown to have little influence on ammonification (Anderson and Drew, 1976; Wainwright, 1978), whereas fumigation generally resulted in an increase in ammonium-N (Rovira, 1976). The reasons for these observations are not clear, however, because a large and diverse group of microorganisms are capable of ammonification. The observed increases in ammonium-N after the addition of fumigants may also result from the release of ammonium-N from microorganisms killed by the compounds.

Nitrification is the process by which ammonium-N, released after the mineralization of N-containing organic compounds, is oxidized to nitrate in two steps, (Eqs. (1) and (2)), most commonly by two genera of autotrophic bacteria:

$$2NH_4^+ + 3O_2 = 4H_2O + 2NO_2^- \quad \text{by } Nitrosomonas \text{ sp.} \tag{1}$$

$$2NO_2^- + O_2 = 2NO_3^- \quad \text{by } Nitrobacter \text{ sp.} \tag{2}$$

Inasmuch as nitrification is usually conducted by a select group of chemoautotrophic microorganisms, it would be expected that this process would be much more sensitive to the action of xenobiotics than processes such as ammonification, which are conducted by a diverse microbial population. Indeed, Parr (1974) suggested that nitrification is one of the soil microbiological transformations most sensitive to pesticides.

Chlorinated hydrocarbons appear to have few acute effects on soil nitrification when applied at low rates. However, long-term chronic effects may result from repeated application of these pesticides. More information is needed concerning the chronic versus acute effects of xenobiotics on microorganisms in soils. In addition, the degradation products of chlorinated compounds may influence nitrification. Corke and Thompson (1970) studied the effects of selected chlorinated aromatic compounds on the oxidation of ammonium and nitrite and found that specific degradation products inhibited each process to a different degree. Their results, using Diuron and Propanil, along with the probable degradation products of each of these compounds are illustrated in Fig. 5. The principle illustrated in Fig. 5 is that degradation products of xenobiotics can affect specific microbial transformations, even if the parent xenobiotic has no effect. Therefore, a more critical evaluation of xenobiotic compounds in this regard is advisable. In addition, the ramifications of inhibiting ammonium oxidation are much less than inhibiting nitrite oxidation. Nitrite is highly toxic to many plant and animal species, including humans, and therefore processes that inhibit its oxidation without inhibiting ammonium oxidation could have adverse health effects.

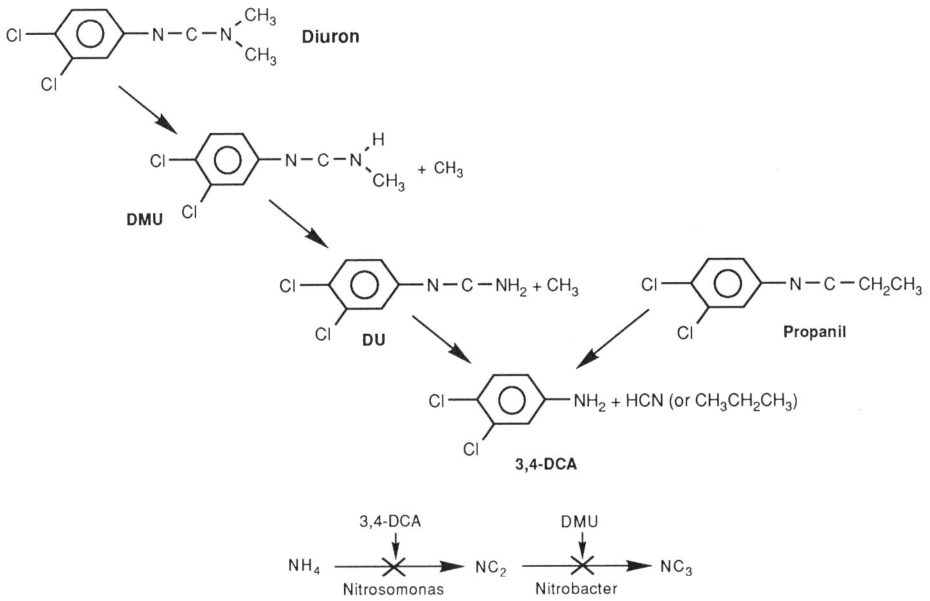

FIG. 5. Inhibition of soil-nitrifying bacteria by degradation products of Diuron and Propanil. From Corke and Thompson (1970).

Fungicides appear to have a greater initial and longer lasting effect on nitrification than most other classes of xenobiotics even though their target (e.g., fungi) is not considered important to the nitrification process (Parr, 1974). The reasons appear to be 2-fold. First, these compounds are usually applied at much higher rates than other compounds, and second, they are designed and selected to affect microorganisms. In addition to the inhibitory effects of organic compounds, nitrification processes are, in general, also inhibited by the action of heavy metals (Giashuddin and Cornfield, 1979; Rother et al., 1982; Chang and Broadbent, 1982; Bewley and Stotzky, 1983). The comparative toxicity of metals to nitrification follows the sequence, $Hg > Cr > Cd > Ni > Cu > Zn > Pb$ (Liang and Tabatabai, 1978).

An interesting observation is the effect of xenobiotics on nitrification processes in soils differing in their "inherent nitrifying capacity." Dubey (1969) defined inherent nitrifying capacity as a soil's ability to support and maintain a large population of active nitrifiers. Both Dubey and Rodriguez (1970) and Parr (1974) observed that soils of high nitrifying capacity are affected less by xenobiotics than soils of low nitrifying capacity. These results may explain some of the discrepancies between experiments and emphasize the need to understand thoroughly the environmental parameters of the soils being evaluated.

Although the effects of xenobiotics on denitrification processes in soils have not been investigated thoroughly, Wainwright (1978) suggested that changes in nitrate-N caused by additions of xenobiotic compounds may result from changes in denitrification processes. For example, Saive (1974) found that although treatment of soil with Benomyl and Fentin hydroxide resulted in similar increases in nitrate levels, different processes were affected. Benomyl appeared to enhance the oxidation of nitrite to nitrate, whereas Fentin hydroxide appeared to decrease the rate of denitrification. The inhibitory effect of a number of other organic compounds, including many chlorinated compounds, on denitrification processes has also been demonstrated (Bollag and Henninger, 1976; Bollag and Nash, 1974). Both nitrification and denitrification are also sensitive to heavy metals: the sequence of toxicity is $Cd > Zn > Cu > Pb$ (Babich and Stotzky, 1985a). From an agronomic perspective, the loss of gaseous nitrogen caused by microorganisms under anaerobic conditions is undesirable; therefore, inhibition of denitrification may not be detrimental.

The atmosphere is the principal source of nitrogen for plant growth. The microorganisms involved in fixing N_2 are either free-living, such as cyanobacteria, *Azotobacter* sp., *Azospirillum* sp., and *Clostridium* sp., or exist in symbiosis with higher plants, such as *Rhizobium* sp. and *Frankia* sp. The effects of xenobiotics on nitrogen fixation appears to be species-specific. For example, symbiotic nitrogen-fixing bacteria appear to be more sensitive to the action of added compounds than free-living bacteria such as *Azotobacter* sp. (Grossbard, 1976). Intraspecies sensitivity is also variable. For example, some strains of the blue-gree alga *Nostoc* sp. can tolerate up to 2000 ppm Propazine, whereas other strains are killed by 1 ppm (Grossbard, 1976).

Nitrogen fixation is often assessed by relatively nonselective methods, such as the acetylene reduction assay (Weaver and Frederick, 1982). In the case of symbiotic nitrogen fixation, distinction must be made between the response of the bacterium and of the host to a xenobiotic. For example, a compound that does not affect nitrogen-fixing bacteria directly may do so indirectly by inhibiting the host. This is reflected in the conclusions of Greaves et al. (1980) that the most reliable indications of the effects of xenobiotic compounds on symbiotic nitrogen fixation were obtained from measurements of plant growth and yield rather than specific nitrogen fixation assays. However, it was suggested that if effects were found using plant growth and yield measurements, further investigations using acetylene reduction methods should be undertaken. Nevertheless, it should be emphasized that the examination of the effects of xenobiotics on symbiotic nitrogen-fixing bacteria independent of the host is important, because the bacteria may exist in soil for long periods of time in the absence of the host.

4. *Transformation of Sulfur and Phosphorus*

Sulfur enters soil primarily in the form of plant residues, animal wastes, chemical fertilizers, and rainwater. A large part of the sulfur in the soil profile is present in organic matter. The importance of microbially mediated transformations of sulfur has become increasingly apparent in recent years (Alexander, 1977; Granat et al., 1976). Sulfate is the principal plant-available source of sulfur, and the oxidation of sulfur to sulfate and the reduction of sulfate are particularly important. The oxidation of sulfur involves principally the activity of specialized chemoautotrophic bacteria, although heterotrophic microorganisms may be important under certain conditions.

Because of the role of microorganisms in the transformation of sulfur and the recognized deficiencies of sulfur in various parts of the world (Coleman, 1966), the effects of xenobiotics on sulfur transformation have become an important issue. Despite recognition of this importance, there have been few studies on the influence of xenobiotics on sulfur transformations. Certain pesticides have been shown to decrease sulfur oxidation when added to soils. Tu and Miles (1976) reported that 2000 ppm Aldrin and Dieldrin decreased the rate of sulfur oxidation for 2 months, whereas Audus (1970) reported no effect at this concentration. Herbicides, such as Paraquat and 2,4-D, have been shown to decrease the oxidation of sulfur, although it is not known if the decrease was the result of a direct action on the principal organisms responsible for oxidation or an indirect effect caused by the loss of plant exudates after the death of the plant (Tu and Bollen, 1968).

Another major nutrient required by both plants and microorganisms is phosphorus. Phosphate exists in soils as inorganic forms and as organic forms that undergo mineralization (Alexander, 1977). Microorganisms, especially mycorrhizal fungi, are important in the uptake of phosphorus by plants (Agrios, 1978). Because of the importance of mycorrhizal fungi, it would be expected that fungicides decrease the amount of soluble phosphorus. However, Wainwright and Snowden (1977) showed that fungicides increased slightly the level of $CaCl_2$-extractable phosphorus in soils, resulting in increased solubilization of added insoluble phosphates. These increases were associated with an increase in the population of phosphorus-solubilizing bacteria after soil treatment. The application of insecticides and herbicides has been shown to have little effect on either phosphorus mineralization from organic matter or solubilization from inorganic forms (Smith and Weeraratna, 1974; Tyunyayeva *et al.*, 1974).

Although few studies have investigated the influence of heavy metals on microbially mediated cycling of inorganic phosphorus, these processes appear, in general, to be inhibited in the presence of heavy metals (Juma and Tabatabai, 1977; Capone *et al.*, 1983).

5. *Cell-Free Enzymes*

Many reactions involving the transformation of inorganic and organic nutrients in soils are catalyzed by enzymes that exist free in the soil solution or in association with clays or organic matter (Burns, 1983). Although studies on the effect of xenobiotic compounds on soil enzymatic activity have been limited, the results show considerable variability with respect to the effect of a compound on the activity of soil enzymes.

Many xenobiotic compounds either stimulate or inhibit the activity of soil enzymes (Table IX), although some may exert only a negligible effect. Some enzymes, such as dehydrogenase and urease, appear to be more sensitive to xenobiotics than others. Grossbard (1976) suggested that the inhibition of the activity of these enzymes may result from the loss of vegetative cover after application of the selected herbicides rather than from a direct antimicrobial effect. However, Cole (1976) found that compounds such as the triazines had a direct bacteriostatic effect. Also, even though most xenobiotics do not appear to inhibit mineralization of nitrogen in soils, they do appear to inhibit urease and protease activity, which is essential to the mineralization of nitrogen. Another interesting observation is that while most enzymes studied appear to be inhibited by most xenobiotics studied, phosphatase activity is often stimulated.

6. *Microbial Populations*

a. Free-Living Populations. Table X summarizes reported effects of some xenobiotic compounds on selected microbial populations. It is difficult to reach any generalizations concerning the effects of xenobiotics on specific microbial populations; although some trends are evident, additional research into documenting the effects by chemical class is needed. In general, algae and photosynthetic bacteria appear to be more susceptible to xenobiotics than other groups, probably because many of the compounds that have been tested inhibit photosynthesis. Actinomycetes and saprophytic fungi appear to be more resistant to the action of xenobiotics, and for many of the compounds tested, an increase in

TABLE IX

Effects of Selected Xenobiotics on Soil Enzymatic Activity [a]

Compound class	Enzyme effect [b]							
	Dehydrogenase	Urease	Catalase	Phosphatase	Protease	Invertase	Asparaginase	Proteinase
Substituted ureas	I	I	IS	S	I	I	IS	?
Triazines	I	I	IS	S	I	S	S	I
Nitrophenols	ISN	IS	I	?	?	N	?	?
Benzoic and phenylacetates	I	I	S	S	?	S	?	I
Heterocyclic compounds	I	I	?	S	I	I	?	I
Carbamates	N	I	?	S	I	I	?	?
Heavy metals	I	I	I	I	I	I	I	I

[a] Data from Grossbard (1976), Bonmati et al. (1985), Wainwright (1978), and Litchfield and Huben (1973).
[b] I, Inhibition; S, stimulation; ?, effects unknown; N, no effect.

TABLE X

Effects of Selected Xenobiotics on Soil Microbial Populations [a]

Microbial group	Xenobiotic effect [b]									
	Chlorinated aromatics	Fumigants	Fungicides	Substituted ureas	Surfactants	Triazines	PCBs	Cadmium	Chromium	Mercury
Chemoheterotrophic bacteria	ISN	I	SN	ISN	I∧S*	N	ISN	IN	IN	IN
Chemoautotrophic bacteria	IN	I	N	?	?	IN	IN	I	IN	IN
Actinomycetes	N	I	SN	SN	S	SN	N	IN	N	IN
Rhizobia	I	I	I	I	?	IN	?	I	IN	I
Photosynthetic bacteria	I	I	IN	?	I	I	?	I	IN	IN
Saprophytic fungi	NS	I	I	IN	S	N	NS	IN	I	I
Mycorrhizal fungi	IN	I	I	I	?	IN	?	IN	I	?
Algae	I	I	I	I	I	I	IN	I	I	I
Protozoa	I	I	I	?	?	IN	IN	?	IN	?

[a] Adapted from Babich and Stotzky (1982), Simon-Sylvestre and Fournier (1979), Lal and Saxena (1982), Grossbard (1976), and Parr (1974).
[b] I, Inhibition; S, stimulation; N, no effect; ?, unknown; ∧, high concentrations only; * low concentrations only.

their numbers was detected (Simon-Sylvestre and Fournier, 1979). This trend, which was also seen with heterotrophic bacteria, may be related to two phenomena. First, the xenobiotic in question served as a carbon or energy source for these microorganisms, in which case an increase in population density would occur. Second, the xenobiotic suppressed the population density of predatory microorganisms, such as protozoa. Another observation is that, in general, compounds such as fungicides have a broad inhibitory effect, causing reduced population densities among all microbial groups. For certain groups, such as the heterotrophic bacteria, this effect is usually only temporary, and populations generally recover to or above pretreatment population densities. As mentioned earlier, this increase is usually attributed to the utilization, by surviving bacteria, of dead microbial cells killed by the xenobiotic.

b. *Root-Associated Microbial Populations.* Rhizosphere populations constitute a particularly important soil microbial community. Because of their unique relationship to the plant root zone that they colonize, rhizosphere microbial populations differ from those in soil not directly associated with roots (Gerhardson and Clarholm, 1986). A small fraction of the carbon compounds resulting from photosynthesis are directly released by the roots into the rhizosphere. These organic compounds stimulate the microbial degradation of organic matter in soil adjacent to the roots, liberating inorganic nutrients that are available for direct uptake by the roots (Lynch, 1983). The release of organic compounds by the plant also maintains the microbial interactions that may restrict the activities of plant growth-inhibiting microorganisms while sustaining the activities of plant growth-stimulating microorganisms (Schippers et al., 1986). However, very little is known about the effects of xenobiotics on release of root exudates and on plant pathogens.

Schippers et al. (1986) divided rhizosphere microorganisms into two groups with respect to their influence on plant development. The first group includes those organisms that, through their metabolic activities, stimulate the growth of plants. Included in this group are N_2-fixing rhizobia, mycorrhizal fungi, free-living N_2-fixers, and microorganisms that mineralize organic matter, liberating inorganic nutrients for plant use. The second group includes those organisms that inhibit plant growth, either directly, as in the case of plant pathogens, or indirectly, by inhibiting plant growth as a result of their metabolic activities. The loss of the first group of rhizosphere microorganisms through the action of xenobiotic chemicals can be detrimental to plant growth. However, it is the effect of xenobiotics on plant pathogens that has the greatest economic impact (Lynch, 1983; Trappe et al., 1984; Roslycky, 1985; Sanders, 1986; Schippers et al., 1986).

Because rhizobia fix atmospheric N_2 in symbiosis with leguminous plants, they are the most important among agriculturally beneficial soil bacteria (Roslycky, 1985). The susceptibility of rhizobia to xenobiotics seems to vary with species, strain, type of chemical, and soil (Greaves et al., 1976). Once again fungicides appear to have the greatest detrimental effect on rhizobia (Wainwright, 1978). Although some herbicides have also been shown to inhibit rhizobia (Roslycky, 1985; Barkay et al., 1986), rhizobia appear to be rather resistant to most herbicides (Greaves et al., 1976). Insecticides appear to have little inhibitory effect on rhizobia and, in some cases, can stimulate growth (Barkay et al., 1986). Heavy metals, particularly Cd, Hg, Pb, Cu, and Zn, also have a detrimental effect of rhizobia (Babich and Stotzky, 1982, 1985a).

Certain fungi also form symbiotic relationships with plant roots. These mycorrhizal fungi may be ectotrophic forms, in which the bulk of the fungus is outside the plant root, or endotrophic, in which most of the fungus is inside the plant tissue. The endotrophic forms of the vesicular-arbuscular (VA) types are increasingly recognized as being of significant importance in agricultural crops, as a result of their wide distribution and high efficiency in transporting phosphorus and other nutrients to the tissues of the host plant (Sanders, 1986). Consequently, xenobiotics that affect mycorrhizal fungi will affect crop productivity. Similarly, xenobiotics that affect plant growth will affect formation of mycorrhizal associations.

Trappe et al. (1984) reviewed the effects of pesticides on mycorrhizal fungi. Not surprisingly, fungicides profoundly affect mycorrhizal fungi, as they are intended to kill fungi. However, most of the fungicides selectively affected some fungi more than others. For example, thiazoles were found to be particularly inhibitory to Zygomycotina and less inhibitory to most Basidiomycotina or Ascomycotina. The dicarboximide fungicides do not appear to inhibit mycorrhizal fungi, and some even stimulate these fungi. On the other hand, the dithiocarbamates seem to inhibit most mycorrhizal fungi.

The effects of herbicides on mycorrhizal fungi can be 2-fold. First, mycorrhizal fungi can be drastically reduced by the herbicide directly. However, stimulatory effects have also been reported (Schwab et al., 1982). Second, because herbicides affect the growth and development of the host plant, their action on mycorrhizal populations can also be indirect. Schwab et al. (1984) suggested that the increase in mycorrhizae on plants treated with the herbicide, Simazine, resulted from the increased release of sugars and amino acids by the host. Similarly, inhibition of mycorrhizal fungi has been suggested to result from the reduction of sugar release by herbicide-treated plants (Trappe et al., 1984).

Effects of insecticides on mycorrhizal fungi have been studied least. As with other pesticides, the effects appear to depend on the type of compound as well as on the type of mycorrhizal fungi. The effects of heavy metals on mycorrhizae are also relatively unknown. However, chromium and cadmium have been shown to be inhibitory (Simon-Sylvestre and Fournier, 1979; Babich and Stotzky, 1985a).

It is difficult to quantify accurately microbial populations in soil because the ecological and physiological factors that control the growth of microorganisms in soil are not well understood. Therefore, a completely accurate environmental risk assessment of the effects of xenobiotics on microbial populations and communities is not currently possible. Consequently, the quantification of microbial populations in soil as a measure of the effect of xenobiotics on microorganisms is often disregarded (Greaves, 1982). Changes in microbial populations, if detectable, can serve as a guide in the interpretation of metabolic data, such as respiration or nitrogen transformations (Grossbard, 1973b). In addition, results obtained from changes in species composition caused by the action of a particular xenobiotic may help to elucidate physiological factors involved in inhibition or stimulation.

B. Mode of Action

Present knowledge of the mechanisms whereby xenobiotics affect microorganisms is fragmentary. Specific cellular or biochemical responses will depend on the type of xenobiotic, as well as on the specific microorganism being affected. The types of cellular functions likely to be affected by xenobiotics include cell membrane permeability, photosynthesis, oxidative metabolism, and the synthesis of nucleic acids and proteins, as well as cellular function.

The interactions of xenobiotics (either through adsorptive or absorptive processes) with cell membranes appear to be important in determining the primary target of a xenobiotic. For example, Hicks and Corner (1973) showed that the lethal action of DDT on Gram-positive bacteria was related to the binding of DDT to membranes of these bacteria. Later studies showed that DDT, as well as other chlorinated hydrocarbons, altered both the ratio of phospholipid head groups and the composition of fatty acids (Rosas et al., 1980). These changes might be expected to alter the structure and function of membranes, possibly resulting in bacterial death.

Some xenobiotic compounds have been reported to alter cellular morphology, in addition to biochemical changes. These alterations include abnormalities in nuclear morphology, such as deep incisions, loose chromatin, and fragmented macronuclei (Lal and Saxena, 1980). Changes in the architecture of the plasma membrane, an altered number of cellular organelles, and damaged cell membranes causing leakage of cellular material have also been reported (Parasher et al., 1978).

Interference of xenobiotics with the structure and chemistry of membranes can affect the permeability of cells, which can, in turn, have serious effects on cellular activity. For example, certain chlorinated aromatics decreased the uptake of some amino acids, as well as the synthesis of proteins (Lal and Saxena, 1982). In addition, the synthesis of nucleic acids was also inhibited by some chlorinated aromatics, such as DDT (Lal and Saxena, 1979). Xenobiotics also inhibited the metabolism of a variety of compounds, including pyruvate, citrate, malate, lactate, succinate, and ethanol (Juneja and Dogra, 1978). The reduced metabolic capability has been attributed to the inhibitory effect of xenobiotics on the enzymes responsible for the catabolism of these compounds. For example, Chlordane has been shown to inhibit the activities of succinate dehydrogenase and other oxidative enzymes (Widus et al., 1971).

Various detrimental effects on photosynthetic microorganisms have been reported for certain xenobiotics, such as the chlorinated aromatics. This is particularly important because these microorganisms have an important role in primary food webs as well as in the oxygen balance of the biosphere. Clegg and Koevening (1974) showed that four organochlorine insecticides (DDT, Aldrin, Chlordane, and Dieldrin) significantly reduced the amounts of ATP detected in algae, suggesting that these compounds interfered with photophosphorylation in the light reaction of photosynthesis. In addition,the fixation of atmospheric CO_2 was also shown to be inhibited by these compounds (Wurster, 1968).

VII. Methods of Assessing the Effects of Xenobiotics on Microorganisms

All known types of microorganisms, including bacteria, fungi, protozoa, and algae, occur in the soil (Alexander, 1977; Lynch, 1983). Bacteria and fungi are especially important because they compose the majority of the microbial biomass in soil and are the primary contributors to the processes of decomposition of organic material, degradation of natural and anthropogenic organic chemicals, and cycling of humus (Lynch, 1983). In addition, many of the important processes (e.g., nitrogen fixation, nitrification, sulfur oxidation) that occur in soils are mediated exclusively by specific groups of bacteria and/or fungi.

There is continuing concern that the accidental or deliberate release of xenobiotic compounds into soil may adversely affect the various segments of the soil microbiota, thus disrupting ecologically important microbe-mediated processes (e.g., nutrient cycling). Pesticides are effective in agriculture because they exhibit some degree of toxicity to biochemical processes and consequently have the potential for affecting nontarget microbiota. Many nonpesticidal organic compounds and heavy metals also affect microbial activity.

In assessing the effect of xenobiotics on soil microorganisms, three problems are immediately apparent: (1) what is the effective concentration of the xenobiotic; (2) which microbial processes or properties should be used to assess the effects of the compound in question; and (3) which laboratory test system will allow an accurate estimate of the effect of the xenobiotic on a given process? Most xenobiotics have either an inhibitory or a stimulatory effect, depending on the concentration of the compound (Grossbard, 1973b). However, the specific response to a concentration gradient is not determined by the amount of a xenobiotic applied to soil but by the amount with which the microbiota is in contact. Many xenobiotics often have low solubilities in water; they are adsorbed on mineral and organic surfaces, and their distribution in soil can be heterogeneous. In addition, the composition and density of microorganisms in soil are not uniform. Thus, the concentration of a compound to which different microbial populations are exposed will vary. To assess the effect of a particular compound on microorganisms in soil, an estimate of its concentration in soil is necessary.

It would be desirable to designate an indicator microbial property or process with which the overall effects of xenobiotics could be assessed. However, this is not possible, because different microbial processes are affected to different extents by various xenobiotics. For example, nitrifiers are more susceptible to fumigants than are ammonifiers and denitrifiers (Parr, 1974). Therefore, to assess the effects of xenobiotic compounds on soil microorganisms, it is necessary to examine the effect of a xenobiotic on a variety of microbial processes and/or microbial species. Three categories of relevant indicators of the effects of xenobiotics on soil microorganisms have been suggested (Greaves, 1982; Babich and Stotzky, 1985a; Barkay et al., 1986): (1) assessment of the effects of xenobiotics on microbially mediated processes, such as respiration and transformations of nitrogen and carbon; (2) the effect of xenobiotic compounds on key soil enzymatic activities, such as those of phosphatase and urease; and (3) the influence of xenobiotics on specific microbial populations, especially those of rhizosphere bacteria and fungi because of their important role in increasing the availability and uptake of nutrients by plants.

A. Laboratory Test Systems

A variety of test systems are available for measuring the effects of xenobiotics on microorganisms in soil. Laboratory systems can range in complexity and size from simple batch systems employing domestic canning jars to complex mesocosms that use blocks of cropped soil housed in large greenhouses and are elaborately equipped with instrumentation to monitor a wide variety of microbial processes.

1. Batch Systems

In batch systems, soil is amended with the xenobiotic of interest and incubated in a closed system. An example of a batch system is the biometer flask developed by Bartha and Pramer (1965). This flask is a compact, commercially available unit used for measuring CO_2 produced in soil by microorganisms. Other inexpensive systems that are easy to construct and use for measuring CO_2 evolution from soil are also available for testing the effects of xenobiotics (Anderson, 1982). In addition, by using ^{14}C-labeled substrates and monitoring the evolution of $^{14}CO_2$, the effects of xenobiotics on the mineralization of specific carbon compounds (e.g., cellulose) can be measured. Batch systems offer a simple and inexpensive means of assessing the effects of xenobiotics on microorganisms, which is particularly valuable because of the frequent need to monitor large numbers of samples and replicates, as well as many different xenobiotics. A major criticism of batch systems is that they fail to model adequately real ecosystems.

2. Microcosms

The use of terrestrial "microcosms" (Fig. 6), or integrated laboratory model ecosystems, is an alternative approach to obtaining information about the impacts of xenobiotics on the biota and their interactions, as well as on the influence of abiotic factors on the degradation and residence time of a xenobiotic within a soil.

Simply defined, a microcosm is a small, controlled laboratory system that mimics the processes and interactions in a larger natural ecosystem (Gillet and Witt, 1979; Van Voris et al., 1985a). Its use to study the interactions and relations of organisms in their natural environment dates back to the mid-nineteenth century (Warington, 1851, 1857). Later, Beyers (1964) and Patten and Witkamp (1967), among others, developed the microcosm into a useful laboratory research tool. The application of this type of integrated test system to the questions of the ecological fate and effects of pesticides and complex chemical wastes came next (Metcalf, 1977; Cole et al., 1976; Gillett and Gile, 1976). This

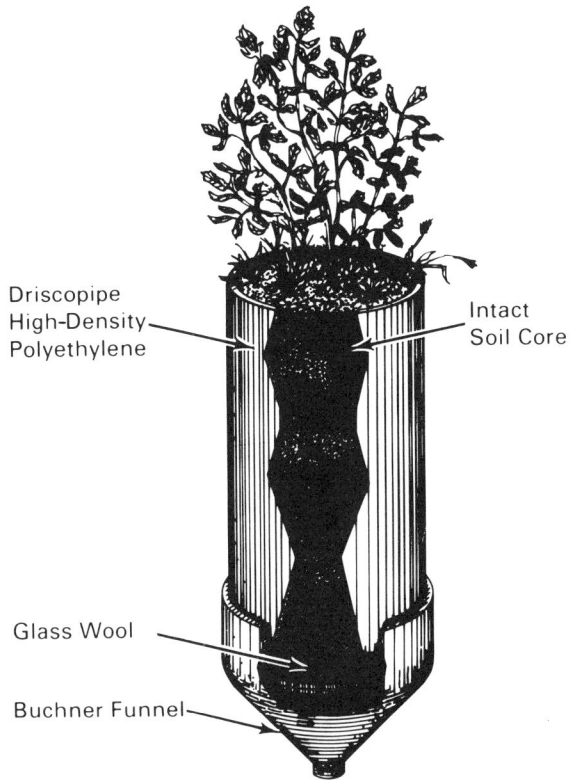

Fig. 6. Terrestrial soil core microcosm test system.

application resulted in a desire to make the test system as identical to the real world as possible and persuaded many scientists to incorporate large and more complex components, such as fish, trees, and small mammals, into their laboratory model ecosystems. These "mesocosms" or "megacosms" outgrew most scientific laboratory buildings (Giesy, 1980). Additionally, these "mesocosms" became too large to replicate to the degree required to obtain reasonable estimates of variance, and too expensive to operate and maintain.

For the smaller systems (e.g., Cole et al., 1976; Ausmus et al., 1979; Gile et al., 1979), the criticisms were that (1) the microcosm did not represent its analog ecosystem, (2) a very large number of test units would be necessary to detect even large ecological effects. (3) the test system could not be used for chemicals that were environmentally persistent, and (4) the validity of the microcosm test system had not been demonstrated. The microcosm approach developed by Van Voris et al. (1985a) and described in a detailed protocol in EPA/600/3-85/047, ASTM 1988 Test Guideline E-1197, as well as in *Federal Register* 9/28/87, Vol. 52, No. 187, pp. 36363–36371, overcomes many of the previous criticisms of the terrestrial microcosm test system. Based on research results compiled over a 5-year period on the "Terrestrial Soil-Core Microcosm" (Van Voris et al., 1984, 1985a,b; Tolle et al., 1981, 1983), this microcosm test system can be used to evaluate any of the following:

1. Fate of xenobiotics within the ecosystem:
   Transport between compartments
   Transformation
   Bioaccumulation
   Chemical speciation
2. Ecological effects of xenobiotics:
   Phytotoxic effects
   Microbiological effects
   Invertebrate effects
   Community effects
   Ecosystem effects (e.g., nutrient cycling)
3. Effects of xenobiotics on:
   Physicochemical properties of soils

B. Methods for Assessing Xenobiotic Impacts on Relevant Microbial Processes and Properties

It is beyond the scope of this section to consider in detail the methods available for studying the effect of xenobiotics on microorganisms in soil. Hence, overviews of methods used to measure important microbial processes and properties are presented. The methods most commonly used for determining whether a xenobiotic has an impact on microorganisms in soil include (1) the measurement of microbial processes, such as respiration and the transformation of carbon and nitrogen and of other important elements, such as phosphorus and sulfur; (2) the measurement of soil enzymatic activities; and (3) the measurement of specific microbial populations.

1. Respiration

The uptake of O_2 or the release of CO_2 by bacterial, fungal, algal, and protozoan cells is used to measure the respiratory activity of microorganisms (Anderson, 1982). Respiration is a measure of the overall activity of the soil microbiota and is frequently used to assess the effects of xenobiotics on the soil microbial community in both field and laboratory experiments. However, there are some problems associated with using respiration to measure the effects of xenobiotics on soil microorganisms. Measurement of respiration lacks specificity, because respiratory activities are not confined to microorganisms but occur also in animals and plants and as a result of abiotic chemical reactions. Therefore, it is difficult to assess whether changes in respiratory activities after the addition of a xenobiotic are microbial or nonmicrobial in orgin. In addition, quantitative respiratory data do not always reflect shifts in microbial equilibria (e.g., changes in species composition), particularly if the compound in question is inhibiting to one microbial species but stimulating to another. Despite these problems, soil respiration is often used as an initial screening procedure for measuring the effects of xenobiotics on soil microorganisms because of the relative ease with which it can be measured (Bitton and Dutka, 1986).

2. *Carbon Transformations*

The effects of xenobiotics on the transformations of organic matter is often monitored by measuring the evolution of CO_2 from soils amended with plant material. The use of ^{14}C-labeled substrates enables monitoring the effects of xenobiotics on the degradation of specific carbon compounds. Buried substrate techniques are also used to measure the effects of xenobiotics on the degradation of specific substrates. For example, the effect of xenobiotics on cellulose degradation in soil has been assessed by monitoring changes in the tensile strength of calico strips buried in soil (Grossbard, 1973a). Another buried substrate technique uses fixed, unexposed color film to measure proteolytic activity in soil (Cullimore and Ball, 1978). Proteolysis is measured by the increase in light transmission through the film resulting from degradation of the gelatin layers of the film (Greaves, 1982).

3. *Nitrogen Transformations*

Ammonification and nitrification are often measured to assess the effects of xenobiotics on nitrogen transformations in soil (Greaves, 1982). In one method, ground plant material is added to soil as a source of organic nitrogen, and the NH_4^+, NO_2^-, and NO_3^- produced are measured (Greaves et al., 1980). If the xenobiotic has no effect, NO_3^- will accumulate. If ammonification is affected, the production of NH_4^+, NO_2^-, and NO_3^-, will be less than in the control soil. If nitrification is affected, NH_4^+ accumulates.

The effect of xenobiotics on denitrification can be examined by monitoring the production of N_2O in sealed chambers enriched with acetylene (C_2H_2) Tiedje, 1982). Another method uses gas chromatography to separate N_2, N_2O, and NO_2 (Payne, 1973). The advantages and disadvantages of a variety of other methods, including ^{15}N balance, the rate of NO_3^- disappearance, and the rate of $^{13}NO_3^-$ conversion to $^{13}N_2$, are discussed by Tiedje (1982).

Nitrogen fixation by both symbiotic and free-living microorganisms can be monitored by the C_2H_2 reduction method (Weaver and Frederick, 1982; Knowles, 1982). C_2H_2 is reduced to ethylene (C_2H_4) by the nitrogen-fixing bacteria, and the C_2H_4 is measured by gas chromatography.

4. Transformations of Other Elements

The effects of xenobiotics on the transformation of elements such as phosphorus, sulfur, iron, and manganese have not been studied in detail. However, simple soil incubation techniques with the extraction of the inorganic element of interest can be used to monitor the adverse or beneficial effects of a xenobiotic (Greaves, 1982).

5. Soil Enzymes

A large number of cell-free enzymes that are active in soils have been described (Ladd, 1978). Burns (1983) has separated these extracellular enzymes into seven different categories according to their source and location in soil systems:

1. Enzymes that normally function within the cytoplasm of viable cells, yet remain active in dead cells and cell debris
2. Periplasmic enzymes released into the environment by leakage through damaged cell membranes
3. Enzymes attached to the outer surfaces of cell walls or associated with extracellular polysaccharides
4. Enzymes that are truly extracellular and are secreted into the soil during cell growth
5. Enzymes temporarily associated with substrates as enzyme–substrate complexes
6. Enzymes adsorbed to the clay constituents of soils
7. Enzymes associated with colloidal organic matter

More than 50 enzymes have been detected in soils (see Table II), and many of these enzymes have a critical role in the cycling of nutrients. The sensitivity, accuracy, and ease of several enzyme assays have made them useful indicators of the effects of xenobiotics. In particular, the activities of phosphatases, dehydrogenases, and urease are frequently studied in assessing the effects of xenobiotics (Greaves, 1982; EPA,

1978). However, the use of soil enzymes as a measure of the effects of xenobiotics on soil microorganisms is not without problems. Enzyme activity, like respiration, is not confined to microorganisms; therefore, it is difficult to determine if a measured effect on extracellular enzyme activity correlates with a corresponding effect on microbes (Grossbard, 1973b). In addition, there is a lack of unequivocal methods to determine the activities of many soil enzymes.

6. *Microbial Populations*

There are two fundamental approaches to estimating soil microbial biomass: procedures aimed at enumerating microbial populations, and those aimed at estimating microbial biomass using biochemical approaches (Atlas, 1982).

The quantitative enumeration of different microbial populations in soil is difficult. There are no universal methods that can be applied successfully to all microorganisms and all soils. Three methods most often used are (1) the plate count method, (2) the most probable number (MPN) method, and (3) the direct cell count method (Atlas, 1982). In the plate count method, a diluted soil suspension is inoculated onto a solid medium containing nutrients to support growth. After incubation, the microbial colonies that have formed are counted. The assumption of the technique is that each colony arose from a single microbial cell and that each viable microorganism formed a colony. Therefore, the total number of colonies is equated with the original numbers of a microbial population. By the use of different carbon sources (e.g., cellulose, chitin), the numbers of microbes having unique biochemical capabilities can be estimated.

The MPN method permits the estimation of population size without an actual count of single cells or colonies. The MPN method is seldom used to enumerate the total number of microorganisms in a soil but rather to enumerate specific groups, such as nitrifiers. The technique employs a statistical approach in which successive dilutions are made to reach an extinction point (Alexander, 1982). Replicates of each dilution are inoculated into (usually) a liquid growth medium, and the pattern of positive and negative scores (i.e., growth) is recorded. A statistical table is then used to determine the MPN of viable organisms in the original sample. The plate count and MPN methods require that microorganisms grow and divide during the assay. Therefore, these methods have a tendency to underestimate the true size of the different microbial populations in soil, because it is difficult or impossible to duplicate the exact conditions required for the growth of all microorganisms likely to be present in soil.

Direct observation of microbial cells in soils can be achieved by light or electron microscopy, with or without the use of stains (Schmidt and Paul, 1982). These methods do not rely on culturing of the selected populations and, therefore, do not require a knowledge of the nutritional and environmental requirements of the microbial populations being studied. A limitation of the direct count method for estimating microbial populations in soils is that it is usually not possible to differentiate living microorganisms from dead ones. Hence, the method usually overestimates the viable population. Further, populations of individual species cannot usually be enumerated by direct observation.

7. Rhizosphere Populations

Because of their presumed role in increasing the availability and uptake of nutrients by plants, an understanding of the effects of xenobiotics on those microorganisms associated with plant roots is important (Leach, 1984). Xenobiotics, particularly pesticides, are often present near the root. Chemicals not directly toxic to microorganisms may indirectly affect rhizosphere microbial populations by changing the physiology of the plant and, subsequently, root exudation patterns (Greaves, 1982; Trappe et al., 1984).

Techniques for studying rhizosphere microbial populations are not as numerous as those for microorganisms living in soil away from roots. Rhizobia are most frequently enumerated by a MPN method that depends on the ability of the rhizobial cell to cause nodule formation on the roots of a legume host (Weaver and Frederick, 1982). Mycorrhizal biomass is often estimated from the density of spores of a specific fungal species in soil (Black and Tinker, 1979), stepwise dilution followed by infection of bait plants (Smith and Bowen, 1979), and microscopic examination of root segments for mycorrhizal colonization and infection (Hadley and Williamson, 1972; Black and Tinker, 1979; Powell, 1982).

C. Assessing the Efficacy of Test Methods

The methods mentioned previously, as well as others, have been evaluated by various research groups using a variety of criteria (EPA, 1981; Greaves, 1982; Domsch et al., 1983). The criteria most often used to accept or reject a method to determine the effect of xenobiotics on microorganisms include:

1. Reproducibility
2. Sensitivity

3. Standardization (i.e., its potential for interlaboratory transfer)
4. Cost
5. Time required to perform test
6. Level of training required to perform test

As can be seen in Table XI, no one method has all the desirable attributes necessary such that the results of the test can be accepted unequivocally. However, certain methods have a better overall rating than others. Microbially mediated chemical reactions involved in the nitrogen and carbon cycles appear to be the most sensitive to xenobiotic effects. Because they are sensitive and because they represent processes that determine the availability of nutrients, these reactions are good indicators with which to assess the response of soil microorganisms to xenobiotics (Barkay et al., 1986).

Although no international agreement on methods for measuring the

TABLE XI

EVALUATION OF MAJOR TEST SYSTEMS FOR MEASURING THE EFFECTS OF XENOBIOTICS ON SOIL MICROORGANISMS [a]

Assay method	Test criteria [b]					
	Reproducibility	Sensitivity	Standardization	Cost	Time required	Level of training required
Respiration	2	3	1	2	3	2
Carbon mineralization	4	3	3	1	3	1
Nitrogen mineralization	3	3	3	2	2	3
Nitrification	4	4	2	2	1	3
Nitrogen fixation	4	3	3	2	2	2
Phosphorus transformation	1	1	2	2	3	2
Sulfur transformation	1	0	2	2	3	2
Enzymatic activities	1	0	2	2	3	2
Microbial populations	1	1	1	1	2	1

[a] Data adapted from EPA (1981), Greaves (1982), and Domsch et al. (1983).
[b] Higher numbers indicate an advantage and lower numbers indicate disadvantage.

effects of xenobiotics on soil microorganisms has been achieved, a number of laboratories have developed systems that offer a reasonable approach to the question. The terrestrial soil core microcosm (Fig. 6) is just one such system. Another system, developed by researchers at the Jealott's Hill Research Station in Berkshire, United Kingdom (Anderson, 1973), appears promising because of its multiplicity of tests and its use of *in situ* methods. The different methods employed in these two test systems include:

1. Respiration (both CO_2 evolution and O_2 consumption)
2. ^{14}C-Mineralization studies using plant residues, fresh leaves, sucrose, starch, protein, amino acids, urea lipids, and phenol (expensive and time-consuming)
3. Hydrolysis of pectin, DNA, and chitin
4. Nitrification
5. Activity of phosphatase, peroxidase, and dehydrogenase
6. Population densities of *Azotobacter* sp.
7. Population density for bacteria, fungi, and actinomycetes by standard plate count techniques
8. Direct counts using ultraviolet microscopy
9. ATP assay using the liciferin–luciferase technique

Although these test systems were designed for measuring the effects of pesticides on microbial ecosystems, it seems reasonable that they could be adapted to other, nonpesticidal xenobiotic compounds. These test systems and others (see Bitton and Dutka, 1986) illustrate the need to use several different methodologies to evaluate the toxicity of xenobiotics to soil microorganisms. If the results are integrated in a monitoring scheme, it is highly probable that valid data could be generated that will aid in the assessment of the effects of xenobiotic chemicals.

The results obtained on the effects of a particular xenobiotic on particular microbial processes often depend on the physicochemical properties of the soil used during the test, illustrating again the importance of abiotic factors on microbial effects. Inasmuch as little is known about the influence of most xenobiotic compounds, it is difficult to make meaningful predictions about the effects of xenobiotic compounds in soils different from the one in which the original tests were performed. To remedy this shortcoming, models that adequately predict the abiotic behavior of a compound in any particular soil and sound scientific hypotheses about the subsequent microbiological impact of a compound should be developed.

In addition to developing ecotoxicity systems that meet the six criteria listed in the beginning of this subsection, suitable methods are needed for quantifying the data generated from these tests so that these data can be used easily by the regulatory agencies responsible for formulating criteria for tolerable levels of xenobiotics in the environment (Babich and Stotzky, 1985a). The "ecological dose" (EcD) concept, developed by Babich and Stotzky (see Babich et al., 1981, 1983; Babich and Stotzky, 1983, 1985a) and defined as the dose of a toxicant that decreases a specific microbe-mediated process by some percentage, represents one method for quantifying the effects of xenobiotics on microbe-mediated processes. This method has been used successfully to quantify the effects of some heavy metals on selected microbe-mediated processes (e.g., glucose mineralization, respiration, nitrification) (Babich et al., 1983), and it appears reasonable that the concept could be modified to quantify the results of toxicity tests that measure the effects of other xenobiotics on microbe-mediated processes.

VIII. Conclusions and Recommendations

The intensive use of chemicals in agriculture and the generation of large quantities of chemical wastes from industry has aroused public concern as to how these compounds might affect humans and the environment. Because soil is the main sink and repository for many of these chemicals and thus has a major role in determining their ultimate fate, it is critical that the capability to forecast the potential impacts of xenobiotics on soil microorganisms be developed as soon as possible. Once released, the distribution of a xenobiotic between environmental compartments depends on the chemodynamic properties of the compound and on the physicochemical properties of the soil, and it occurs across soil–water and soil–air interfaces and across biological membranes. Abiotic and biotic processes can transform the chemical compound. Ideally, the conversion of the xenobiotic is to carbon dioxide, water, and mineral elements, or at least to some harmless substance. However, intermediate transformation products can be formed that become toxic pollutants in their own right.

The biological activity of soil as a living environment is the result of the activities of a host of resident microorganisms, invertebrates, and plants. The microbial components, mainly bacteria, fungi, algae, and protozoa, are important to the fertility of soils through their role in the degradation of plant and animal matter and in the cycling of the organic and inorganic nutrients contained in soil. Among the more important processes mediated by soil microorganisms are the transformation of carbon, nitrogen, phosphorus, and sulfur. Anything that disrupts the activity of the microorganisms could be expected to affect the nutritional quality of soils and would therefore have serious ecological consequences. Because of this, it is important to understand the effects of xenobiotic compounds on soil microorganisma and their activities.

In assessing the effects of xenobiotic compounds on soil microorganisms, it is necessary to decide which microbial processes or properties should be evaluated. The most commonly used criteria are those that measure gross or overall activities, such as respiration or nitrogen mineralization, sometimes accompanied by total counts of microbial density. However, these parameters lack specificity. Another approach is monitoring changes in physiological activities, such as nitrification, nitrogen fixation, denitrification, decomposition of individual fractions of organic matter, and the metabolism of phosphorus and sulfur.

When exposed to xenobiotic compounds, various segments of the soil microbial community are affected to different extents. The degree to which a xenobiotic affects microbial activities is largely dependent on the chemical, its dosage and method of application, and the particular physicochemical characteristics of the soil, such as soil type, temperature, water content, and pH. Soil physicochemical factors are particularly important and probably account for many of the variations in effects seen with the same compound. A correlation between a compound class and its effects on soil microorganisms is not possible at this time. However, a few generalizations have emerged. Broad-range biocidal compounds, such as soil fungicides, appear to affect all microbial processes, at least temporarily. This is also observed with most heavy metals. Compounds such as herbicides and insecticides are less active against microbial activity and, under certain conditions, may stimulate their activity.

Although all soil microbial processes can be affected to a limited extent by any particular xenobiotic compound, some processes appear more sensitive to xenobiotics than others. For example, nitrification appears to be highly sensitive to xenobiotics, whereas nitrogen mineralization is somewhat insensitive. These results reflect the differences in the microorganisms that mediate these processes. Nitrification is conducted by a select group of chemoautotrophic bacteria, whereas nitrogen mineralization is conducted by a large and diverse group of heterotrophic microorganisms.

The present state of knowledge of the cellular and biochemical effects of xenobiotics is insufficient to delineate their mode of action. However, available data suggest that xenobiotics may interfere with photosynthesis, oxidative metabolism, and the synthesis of cellular constituents. In addition, certain compounds, such as chlorinated aromatics and heavy metals, have been shown to alter cellular membrane composition, thereby changing cell membrane permeability and cellular physiology.

Numerous methods exist for measuring different microbial processes, microbial populations, and soil enzymes, and most of these methods can be applied to assessing the effects of xenobiotics. The most pressing issue appears to be deciding which of the available methods are the most valid for investigating xenobiotic effects and how they can be applied for the best results. Additional studies are needed to identify and select the microbially mediated ecological processes that can best be used to generate meaningful data of the short- and long-term effects of xenobiotics before selection and standardization of techniques can be accomplished.

In the interim, however, it is recommended that agencies responsible for monitoring xenobiotic effects use a suite of test methods rather than only one or two. To use time and funds efficiently, these testing programs should have a multitiered design, in which the methods used would increase in complexity as warranted by the potential hazard of the compound. For example, simple, less expensive tests, such as respiration studies, can be used initially to monitor the effect of a xenobiotic compound on soil microorganisms. If the results indicate a strong microbe–xenobiotic interaction, then additional, more complex tests, such as the terrestrial soil core microcosm test protocol may be indicated.

Although the answer to the question, "Do xenobiotic compounds impact soil microorganisms?" is yes, the attention that has been given to the effects of xenobiotics on soil microorganisms does not yet provide a systematic approach by which environmental risk assessments can be performed for all classes of xenobiotics with any degree of accuracy or confidence.

Agricultural chemicals do not appear to have any long-term harmful effects on soil microbial activity when applied at recommended field levels (Greaves et al., 1976; Smith, 1982; Wainwright, 1978; Grossbard, 1976; Parr, 1974). However, when applied at higher levels than recommended, these compounds do appear to have an adverse short-term effect, particularly on nitrogen and carbon transformations (Wainwright, 1978; Barkay et al., 1986; Grossbard, 1976). Furthermore, contradictory results are often observed for the same chemical applied at the same levels (Barkay et al., 1986). Although these contradictory results are often explained on the basis of differences in laboratory procedures or soil types, no generally accepted explanations exits.

These contradictory results illustrate the basic problems that must be resolved before an adequate predictive environmental risk assessment capability can be developed for the effects of xenobiotics on microorganisms in soil. First, a standardized system (i.e., test method and defined test protocol) for evaluating the effects of xenobiotics on soil microorganisms must be established. Such a system would give both industry and regulatory agencies the capability to assess adequately the effects of different compounds on microbial processes. Second, information correlating the effects of certain groups of compounds (e.g., the chlorinated or N-containing aromatics) on the growth and activities of soil microorganisms and the specific physiological effects of the different compound classes must be obtained before an addquate predictive environmental risk assessment capability can be developed. To resolve these two problems, the following research is recommended:

1. A comprehensive system for evaluating the effects of xenobiotics on soil microbial activity should be established, involving:
   a. Identification and selection of the microbially mediated processes that can best be used to generate quantitative data concerning the long- and short-term effects of xenobiotics.
   b. Review and selection of methods currently used to measure the selected processes. In reviewing current ecotoxicity test systems, the following criteria are recommended: (1) the results should be capable of being reproduced in other laboratories; (2) the methods should be amenable to standardization techniques; (3) the tests should be sensitive to low levels of xenobiotics that are often found in soil; (4) the test systems should be as realistic as possible with respect to the environmental conditions, the environmental dose, and the form of the xenobiotic; and (5) the methods should be both economical and uncomplicated without losing reliability.
   c. Further development of methods to quantify and evaluate data of the short- and long-term effects of xenobiotics on microorganisms in soil, for example, the EcD concept (Babich et al., 1981, 1983; Babich and Stotzky, 1983, 1985a). Such quantification of the effects of xenobiotics on microbe-mediated processes could be utilized by the regulatory agencies to formulate criteria for acceptable levels of xenobiotics in soil.
2. A systematic evaluation of the various compound classes is needed to establish whether a correlation exists between compound structure and inhibitory effects on microbial processes. At present, no unequivocal correlations between compound structure and effects have been established. In addition, there is a paucity of information concerning the effects of nonagricultural xenobiotic organic compounds on soil microorganisms. Although this task might be monumental in both effort and cost, it would provide the basic scientific information needed to predict, based on the chemical properties of a compound, its ecotoxicological behavior with respect to microorganisms in soil. Such an evaluation should include:

a. Establishment of a comprehensive data base of the known effects of xenobiotics on soil microbial activity to determine if a statistical correlation between compound structure and effects is currently available.
b. Performance of laboratory tests (using selected microbially mediated processes) on compound classes that have not been previously evaluated.

AKNOWLEDGMENTS

This work was supported by the United States Environmental Protection Agency, Corvallis Environmental Research Laboratory, under a Related Services Agreement with the United States Department of Energy, contract DE-AC06-76RLO 1830.

REFERENCES

Agrios, G. N. (1978). "Plant Pathology," 2nd Ed. Academic Press, New York.
Alexander, M. (1977). "Introduction to Soil Microbiology," 2nd Ed. Wiley, New York.
Alexander, M. (1981). Science 211, 132–138.
Alexander, M. (1982). In "Methods of Soil Analysis. Part 2: Chemical and Microbiological Properties" (A. L. Page, R. H. Miller, and D. R. Keeney, eds.), 2nd Ed., pp. 815–820. Am. Soc. Agron., Madison, Wisconsin.
Anderson, J. P. E. (1982). In "Methods of Soil Analysis. Part 2: Chemical and Microbiological Properties" (A. L. Page, R. H. Miller, and D. R. Keeney, eds.), 2nd Ed., pp. 831–871. Am. Soc. Agron., Madison, Wisconsin.
Anderson, J. R. (1973). Bull. Ecol. Res. Comm. NFR No. 17, 473–474.
Anderson, J. R., and Drew, E. A. (1976). Zentrabl. Bakteriol., Parasitenkd., Infektionskr. Hyg., Abt. 2 131, 136–147.
Atlas, R. M. (1982). In "Experimental Microbial Ecology" (R. G. Burns and J. H. Slater, eds.), pp. 84–102. Blackwell, London.
Atlas, R. M., and Bartha, R. (1987). "Microbial Ecology: Fundamentals and Applications." Cummings, Menlo Park, California.
Audus, L. J. (1970). Meded. Fac. Landbouwwet. Rijksuniv. Gent. 35, 465–492.
Ausmus, B. S., Kimbrough, S. S., Jackson, D. R., and Lindberg, S. (1979). Environ. Pollut. 20, 103–111.
Babich, H, and Stotzky, G. (1977). Appl. Environ. Microbiol. 33, 696–705.
Babich, H., and Stotzky, G. (1978). Adv. Appl. Microbiol. 23, 55–117.
Babich, H., and Stotzky, G. (1979). Appl. Environ. Microbiol. 38, 506–513.
Babich, H., and Stotzky, G. (1982). In "Experimental Microbial Ecology" (R. G. Burns and J. H. Slater, eds.), pp. 631–670. Blackwell, London.
Babich, H., and Stotzky, G. (1983). Environ. Health Perspect. 49, 247–260.
Babich, H., and Stotzky, G. (1985a). Environ. Res. 36, 111–137.
Babich, H., and Stotzky, G. (1985b). Arch. Environ. Contam. Toxicol. 14, 409–415.
Babich, H., Davis, D. L., and Trauberman, J. (1981). Environ. Manage. 5, 191–205.
Babich, H., Bewley, R. J. F., and Stotzky, G. (1983). Arch. Environ. Contam. Toxicol. 12, 421–428.
Barkay, T., Shearer, D. F., and Olson, B. H. (1986). In "Toxicity Testing Using Microor-

ganisms" (B. J. Dutka and G. Bitton, eds.), Vol. 2, pp. 133–155. CRC Press, Boca Raton, Florida.
Bartha, R., and Pramer, D. (1965). *Soil. Sci.* **100,** 68–70.
Bartha, R., Lanzillota, R. P., and Pramer, D. (1967). *Appl. Microbiol.* **15,** 67–75.
Baver, L. D., Gardner, W. H., and Gardner, W. R. (1972). "Soil Physics." Wiley, New York.
Beste, C. E. (1983). "Herbicide Handbook." Weed Sci. Soc. Am., Champaign, Illinois.
Bewley, R. J. F., and Stotzky, G. (1983). *Sci. Total Environ.* **31,** 57–69.
Beyers, R. J. (1964). *Am. Biol. Teacher* **26,** 491–498.
Bitton, G. and Dutka, B. J. (1986). *In* "Toxicity Testing Using Microorganisms" (B. J. Dutka and G. Bitton, eds.), Vol. 2, pp. 1–8. CRC Press, Boca Raton, Florida.
Black, R. L. B., and Tinker, P. B. (1979). *New Phytol.* **83,** 401–413.
Bollag, J.-M., and Henninger, N. M. (1976). *J. Environ. Qual.* **5,** 15–18.
Bollag, J.-M., and Nash, C. L. (1974). *Bull. Environ. Contam. Toxicol.* **12,** 241–248.
Bonmati, M., Pujola, M., Sana, J., Soliva, M., Felipo, M. T., Garan, M., Ceccanti, B., and Nannipieri, P. (1985). *Plant Soil* **84,** 79–91.
Boyd, S. A., and Shelton, D. R. (1984). *Appl. Environ. Microbiol.* **47,** 272–277.
Burns, R. G. (1979). *In* "Adhesion of Microorganisms to Surfaces" (D. C. Ellwood, J. Melling, and P. Rutter, eds.), pp. 107–138. Academic Press, London.
Burns, R. G. (1983). *In* "Microbes in Their Natural Environment" (J. H. Slater, R. Whittenbury, and J. W. T. Wimpenny, eds.), pp. 249–298. Cambridge Univ. Press, London.
Burns, R. G. (1986). *In* "Interactions of Soil Minerals with Natural Organics and Microbes" (P. M. Huang and M. Schnitzer, eds.), pp. 429–451. Soil Sci. Soc. Am., Madison, Wisconin.
Cairns, J., Buikema, A. L., Heath, A. G., and Parker, B. C. (1978). "Effects of Temperature on Aquatic Organism Sensitivity to Selected Chemicals," Bull. 106. Virginia Water Resour. Res. Cent. Virginia Polytech., Blacksburg..
Capone, D. G., Reese, D. D., and Kiene, D. P. (1983). *Appl. Environ. Microbiol.* **45,** 1586–1591.
Chang, F. H., and Broadbent, F. E. (1982). *J. Environ. Qual.* **11,** 1–4.
Clegg, T. J., and Koevening, J. L. (1974). *Bot. Gaz.* **135,** 368–372.
Cole, L. K., Metcalf, R. L., and Sandborn, J. R. (1976). *Int. J. Environ. Stud.* **10,** 7–14.
Cole, M. A. (1976). *Weed Sci.* **24,** 473–476.
Coleman, R. (1966). *Soil Sci.* **101,** 230–239.
Corke, C. T., and Thompson, F. R. (1970). *Can. J. Microbiol.* **16,** 567–571.
Cullimore, D. R., and Ball, L. (1978). *Appl. Environ. Microbiol.* **36,** 959–961.
Debosz, K., Babich, H., and Stotzky, G. (1985). *Bull. Environ. Contam. Toxicol.* **35,** 517–524.
Dixon, J. B., and Weed, S. B. (1977). "Minerals in Soil Environments." Soil Sci. Soc. Am., Madison, Wisconsin.
Doelman, P., and Haanstra, L. (1984). *Plant Soil* **79,** 317–327.
Doetsch, R. N., and Cook, T. M. (1974). "Introduction to Bacteria and Their Ecobiology." Univ. Park Press, Baltimore, Maryland.
Domsch, K. H., Jagnow, G., and Anderson, T. H. (1983). *Res. Rev.* **86,** 65–105.
Dubey, H. D. (1969). *Soil Sci. Soc. Am. J.* **33,** 893–896.
Dubey, H. D., and Rodriguez, R. L. (1970). *Soil Sci. Soc. Am. J.* **34,** 435–439.
EPA (1978). *Fed. Regist.* **43,** Part II, 29696–29741.
EPA (1981). *Ecotoxicol. Test Sys.: Proc. Ser. Workshops* EPA-560/6-81-004.
Farmer, V. C. (1978). *In* "The Chemistry of Soil Constituents" (D. J. Greenland and M. H. B. Hayes, eds.), pp. 405–408. Dekker, New York.
Farrah, H., and Pickering, W. F. (1978a). *Water Air Soil Pollut.* **8,** 189–197.
Farrah, H., and Pickering, W. F. (1978b). *Aust. J. Chem.* **31,** 1510–1509.

Frankenberger, W. T., Troeh, F. R., and Dumenil, L. C. (1979). *Soil Sci. Soc. Am. J.* **43,** 333–338.
Gaur, A. C., and Misra, K. C. (1977). *Plant Soil* **46,** 5–15.
Gerhardson, B., and Clarholm, M. (1986). In "Microbial Communities in Soil" (V. Jensen, A. Kjoller, and L. H. Sorensen, eds.), pp. 19–34. Elsevier, New York.
Ghiorse, W. C. (1984). *Annu. Rev. Microbiol.* **38,** 515–550.
Giashuddin, M., and Cornfield, A. H. (1979). *Environ. Pollut.* **19,** 67–70.
Gibson, D. T. (1984). "Microbial Degradation of Organic Compounds" Dekker, New York.
Giesy, J. P., Jr., ed. (1980). "Microcosms in Ecological Research," CONF-7891101. Tech. Inf. Cent., U.S. Dep. Energy, Washington, D.C.
Gile, J. D., Gillett, J. W., and Collins, J. C. (1979). "The Soil-Core Microcosm—Potential Screening Tool," Ecol. Res. Ser., EPA 600/3-78-0010. U.S. Environ. Prot. Agency, Corvallis, Oregon.
Gillett, J. W., and Gile, J. D. (1976). *Int. J. Environ. Stud.* **10,** 15–22.
Gillett, J. W., and Witt, J. M., eds. (1979). "Terrestrial Microcosms," Proceedings of the Workshop on Terrestrial Microcosms, NSF/RA 79-0027. Nat. Sci. Found., Washington, D.C.
Granat, L., Hallberg, R. O., and Rodhe, H. (1976). In "Nitrogen, Phosphorus and Sulfur—Global Cycles" (B. H. Svensson and R. Soderlund, eds.), SCOPE Rep. 7. *Ecol. Bull.* No. 22, 23–73.
Greaves, M. P. (1982). In "Experimental Microbial Ecology" (R. G. Burns and J. H. Slater, eds.), pp. 613–630. Blackwell, London.
Greaves, M. P., Davies, H. A., Marsh, J. A. P., and Wingfield, G. I. (1976). *CRC Crit. Rev. Microbiol.* **5,** 1–38.
Greaves, M. P., Poole, N. J., Domsch, K. H., Jagnow, G., and Verstraete, W. (1980). "Recommended Tests for Assessing the Side-Effects of Pesticides on the Soil Microflora," Tech. Rep. No. 59. Agric. Res. Counc. Weed Res. Organ., Oxford.
Grossbard, E. (1973a). *Bull. Ecol. Res. Comm., NFR* No. 17, 473–474.
Grossbard, E. (1973b). *Bull. Ecol. Res. Comm., NFR* No. 17, 457–463.
Grossbard, E. (1976). In "Herbicides: Physiology, Biochemistry, and Ecology" (L. J. Audus, ed.), pp. 99–147. Academic Press, New York.
Grossbard, E., and Davies, H. A. (1976). *Weed Res.* **16,** 163–169.
Hadley, G., and Williamson, B. (1972). *New Phytol.* **71,** 445–455.
Hartley, D., and Kidd, H. (1983). "The Agrochemical Handbook," R. Soc. Chem. Unwin Brothers, Ltd., Old Woking, Surrey, England.
Hicks, G. F., and Corner, T. R. (1973). *Appl. Microbiol.* **25,** 381–385.
Hutzinger, O., and Veerkamp, W. (1981). In "Microbial Degradation of Xenobiotics and Recalcitrant Compounds" (T. Leisinger, R. Hutter, A. M. Cook, and J. Nuesch, eds.), pp. 3–45. Academic Press, London.
Jenkinson, D. S., and Powlson, D. S. (1976). *Soil Biol. Biochem.* **8,** 209–213.
Juma, N. G., and Tabatabai, M. A. (1977). *Soil Sci. Soc. Am. J.* **41,** 343–346.
Juneja, S., and Dogra, R. C. (1978). *J. Appl. Bacteriol.* **49,** 107–115.
Keswick, B. H. (1984). In "Groundwater Pollution Microbiology" (G. Bitton and C. P. Gerba, eds.), pp. 39–64. Wiley, New York.
Kahn, D. H., and Frankland, B. (1984). *Environ. Pollut.* **33,** 63–74.
Knowles, R. (1982). In "Methods of Soil Analysis. Part 2: Chemical and Microbiological Properties" (A. L. Page, R. H. Miller, and D. R. Keeney, eds.), 2nd Ed., pp. 1071–1092. Am. Soc. Agron., Madison, Wisconsin.
Kobayashi, H., and Rittmann, B. E. (1982). *Environ. Sci. Technol.* **16,** 170A–183A.
Ladd, J. N. (1978). In "Soil Enzymes" (R. G. Burns, ed.), pp. 51–96. Academic Press, London.

Lal, R., and Saxena, D. M. (1979). *Res. Rev.* **73,** 49–86.
Lal, R., and Saxena, D. M. (1980). *Arch. Protistenkd.* **122,** 382–386.
Lal, R., and Saxena, D. M. (1982). *Microbiol. Rev.* **46,** 95–127.
Leach, F. R. (1984). *In* "Groundwater Pollution Microbiology" (G. Bitton and C. P. Gerba, eds.), pp. 303–351. Wiley, New York.
Liang, C. N., and Tabatabai, M. A. (1978). *J. Environ. Qual.* **7,** 291–293.
Litchfield, and Huben. (1973) *Bull. Ecol. Res. Comm.* **17,** 464–466.
Low, P. F. (1961). *Adv. Agron.* **13,** 269–327.
Low, P. F. (1979). *Soil Sci. Soc. Am. J.* **43,** 651–658.
Lynch, J. M. (1983). "Soil Biotechnology: Microbiological Factors in Crop Productivity." Blackwell, London.
Martin, M. H., Duncan, E. M., and Coughtrey, P. J. (1982). *Environ. Pollut., Ser. B* **3,** 147–157.
Metcalf, R. L. (1977). *Annu. Rev. Entomol.* **22,** 241–262.
Ogram, A. V., Jessup, R. E., Ou, L. T., and Rao, P. S. C. (1985). *Appl. Environ. Microbiol.* **49,** 582–587.
Papendick, R. I., and Campbell, G. S. (1980). *In* "Water Potential Relations in Soil Microbiology" (D. M. Kral, ed.), pp. 1–22. Soil Sci. Soc. Am., Madison, Wisconsin.
Parasher, C. D., Ozel, M., and Geike, F. (1978). *Chem.–Biol. Interact.* **20,** 89–95.
Parr, J. F. (1974). *In* "Pesticides in Soil and Water" (W. D. Guenzi, ed.), pp. 315–340. Soil Sci. Soc. Am., Madison, Wisconsin.
Patten, B. C., and Witkamp, M. (1967). *Ecology* **48,** 813–824.
Payne, W. J. (1973). *In* "Modern Methods in the Study of Microbial Ecology" (T. Rosswall, ed.), Ecol. Bull., pp. 262–268. Swed. Nat. Sci. Res. Counc., Stockholm.
Powell, C. L. (1982). *In* "Experimental Microbial Ecology" (R. G. Burns and J. H. Slater, eds.), pp. 447–471. Blackwell, London.
Reineke, W. (1984). *In* "Microbial Degradation of Organic Compounds" (D. T. Gibson, ed.), pp. 319–360. Dekker, New York.
Rosas, S. B., Secco, M. D. C., and Ghittoni, N. E. (1980). *Appl. Environ. Microbiol.* **40,** 231–234.
Roslycky, E. B. (1985). *Can. J. Soil Sci.* **65,** 667–675.
Rother, J. A., Millbank, J. W., and Thornton, I. (1982). *J. Soil Sci.* **33,** 101–113.
Rovira, A. D. (1976). *Soil Biol. Biochem.* **8,** 241–247.
Russell, E. W. (1973). "Soil Conditions and Plant Growth," 10th Ed. Longman, London.
Saive, R. (1974). *Ann. Gembloux* **80,** 55–77.
Sanders, F. E. (1986). *In* "Microbial Communities in Soil" (V. Jensen, A. Kjoller, and L. H. Sorensen, eds.), pp. 61–72. Elsevier, New York.
Schippers, B., Bakker, P. A. H. M., Bakker, A. W., Weisbeek, P. J., and Lugtenberg, B. (1986). *In* "Microbial Communities in Soil" (V. Jensen, A. Kjoller, and L. H. Sorensen, eds.), pp. 35–48. Elsevier, New York.
Schlegel, H. G., and Jannasch, H. W. (1981). *In* "The Prokaryotes" (M. P. Starr, H. Stolp, H. G. Truper, A. Balows, and H. G. Schlegel, eds.), pp. 43–82. Springer-Verlag, New York.
Schmidt, E. L., and Paul, E. A. (1982). *In* "Methods of Soil Analysis. Part 2: Chemical and Microbiological Properties" (A. L. Page, R. H. Miller, and D. R. Keeney, eds.), 2nd Ed., pp. 803–814. Am. Soc. Agron., Madison, Wisconsin.
Schwab, S. M., Johnson, E. L. V., and Menge, J. A. (1982). *Plant Soil* **64,** 238–287.
Simon-Sylvestre, G., and Fournier, J. C. (1979). *Adv. Agron.* **31,** 1–92.
Smith, A. E. (1982). *Can. J. Soil Sci.* **62,** 433–460.
Smith, M. S., and Weeraratna, C. S. (1974). *Pestic. Sci.* **5,** 721–729.

Smith, S. E., and Bowen, G. D. (1979). *Soil Biol. Biochem.* **11,** 469–473.
Stevenson, F. J. (1985). "Humus Chemistry." Wiley, New York.
Stevenson, F. J., and Finch, A. (1986). In "Interactions of Soil Minerals with Natural Organics and Microbes" (P. M. Huang and M. Schnitzer, eds.), pp. 29–58. Soil Sci. Soc. Am., Madison, Wisconsin.
Stotzky, G. (1974). In "Microbial Ecology" (A. I. Laskin and H. Lechevalier, eds.), pp. 57–135. Chem. Rubber Publ., Co., Cleveland, Ohio.
Stotzky, G. (1986). In "Interactions of Soil Minerals with Natural Organics and Microbes" (P. M. Huang and M. Schnitzer, eds.), pp. 305–428. Soil Sci. Soc. Am., Madison, Wisconsin.
Stotzky, G., and Burns, R. G. (1982). In "Experimental Microbial Ecology" (R. G. Burns and J. H. Slater, eds.), pp. 105–133. Blackwell, London.
Stotzky, G., and Schenck, S. (1976). *CRC Crit. Rev. Microbiol.* **4,** 333–382.
Tiedje, J. M. (1982). In "Methods of Soil Analysis. Part 2: Chemical and Microbiological Properties" (A. L. Page, R. H. Miller, and D. R. Keeney, eds.), 2nd Ed., pp. 1011–1026. Am. Soc. Agron., Madison, WI.
Tinsley, I. J. (1979). "Chemical Concepts in Pollutant Behavior." Wiley, New York.
Tolle, D. A., Van Voris, P., Arthur, M. F., Morris, J. P., and Larson, M. (1981). *Bull. Ecol. Soc. Am.* **62,** 141–142.
Tolle, D. A., Arthur, M. F., and Van Voris, P. (1983). *Sci. Total Environ.* **31,** 243–261.
Trappe, J. M., Molina, R., and Castellano, M. (1984). *Annu. Rev. Phytopathol.* **22,** 331–359.
Tu, C. M., and Bollen, W. B. (1968). *Weed Res.* **8,** 28–31.
Tu, C. M., and Miles, J. R. W. (1976). *Res. Rev.* **64,** 5–65.
Tyler, G. (1975). *Proc. Symp. Int. Conf. Heavy Met. Environ. Toronto* **2,** 217–226.
Tyunyayeva, G. N., Minenko, A. K., and Ponkov, L. A. (1974). *Sov. Soil Sci.* **6,** 320–324.
Van Voris, P., Tolle, D. A., Arthur, M. F., Chesson, J., and Zwick, T. C. (1984). "Development and Validation of a Terrestrial Microcosms Test System for Assessing Ecological Effects of Utility Wastes," EA-3672, Proj. 1224-5. Electr. Power Res. Inst., Palo Alto, California.
Van Voris, P., Tolle, D. A., and Arthur, M. F. (1985a). "The Experimental Terrestrial Soil-Core Microcosm Test Protocol," EPA-600/3-85/047. U.S. Environ. Prot. Agency, Corvalis, Oregon.
Van Voris, P., Tolle, D. A., Arthur, M. F., and Chesson, J. (1985b). In "Multispecies Toxicity Testing" (J. Cairns, Jr., ed.), pp. 117–143. Pergamon, New York.
Wainwright, M. (1978). *J. Soil Sci.* **29,** 287–298.
Wainwright, M., and Snowden, F. J. (1977). *Plant Soil* **48,** 335–345.
Warington, R. (1851). *Q. J. Chem. Soc.* **3,** 52–54.
Warington, R. (1857). *Proc. R. Inst. London* **2,** 403–408.
Weaver, R. W., and Frederick, L. R. (1982). In "Methods of Soil Analysis. Part 2: Chemical and Microbiological Properties" (A. L. Page, R. H. Miller, and D. R. Kenney, eds.), 2nd Ed., pp. 1043–1070. Am. Soc. Agron., Madison, Wisconsin.
Widus, R., Trudgill, P. W., and Maliszewski, M. J. (1971). *J. Gen. Microbiol.* **69,** 15–22.
Wurster, C. F., Jr. (1968). *Science* **159,** 1474–1475.
Zachara, J. M., Felice, L. J., Riley, R. G., Harrison, F. L., and Mallon, B. (1984). "The Selection of Organic Chemicals for Subsurface Transport Research," DOE/ER-0217. NTIS, Springfield, Virginia.

Disclosure Requirements for Biological Materials in Patent Law

SHUNG-CHANG JONG AND JEANNETTE M. BIRMINGHAM

American Type Culture Collection
Rockville, Maryland 20852

I. Principles of the Patent System
II. History of Depositing Microbial Cultures in Connection with Patent Applications
III. Legal Decisions Affecting Deposits for United States Patents
IV. Deposit of a Living Organism in a Culture Collection for the Purpose of Patent Disclosure
   A. The Deposit and Its Documentation
   B. United States Postal and Quarantine Requirements
   C. Time of Deposit
   D. Availability of the Deposit during Pendency
   E. Nature of the Deposit
   F. Security of the Deposit
   G. Release of the Deposit
V. United States Regulations Concerning the Deposit of Biological Materials
VI. Budapest Treaty
VII. Other International Agreements Relevant to Biological Inventions
   A. Paris Union Convention
   B. Patent Cooperation Treaty (PCT)
   C. European Patent Convention (EPC) and European Patent Office (EPO)
VIII. Patent Law in European Patent Convention Countries
IX. Patent Law in European Countries outside the European Patent Convention
X. Patent Law in Non-European Countries
XI. Summary of Deposit of Biological Materials
   A. European Patent Convention Countries
   B. European Countries Not in the European Patent Convention
   C. Non-European Countries
   D. Time Scale
XII. Patent Protection for New Plant Varieties and New Animals
   A. Protection for New Plant Varieties
   B. Protection for New Animals
XIII. Summary of International Protection for Microorganisms, Plants, and Animals
   A. European Patent Convention Countries
   B. Eastern European Countries
   C. North American Countries
   D. South American Countries
   E. Australia
   F. Japan
   G. People's Republic of China

XIV. Conclusion
References

I. Principles of the Patent System

A patent is a form of protection issued by a government to an inventor of a new product or process who publicly discloses the details of his or her invention and in return is granted for a limited period a legally enforceable right to exclude others from commercially exploiting it. The rationale behind the patent law is simple: to foster innovation, inventors must be guaranteed some degree of exclusivity on their inventions in order to be assured a reasonable profit and to justify the risks of development. In disclosing how the invention works, the inventor makes this knowledge available to the public, and others may build upon that knowledge (U.S. Congress, OTA, 1989).

For over 650 years patents have played a role in the economic policies of countries. Early protection of inventors' privileges was seen with the passage of the British Statute of Monopolies in 1623/1624. The United States Congress, in accordance with a Constitutional mandate in Art. 1, Sec. 8, "to promote the progress of science and useful arts," enacted its first patent law in 1790. France followed in 1791. The laws of these three nations influenced the subsequent development of patent protection in other countries, and most legislation dates from the late 1800s. The United States system is currently guided by the United States Code, Title 35, Patents (35 USC), enacted in 1952. Patent protection continues to promote the technical innovation and industrial development of the countries of the world, and in March 1984 the People's Republic of China joined the long list of countries to offer legal protection to inventors (Beier and Straus, 1985).

There are four universally accepted requirements for the patentability of any invention. The first three concern the invention itself. It must be novel, inventive (not obvious), and useful (industrially applicable). The fourth concerns the specification. The details of the invention must be disclosed by means of a patent specification that describes the invention in sufficient detail to allow a person skilled in the art to reproduce it (35 USC 112). Claims, which define the invention or the scope of protection sought, are appended to the technical description, and together they form the specification as a whole (Crespi, 1985).

A significant number of today's inventions are originating in the relatively new and still developing field of biotechnology. Broadly defined, it includes any technique that uses living organisms (or parts of organisms) to make or modify products, to improve plants or animals, or to develop microorganisms for specific uses. Patents on biotechnological systems are not new, however. Some date from the early days of the United States patent system. The first patent dealing with microorganisms was issued to Louis Pasteur (United States patent 141,072) and included a claim to a biologically pure culture of a microorganism. Acetic acid fermentation and other food patents date from the early 1800s, whereas therapeutic patents were issued as early as 1895 (U.S. Congress, OTA, 1989).

Although the requirement of full disclosure was relatively easy to meet in the earlier days of the patent system, it has become increasingly more difficult with the advances in technology. In general, however, biotechnological inventions can also obtain patent protection as long as the basic criteria for patentability are fulfilled. Biotechnological inventions may include products, compositions, processes, and use or methods of use.

In biotechnological inventions that involve the use of new living biological materials—that is, those not generally known and available to the public or those not readily isolated from known and available sources—a written description is not usually considered sufficient for the purposes of disclosure. No matter how carefully the description may be worded, if the biological material is unavailable, the invention cannot be reproduced. In such a case, the strain or cell line itself forms an essential part of the disclosure (Bousfield, 1988).

II. History of Depositing Microbial Cultures in Connection with Patent Applications

The practice of depositing microorganisms in culture collections other than those of the patent applicants is not noted prior to 1949 (Pridham and Hesseltine, 1975). In that year a United States patent examiner recommended to an employee of Parke-Davis who was filing a patent application for an invention involving a microorganism, that a deposit of the microorganism should be made with a culture collection. Though not a formal requirement, it was recommended that in cases such as this, in which words alone were not sufficient to describe the invention adequately, a deposit was advisable. On July 8, 1949, the applicant on behalf of Parke-Davis deposited a culture of *Streptomyces venezuelae* in the American Type Culture Collection (ATCC) in Rockville, Maryland. It was assigned an accession number (ATCC 10712) and is cited in United States patent 2,483,892, issued October 4, 1949, for the manufacture of chloramphenicol (Brandon, 1989a).

For similar reasons, in August 1949 the American Cyanamid Company deposited a culture of *Streptomyces aureofaciens* with the Agricultural Research Service Culture Collection (NRRL). It was assigned accession number NRRL 2209 and is listed in United States patent 2,482,055, issued Septembert 13, 1949, for the production of aureomycin (Pridham and Hesseltine, 1975).

These two historic deposits for patent purposes, apparently the first in the world, were made voluntarily by the applicants to satisfy demands for complete disclosure of specifications and were the forerunners to the current requirements that patent applications for inventions involving microorganisms be supported by a deposit in a recognized public patent depository.

The deposit of a new strain in a culture collection was initially conceived as a reference point from which the United States Patent and Trademark Office (PTO) examiner could determine the adequacy of the total information provided and resolve any disputes as to the nature, identity, and operability of what was described or claimed in the specification. Moreover, the continued presence of the deposit and its eventual availability to the public would ensure compliance with the requirement for usabiltiy.

In most cases the deposit of a new organism is made, at the latest, by the filing date for a patent. Apart from drafting the specification, this is often the first step taken by the inventor, who is the only one knowledgeable about the organism, the technical difficulties in handling it, how long is needed to grow it, and any legal constraints in respect to its pathogenicity, which might delay matters (Bousfield, 1988).

III. Legal Decisions Affecting Deposits for United States Patents

Since the late 1960s the patent activity in the biotechnology industry has escalated rapidly, and several decisions have been made in the United States that affect the nature of the deposit, the timing of the deposit, and the location of the depository when one applies for a United States patent. A brief summary of some of the most significant decisions follows.

In 1970 the practice of making a deposit, which had begun in 1949, was challenged in the United States Court of Customs and Patent Appeals (CCPA). In re Argoudelis (CCPA, 168 USPQ 99) was a landmark decision. Patent contract requires the inventor to give full disclosure of the invention to the Patent Office including the best mode for practicing the invention. In the microbiological art, this means that the best culture must be disclosed. This invention involved the production of two new antibiotics by a newly isolated organism. In order to comply with disclosure requirements, the inventors deposited a viable subculture of their best *Streptomyces* strain with NRRL. They requested that the depository maintain the culture in their permanent collection and restrict release to the public until a patent was granted. The depository agreed and gave an accession number to the culture, which was disclosed when the application was filed with the PTO.

The PTO argued that this deposit was secret and confidential, and therefore the microorganism had not been made available to the general public at the time of the patent application to comply with the full-disclosure requirement. The PTO rejected the procedure, and the rejection was appealed to the CCPA. The court held that the enabling disclosure did not require the organism to be available at the time of filing or that the specification be enabling as of the filing date. The disclosure was "sufficient to permit a thorough examination by the Patent Office and to preclude the possibility that a patent could issue without any person skilled in the art being henceforth enabled to make and use the invention." The CCPA, therefore, approved the *Argoudelis* procedure, and there was established in the United States the legal criterion for a full disclosure in a microbiological invention. The procedure was adopted by the PTO as complying with the statutory requirements of 35 USC 112 for an adequate disclosure of the microorganisms required to carry out the invention. This decision was generally accepted by the major patent offices throughout the world (Beier et al., 1985). The first published guidelines on the deposit of microorganisms for patent purposes appeared in the *Official Gazette* in 1971 (Wahl, 1971).

In 1975 the CCPA (*Feldman vs Aunstrup*, 186 USPQ 108) expanded the scope of the type of depository that the PTO would accept—private, nongovernmental, non-United States, or even profit-making types—when it agreed that a deposit made with the well-known Centraalbureau voor Schimmelcultures (CBS) in the Netherlands, in conjunction with the filing of a United States patent application, was acceptable (Crespi, 1985). Most countries permit a culture deposit in a recognized public culture repository, even though the repository is not located in the particular country in which a patent application is being filed. This is reasonable because there are many fine, reputable public repositories located througout the world. In this decision the court also held that the use of a theretofore unknown strain in an old process was patentable because of the prior unavailability of the strain (U.S. Congress, OTA, 1989).

Before 1980 PTO would not grant patents for microorganisms or cells *per se*, although patent protection had been granted for many compositions containing living things. In 1980 the United States Supreme Court in *Diamond vs Chakrabarty*, 206 USPQ 193 (Sup. Ct. 1980) upheld a decision of the United States CCPA and found the Chakrabarty invention of a synthetic plasmid-injected *Pseudomonas*, capable of degrading crude oil, patentable, The PTO maintained that Congress had not intended living organisms to be encompassed by the terms "manufacture" or "composition of matter" under patent law (35 USC 101) and argued that microorganisms are products of nature and as such unpatentable. The PTO claimed that the Plant Patent Act (PPA) of 1930 and the Plant Variety Protection Act (PVPA) of 1970, discusssed later, showed the intention of Congress to limit the legal protection for living organisms to those specifically covered by these statutes, and, therefore, to exclude bacteria as patentable subject matter (Beier et al., 1985).

In rejecting the argument of the PTO in *Diamond vs Chakrabarty* the Supreme Court held that Congress intended that the patent laws be broadly construed. The claim was not to an unknown natural phenomenon but to "non-naturally occurring manufacture or composition of matter, a product of human ingenuity having a distinctive known character and use." In rejecting the PTO's product of nature argument the Court also held that the inventor had produced a new bacterium with markedly different characteristics from any found in nature and one having the potential for significant utility (Beier et al., 1985). This decision helped to precipitate increased research and development, assuring the commercialization of biotechnology in the United States. The PTO subsequently issued patents directed to bacteria, viruses, fungi, and yeast, as well as to animal and human cell lines (such as hybridomas).

With regard to the timing of deposit, the 1985 decision, *In re Lundak* (Fed. Cir. 1985), handed down by the Court of Appeals of the Federal Circuit, concluded that the culture, in this case, a cell line, is only required to be made available to the PTO should the PTO so request, as authorized by 35 USC 114, during the pendency of a patent application. Through an oversight Lundak's patent application was made before the culture was deposited. The inventor, a university professor, argued that his cell line was effectively on deposit in the university laboratories and elsewhere and would have been available to the PTO if required. Thus, his deposit of the cell line with ATCC, which was made a few days after filing but prior to issuance of his patent and which is referred to in his specifications, met the statutory requirements (Brandon, 1988; Crespi, 1988).

While the Supreme Court in *Chakrabarty* was careful to limit its holding to claims to microorganisms, the Court used rather sweeping language to arrive at its conclusion. The Supreme Court had quoted the Congressional Committee reports that accompanied the 1952 Patent Act (35 USC), which said that Congress intended statutory subject matter to "include anything under the sun that is made by man." In 1985 the Board of Appeals in *Ex parte Hibberd*, 227 USPQ 443 (Pt. Bd. App. and Int. 1985), when considering the patentability of a sexually reproducing tryptophan-containing corn plant obtained by tissue culture and selection techniques, relied on that same statement to justify utility patent protection for plants (Hoscheit, 1988).

In 1987 the PTO Board of Patent Appeals and Interferences (BPAI) applied the same logic in Ex parte Allen, 2 USPQ 2d 1425 (Pt. Bd. App. and Int. 1987). The application concerned a polyploid Pacific oyster produced by applying hydrostatic pressure to the zygotes (United States serial no. 06/647,963, filed September 6, 1984). The board held that polyploid oysters are nonnaturally occurring manufactures of compositions of matter within the meaning of 35 USC 101. Various groups violently objected to this decision and urged that animal patents be banned. Adopting the Allen precedent, the PTO commissioner issued a notice stating that the PTO would examine applications for nonnaturally occurring nonhuman multicellular living organisms including animals and would consider them patentable subject matter (Hoscheit, 1989).

The most recent decision to affect biotechnology was the issuance of the first animal patent ("nonhuman mammal") in 1988. Leder and Stuart from Harvard University were awarded a United States patent for an animal (mouse) whose cells carry oncogenes (United States patent 4,736,866, issued April 12, 1988). The claim was for "a transgenic non-human mammal all of whose germ cells and somatic cells contain a recombinant activated oncogene sequence introduced at an embryonic stage." This patent intensified the debate calling for a possible moratorium or modification of animal patents following Ex parte Allen. It is the position of the PTO, however, that previous court cases support the patentability of multicellular animals (Hoscheit, 1989).

In 1949 only bacteria and fungi were accepted by culture collections for patent purposes. Now, in response to the needs of the "patent community," many other types of genetic material are accepted. Plasmids, vectors, cells, plant tissues, seed and other similar types of material, which are newly isolated, novel, synthetic, or not generally available to the public on a long-term basis, have been deposited. From the beginning, cultures deposited as a part of the patent disclosure have occupied a unique position within culture collections because of the need for confidentiality, special records, and special handling. Today, most major depositories have a "patent culture collection." The ATCC, for example, now holds >9,000 deposits for patent purposes (Brandon, 1989a).

IV. Deposit of a Living Organism in a Culture Collection for the Purpose of Patent Disclosure

A culture collection must retain and be a convenient source of the inventor's deposit while remaining independent of both the patent applicant and the PTO. The role it plays in the application for a biotechnological patent and the protection it offers may be discussed using the policies and services of the ATCC as an example. Practices and policies may differ in other collections. It is not the role of the depository to provide legal advice or to know about the legal requirements of the patenting system, but it is in a good position to assist patent applicants in choosing the best deposit procedure to meet their individual requirements.

A. The Deposit and Its Documentation

The ATCC currently accepts algae, animal embryos, animal viruses, bacteria, cell lines, fungi/yeasts, hybridomas, oncogenes, phages, plant tissue cultures, plant viruses, plasmids, protozoa, and seeds for the purpose of patent disclosure.

In the form that accompanies the deposit, all materials must be clearly identified with the acronym of the depositor's collection, a number designation in that collection, the name of the organization, and the name of the organism. A reasonable description of the microorganism is needed to recognize whether the culture is contaminated or mixed and for culturing purposes, if sufficient material has not been included. If the deposit is made to meet requirements set by the Budapest Treaty, which will be discussed later, additional forms must be completed.

B. United States Postal and Quarantine Requirements

Depositors should be aware of the requirements for shipping their material to avoid delays in making patent deposits. Both national and international regulations for the packaging and shipping of biological materials are well defined. They include the use of permits, testing at designated facilities, specific packaging and labeling, and adherence to regulations established by postal authorities and international carriers. International depositories may receive cultures from numerous countries, but their movement into the depository country requires observance of that country's regulations. Potential depositors in an American collection must contend with a variety of regulations imposed by agencies of the United States government depending on the nature of the biological material and its intended use.

The United States Public Health Service (USPHS) requires a permit for entry of certain animal viruses and pathogenic microorganisms, primarily those not indigenous to the United States that could have an impact on public health. The regulation is administered by the Centers for Disease Control (CDC) through its Foreign Quarantine Program. Required import permits can delay receipt of cultures from outside the United States. This service also regulates the packaging and shipping of etiological agents for interstate transport.

The Animal and Plant Health Inspection Service (APHIS) of the United States Department of Agriculture (USDA) regulates all plant pathogens and potential sources of those pathogens. All potential plant pathogens require a permit from the Plant Quarantine Division of the USDA. All cell lines grown on media containing animal serum, a potential source of hoof and mouth disease and other viruses, must be tested in designated USDA laboratories.

The United States Customs Service determines if materials are admissible into the United States and if they should be referred to other government agencies for examination, permits, and release. It also decides if import duties should be imposed and if the packages contain the goods that are declared.

The United States Department of Transportation is responsible for establishing and enforcing regulations for the safe shipment of etiological agents and other restricted materials in domestic transport. Shipping of >50 ml of an etiological agent requires special testing of the container, which must then be shipped by cargo aircraft only. The International Civil Aviation Organization (ICAO) is responsible for the international shipping of dangerous goods, and the International Air Transport Association (IATA) also publishes regulations in agreement with those of the ICAO.

The exportation of cultures from the United States is regulated by the United States Department of Commerce. Many biological materials require a validated export license, which depends on the country of destination and the culture being shipped. Application for a license by a depository may delay a request by 2–3 weeks. The United States Postal Service has very specific packaging requirements for the shipment of biological materials, particularly infectious material.

As a part of its patent deposit service the ATCC assists the depositor to assure compliance with the proper packaging, labeling, and permit and testing requirements for the patent culture.

C. Time of Deposit

Cultures are accessioned on the day the deposit material is received by the culture collection. An accesssion number is assigned, and the depositor is sent written notification of the receipt of the deposit. The deposit is tested for viability. If the culture is viable, the date of receipt becomes the deposit date. Bacteria, fungi, and cell lines usually require only a few days for testing, but viruses require 3–4 weeks. Viability certificates are provided to depositors. The ATCC recommends that all strains be tested for viability upon deposit, and such testing is now a requirement for United States patent procedures. A deposit for United States patent procedure purposes may be converted to a deposit for international purposes at a later date.

D. Availability of the Deposit during Pendency

The availability of a patent culture has concerned both depositors and the United States PTO. If a depositor chooses not to restrict the availability of the strain during the patent application process, that strain will be made available upon request. For those who do not specify, it is assumed that the culture is to be restricted (progeny of the strain will not be sent to anyone other than the depositor, the depositor's designee, or the United States PTO) until the patent is issued.

Patent culture availability during the pendency of a patent application is determined by the commissioner of patents and the depositor. Because of the confidential nature of applications, the only persons with access to the name, acronym, and strain number of microorganisms are the depositors and their attorneys, certain personnel in the United States PTO, and the curators of the collection where the culture is deposited. Except for actions taken by these individuals, no requests for the culture can be initiated until the patent issues. Prior publication of the name, acronym, and strain number does sometimes occur when foreign patents issue or foreign application are published before the United States patent is granted. Distribution is restricted until either the depositor or the United States PTO communicates with the ATCC and specifically requests in writing that the deposit be released for distribution.

E. Nature of the Deposit

A depository can specify the form and quantity of material it will accept. The ATCC requires 6 ampules of bacteria, fungi, and other microorganisms; 25 ampules of cell cultures (including hybridomas), plasmids (when not in a host), and viruses; 25 embryo samples (12 animal embryos per sample); and 2500 seeds for a patent deposit. Except for microorganisms, the depositor must furnish a sufficient quantity of material for the duration of deposit or contract with the collection to do so.

The ATCC accepts material in nearly any form, but frozen or freeze-dried is preferred. Frozen materials are placed directly into a liquid-nitrogen refrigerator (−196°C) or a freezer (−60°C), depending on the temperature recommended by the depositor for storage. To check viability an ampule is thawed and the culture grown. Freeze-dried material is stored at 5°C or −60°C; one ampule may be used for a viability test.

Test tube cultures of microorganisms are generally scraped to harvest the cells, added to a cryoprotective agent, then frozen and placed directly into liquid nitrogen without subculturing. If the culture appears to be in poor condition, it is subcultured once, frozen, and placed into liquid nitrogen. If ATCC handles the culture in any way by freezing it, freeze-drying it, or growing any additional material, a sample of the material is returned to the depositor for verification of the characteristics of the strain.

Plant cell lines provided frozen are simply placed in liquid nitrogen. Plant cells deposited in the callus state are frozen and stored in liquid nitrogen. Frozen animal embryos are also stored in liquid nitrogen.

F. Security of the Deposit

Culture collections must provide for the safety and security of their patent cultures. All restricted patent deposits at the ATCC are kept under double locks, and only pertinent staff members have access to them. Data are also kept in locked files and are available only to authorized staff.

The refrigerators or freezers in which material is stored at the ATCC are under constant surveillance. In addition, sensors connected to the equipment sound external alarms should the temperature rise above an acceptable level. Backup equipment is available should the material need to be tranferred to another refrigerator or freezer. An auxiliary generator is available in case of electrical failure.

The ATCC also offers a safe deposit service for those valuable cultures for which patent protection has not been sought. Cultures are stored in liquid nitrogen and the depositor retains all proprietary rights. If a patent is sought at a later date, the culture can be transferred to the patent depository.

G. Release of the Deposit

Unless the depositor informs the ATCC that a patent has been issued, the ATCC responds to requests for "restricted patent cultures" by noting that the culture is "unavailable pending issuance of pertinent patent." If the United States patent number is included in the request, the ATCC requests a copy of the patent from the PTO to verify the citation of the culture and then the culture is released. Release for distribution of many strains that are subjects of United States patents have been based on routine scanning of the *Official Gazette* of the United States PTO. After the United States patent issues, the culture must be made available to investigators. The ATCC lists the culture in its collection catalog, and any information provided is made available upon request.

If notified that a patent application has been abandoned, ATCC will return the strain to the depositor provided there is no agreement to the contrary such as under the Budapest Treaty (Brandon, 1989b).

V. United States Regulations Concerning the Deposit of Biological Materials

Regulations concerning the deposit of cultures as part of a patent disclosure may vary from country to country. Wherever a culture deposit is made, in order to satisfy current United States patent law (35 USC 112), certain rules must be followed. They prescribe the procedures and conditions for making a deposit, the examining procedures used to address deposit issues, and the procedures concerning access to a deposit once a patent is granted.

Clarification and updating of these rules to assist the inventor and the depository in defining the position of the PTO on deposits were proposed in 1986 and were published in the *Federal Register* Vol. 53, No. 194, October 6, 1988. Most of the proposals reflected existing policy with some changes being formally adopted by PTO. These rules became effective for all applications for patent, reissue of patent, or request for reexamination filed on or after January 1, 1990. Excerpts from the current regulations regarding the deposit of biological materials for patent purposes are taken directly from the *Federal Register* Vol. 54, No. 161, August 22, 1989, and are presented here.

1. *Biological Material*

For the purposes of patents for inventions under 35 USC 101, the term biological material shall include material that is capable of self-replication either directly or indirectly. Representative examples include bacteria, fungi including yeasts, algae, protozoa, eucaryotic cells, cell lines, hybridomas, plasmids, viruses, plant tissue cells, lichens, and seeds. Viruses, vectors, cell organelles and other non-living material existing in and reproducible from a living cell may be deposited by deposit of the host cell capable of reproducing the non-living material.

2. *Need or Opportunity to Make a Deposit*

Where an invention is, or relies on, a biological material, the disclosure may include reference to a deposit of such biological material. Biological material need not be deposited unless access to such material is necessary for the satisfaction of the statutory requirement for patentability under 35 USC 112 . . . It need not be deposited if it is known and readily available to the public or can be made or isolated without undue experimentation.

The reference to a specific organism or other biological material in a specification disclosure does not create any presumption that the specific material is necessary to satisfy 35 USC 112 or that a deposit in accordance with these regulations is or was required.

3. *Acceptable Depository*

A deposit may be made in any International Depository Authority (IDA) as established under the Budapest Treaty . . . or any depository recognized to be suitable by the Office. Suitability will be determined by the Commissioner . . . The depository must

1. Have a continuous existence,
2. Exist independent of the control of the depositor,
3. Possess the staff and facilities sufficient to enable it to examine the viability of a deposit and store the deposit in a manner which ensures that it is kept viable and uncontaminated,
4. Provide for sufficient safety measures to minimize the risk of losing biological material deposited with it,
5. Be impartial and objective,
6. Furnish samples of the deposited material in an expeditious and proper manner,
7. Promptly notify depositors of its inability to furnish samples and the reasons why.

4. *Time of Making an Original Deposit*

An original deposit may be made at any time before filing an application for patent or . . . during pendency of the application . . .

When the original deposit is made after the effective filing date of an application for patent, the applicant shall promptly submit a verified statement from a person in a position to corroborate the fact . . . that the biological material which is deposited is a biological material specifically identified in the application . . .

5. *Replacement or Supplement of a Deposit*

A depositor, after receiving notice . . . that the depository . . . cannot furnish samples or the deposit has been contaminated or has lost its capability to function as described in the specification shall notify the Office in writing. A replacement or supplemental deposit . . . shall not be accepted unless it meets the requirements for making an original deposit . . . and a certificate of correction is requested . . . promptly after the replacement or supplemental deposit has been made.

A depositors's failure to replace a deposit, or in the case of a patent, to diligently replace a deposit and promptly thereafter request a certificate of correction . . . shall cause the application or patent involved to be treated in any Office proceeding as if no deposit were made.

6. *Term of Deposit*

A deposit . . . shall be made for a term of at least thirty years and at least five years after the most recent request for the furnishing of a sample of the deposited biological material was received by the depository.

7. Viability of Deposit

A deposit of biological material . . . must be viable at the time of deposit and during the term of deposit. Viability may be tested by the depository. No evidence is necessarily required regarding the ability of the deposited material to perform any function described in the patent application.

A viability statement for each deposit . . . not made under the Budapest Treaty . . . must be filed in the application.

If a viability test indicates that the deposit is not viable upon receipt, or the examiner cannot, for scientific or other valid reasons, accept the statement of viability received from the applicant, the examiner shall proceed as if no deposit has been made . . .

8. Furnishing of Samples

A deposit must be made under conditions that assure that access to the deposit will be available during pendency of the patent application . . . to one determined by the Commissioner to be entitled thereto . . . and all restrictions imposed by the depositor on the availability to the public of the deposited material will be irrevocably removed upon the granting of the patent.

9. Examination Procedures

The examiner shall determine . . . in each application . . . if a deposit is needed . . . and . . . if a deposit actually made is acceptable for patent purposes. If a deposit is needed and has not been made or replaced or supplemented . . . the examiner . . . shall reject the affected claims . . . explaining why a deposit is needed and/or why a deposit actually made cannot be accepted.

The applicant for patent or patent owner shall respond to a rejection . . . by making an acceptable . . . deposit or assuring the Office in writing that an acceptable deposit will be made on or before the date of payment of the issue fee or, in the case of a patent owner, requesting a certificate of correction . . . or . . . arguing why a deposit is not needed . . . or why a deposit actually made should be accepted.

If an application is otherwise in condition for allowance except for the required deposit and the Office has received a written assurance that an acceptable deposit will be made . . . the Office will mail to the applicant a Notice of Allowance and Issue Fee Due together with a requirement that the required deposit be made within three months . . . Failure to make the required deposit . . . will result in abandonment of the application for failure to prosecute.

For each deposit, the specification shall contain:
1. Accession number for the deposit,
2. Date of the deposit,
3. Taxonomic description of the deposit,
4. Name and address of the depository.

VI. Budapest Treaty

Patenting in biotechnology is usually an international activity. To eliminate the problem of multiple deposits to cover each country where patent protection for inventions involving the use of microorganisms was sought, the Budapest Treaty on the International Recognition of the Deposit of Microorganisms for the Purposes of Patent Procedure was drawn up.

The Budapest Treaty is a product of the World Intellectual Property Organization (WIPO), an intergovernmental organization with headquarters in Geneva, Switzerland, which is one of the 16 specialized agencies of the United Nations system of organizations. The origins of WIPO date back to 1883, and it is responsible for promoting the protection of intellectual property throughout the world through cooperation among member states.

In 1974 the director general of WIPO convened a Committee of Experts to discuss international cooperation over the deposit of microorganisms for patent purposes. The committee concluded that a treaty would be necessary to put the proposed solution into effect. Drafts prepared by the International Bureau of WIPO and examined by the committee served as the basis for the deliberations of a diplomatic conference convened by the director of WIPO, organized by him in cooperation with the government of Hungary, and held in Budapest from April 14 to 28, 1977. The treaty was signed on April 28, 1977, and came into force on August 19, 1980.

Under this treaty a single microorganism deposit with a recognized depository authority satisfies the requirement of all the countries selected in multicountry filings under the Patent Cooperation Treaty (PCT) or the European Patent Convention (EPC). No signatory country may require anything different from or additional to those requirements that are provided in the treaty. As of June 30, 1988, there are 22 signatory countries to the Budapest Treaty:

Australia	Liechtenstein
Austria	Netherlands
Belgium	Norway
Bulgaria	Philippines
Denmark	Republic of Korea
Federal Republic of Germany	Soviet Union
Finland	Spain
France	Sweden
Hungary	Switzerland
Italy	United Kingdom
Japan	United States

All countries party to the Budapest Treaty must recognize a deposit made in an IDA, but not all require deposits to be made in an IDA. An IDA is a scientific institution, typically a culture collection, that is capable of storing microorganisms. France, the Federal Republic of Germany, Switzerland, the United Kingdom, the United States, and the EPC will recognize other culture collections that can comply with their particular requirements. Hungary accepts deposits made in collections on its own soil, but the only deposits it will recognize elsewhere are those in IDAs. The Japanese Patent Office (JPO) will recognize deposits outside Japan only if they have been made under the Budapest Treaty or have been "converted" to Budapest Treaty deposits, regardless of their previous public availability. The European Patent Office (EPO) accepts Budapest Treaty deposits as meeting its deposit requirements. Deposits under the treaty also meet United States PTO requirements. Countries not signatory to the treaty will generally accept deposits in an IDA.

Requirements for a culture collection to be designated an IDA under the Budapest Treaty are as follows:

1. Location on territory of contracting/member state
2. Continuous existence with provision to relocate cultures in event of cessation of activities
3. Impartial policy concerning all depositors
4. Appropriate staff and facilities for both scientific and administrative duties
5. Acceptance of specified types of microorganisms or cell lines
6. Determination of viability of deposit and issuance of any required viability statements
7. Storage of deposited strains for at least 30 years
8. Provision of sufficient safety measures to minimize loss
9. Maintenance of secrecy of deposit as required under the treaty
10. Furnishing of samples in an appropriate and timely manner as required by the treaty.

On January 31, 1981, WIPO approved the ATCC as the first international depository for patent purposes (Brandon, 1989b). There are currently 19 IDAs worldwide, of which three are in the United States, six in the United Kingdom, and three in the Soviet Union.

Agricultural Research Culture Collection (NRRL), United States
American Type Culture Collection (ATCC), United States
Australian Government Analytical Laboratories (AGAL), Australia
Centraalbureau voor Schimmelcultures (CBS), the Netherlands
Collection Nationale de Cultures de Micro-organismes (CNCM), France
Commonwealth Agricultural Bureau (CAB), International Mycological Institute, United Kingdom

Culture Collection of Algae and Protozoa (CCAP), United Kingdom
Deutsche Sammlung von Mikroorganismen (DSM), Federal Republic of Germany
European Collection of Animal Cell Cultures (ECACC), United Kingdom
Fermentation Research Institute (FRI), Japan
Institute of Microorganism Biochemistry and Physiology of the USSR Academy of Science (IBFM), Soviet Union
In Vitro International, Inc. (IVI), United States
Mezogazdasagi Es Ipari Mikroorganizmusok Magyar Nemzeti Gyujtemenye (MIMNG), Hungary
National Bank for Industrial Microorganisms and Cell Cultures (NBIMCC), Bulgaria
National Collection of Industrial Bacteria (NCIB), United Kingdom
National Collection of Type Cultures (NCTC), United Kingdom
National Collection of Yeast Cultures (NCYC), United Kingdom
USSR Research Institute for Antibiotics of the USSR Ministry of the Medical and Microbiological Industry (VNIAA), Soviet Union
USSR Research Institute for Genetics and Industrial Microorganism Breeding of the USSR Ministry of the Medical and Microbiological Industry (VNII Genetika), Soviet Union.

Any cultures already on deposit at an IDA may be converted to meet requirements of the Budapest Treaty, but, if they were deposited before the collection became an IDA, the date of deposit of the "conversion" is the date on which the depository acquired IDA status. Otherwise, the date of deposit is the date on which the collection physically received the culture. Only the original depositor may make the conversion. A form is required and a viability test must be performed (Bousfield, 1988).

Requirements of a depositor to deposit a culture in an IDA under the Budapest Treaty include the following:

1. Indication that the deposit is being made under the treaty
2. Agreement not to withdraw the deposit for 30 years (even if application is abandoned)
3. Name and address of the depositor
4. Assignment of strain designation
5. Indication if deposit is a mixture
6. Indication of details necessary to grow and test viability and, if a mixture, description of components
7. Indication of the properties that are or may be dangerous

8. Submission of the deposit in the form and quantity required by the IDA
9. Payment of a one-time fee

The treaty permits the IDA to set certain conditions of its own. These overlap with the requirements of the depositor just noted.

1. The microorganism must be deposited in the form and quantity necessary for the purpose of the treaty.
2. A form established by such authority and completed by the depositor for administrative procedures of such authority must be furnished.
3. The written statement must be drafted in the language, or in any of the languages, specified by such authority.
4. The fee for storage must be paid.
5. To the extent permitted by the applicable law, the depositor must enter into a contract with such authority defining the liabilities of the depositor and the authority.

These conditions allow the IDA to make a contract with the depositor that would be usual under the laws of contract of the IDA's own country. Without this provision, some culture collections would have been unwilling to become IDAs. It is entirely up to the IDA whether it requires any or all of the foregoing from the depositor, but if it does, then the depositor has no option but to comply (Bousfield, 1988).

For its part the IDA must also fulfill certain obligations under the treaty:

1. Issue to the depositor an official receipt.
2. Test the viability of the culture deposited and issue an official statement to the depositor disclosing the result.
3. Keep the deposit secret from all except those entitled to receive samples.
4. Maintain the deposit for ≥ 30 years, checking the viability at reasonable intervals or at any time on the demand of the depositor.
5. Supply cultures to anyone entitled under the relevant law to receive them provided the IDA has been given proof of entitlement.
6. Inform the depositor when and to whom it has released samples.
7. Be impartial and available to any depositor under the same conditions (Bousfield, 1988).

If by some chance a microorganism that was viable when deposited dies during storage, or if for any reason the IDA can no longer supply cultures of it, then the IDA must notify the depositor immediately. The latter then has the option of replacing it, and provided this is done within 3 months, the date on which the original deposit was made still stands. When making a new deposit, the depositor must provide the IDA with the following (Bousfield, 1988):

1. A signed statement that the culture being submitted is of the same microorganism as deposited previously
2. An indication of the date on which the depositor received notification from the IDA of its inability to supply cultures of the previous deposit
3. The reason for making the new deposit
4. A copy of the receipt and the last positive viability statement in respect of the previous deposit
5. A copy of the most recent scientific description and/or taxonomic designation submitted to the IDA with respect to the previous deposit
6. If a new deposit is being made with a different IDA, all the indications required for the original deposit apply.

A new deposit can be made with a different IDA if the original IDA is no longer operating as such (either entirely or just in the respect of the particular kind of microorganism) or if import–export regulations render the original IDA inappropriate for that particular deposit. Provisions for making a new deposit cannot be applied to a microorganism that was shown by the IDA to be nonviable when it was originally deposited (Bousfield, 1988). Material that is contaminated may not be replaced. If the contaminant can be eliminated, it would still be an acceptable deposit (Brandon, 1989b).

Samples of the deposited culture can be released to industrial property offices, to the depositor or a party authorized by the depositor, and to parties legally entitled. In the latter connection the treaty recognizes differences in national law. One procedure for releasing a sample to a third party requires a certificate from an industrial property office that has published a patent application referring to the deposited strain and identifying it by accession number. This certificate must inform the culture collection that the requesting party has the right to receive the strain under applicable national law. A certificate procedure of this kind is relevant to national laws that involve a dual publication system such as in European law but is unnecessary where the patent is published only upon being granted, as in the United States.

The Budapest Treaty provides an internationally uniform system of deposit and lays down the procedures that depositor and depository must follow, the duration of deposit, and mechanisms for the release of samples. The treaty does not deal with the timing of deposit, nor, generally, of release. These are determined by the relevant national laws. Likewise, the recipients of samples (other than patent offices and people with the depositor's authorization) are referred to merely as "parties legally entitled." Exactly who such parties are and under what conditions they may obtain samples are determined by national law. For purely national purposes, deposit under the Treaty is often not necessary. However, for the international recognition of a single deposit, using the Budapest Treaty is by far the safest course of action (Bousfield, 1988).

VII. Other International Agreements Relevant to Biological Inventions

A number of differences exist between nations regarding protection for biotechnological inventions including what constitutes patentable subject matter. Several international agreements have been made concerning basic intellectual property rights and the procedural mechanisms involved in the patenting process. The Budapest Treaty has already been discussed, and the Union for the Protection of New Varieties of Plants will be considered in the section on the patenting of new plants and animals. The accompanying tabulation lists the most significant agreements, the date on which they came into force, and the current number of signatory countries (U.S. Congress, OTA, 1989).

Agreement	Date Enforced	Number of Signatories
Paris Union Convention	July 7, 1884	97
Union for the Protection of New Varieties of Plants (UPOV)	August 10, 1968	17
European Patent Convention (EPC)	October 7, 1977	13
Patent Cooperation Treaty (PCT)	January 24, 1978	40
Budapest Treaty	August 19, 1980	22

A. Paris Union Convention

The Paris Union Convention is a universal treaty establishing certain basic rights for residents and nationals of its member countries to protect industrial property rights including patents under the laws of the other member countries. The original convention was signed in 1883 by 11 countries. Nine revision conferences have been held since then, and >90 nations were members in 1988. The union is administrated by WIPO.

According to the convention, nationals of any country of the union may enjoy in all other member countries the advantages that their respective laws grant to their own nationals. A practical and important right granted by the convention is the right of priority, which enables any resident or national of a member country first to file a patent application in any member country and thereafter to file a patent application for the same invention in any of the other member countries within 12 months of the original filing. The effect is that the subsequently filed applications will enjoy the right of priority established by the first filing date.

The convention does not place an obligation of working the invention. It only limits the extent national law may provide for not working the patented invention. For owners of biotechnology inventions, working requirements represent perhaps the most serious loss of effective patent protection in foreign countries. If, because of the obligation for a patentee to make freely available a sample of the deposited organism, it proves to be easier for competitors within such foreign countries to practice certain biological inventions without technological assistance from the patentee, there may be more of a temptation for the competition to seek a compulsory license or revocation or forfeiture of the patent.

Article 19 of the Paris Union Convention permits member nations to enter into separate agreements for the protection of industrial property, as long as those agreements are not contrary to the provisions of the convention. Under this provision several multinational agreements have been concluded (U.S. Congress, OTA, 1989).

B. Patent Cooperation Treaty (PCT)

Membership in this convention is open to any Paris Union Convention country. It came into force in 1978, and as of 1988 included 40 signatories. The PCT does not deal with substantive requirements for patenting, as each signatory determines patentability under its own domestic law. Instead, it relates to procedural requirements to simplify the filing, searching, and publication of international patent applications in order to eliminate multiple filings and duplicate filing costs.

Procedural steps are carried out in essentially two stages. The international stage begins when an applicant files the international patent application with one of the receiving offices, generally the national patent office of the country in which the applicant is a resident or national. An international search is then conducted by an appropriate international searching authority (ISA). In the case of United States-initiated applications, the ISA is the PTO or the EPO. Following the international search, the application is sent to the international bureau, WIPO in Geneva, which then publishes the application and provides copies to each of the designated offices in the countries where protection is sought. The applicant provides a translation as necessary and any required national fee to begin the national stage. The application is then subjected to national procedures in each of the designated countries.

The PCT does not contain any definition of patentable subject matter and does not contain any requirements regarding the deposit of microorganisms or the description of the characteristics of a deposited microorganism (U.S. Congress, OTA, 1989).

C. European Patent Convention (EPC) and European Patent Office (EPO)

With the need for free movement of goods and against anticompetitive acts, 14 countries of the European Common Market signed the Convention on the Grant of European Patents in October 1973. To date 13 countries are signatories of that convention, which came into force in 1977. The EPC is a system of law, common to all of the member countries, established for the granting of so-called European patents. Primarily, the convention establishes a single supranational European Patent Office (EPO) with a uniform procedural system for the centralized filing, searching, examination, and opposition with respect to a single European patent application. If granted, a patent matures into a bundle of individual European patents, one for each of the countries designated by the applicant.

Article 52(1) of the EPC defines patentable subject matter as inventions that are susceptible of industrial application, new, and involve an inventive step. This very general and broad definition is limited by certain exclusions in Art. 52(2). Article 52(4) further excludes "methods for the treatment of the human or animal body by surgery or therapy and diagnostic methods practiced on the human or animal body." This does not apply to products, and in particular to substances or compositions, for use in any of the excluded methods.

Although plant and animals varieties and "essentially biological" processes are specifically excluded from patentable subject matter, the EPC does not appear, in principle, to exclude entirely the patenting of microbiological inventions in any of the major classes: microorganisms per se, processes for producing microorganisms, processes for using microorganisms, products obtained from microbiological processes, and DNA/RNA molecule or subcellular units (U.S. Congress, OTA, 1989).

The EPC made provision for the deposit of a new microorganism in Rule 28 of its regulations. This required an applicant to make the deposit in a culture collection not later than the European patent application date, to include identifying details of the deposit with the accession number, and to make the deposited microorganism available from the culture collection to any person from the date of first publication of the patent application. The European patent system has a policy of dual publication in which the first publication of the unexamined application takes place 18 months after the application date or priority date followed by republication of the finally accepted text upon grant of the patent.

Rule 28 was amended in June 1980. The changes were designed to bring the original rule more into line with the terminology and procedures of the Budapest Treaty and provide some control over the availability of the deposited culture in the interim between the early publication of the application and the grant of the European patent. The EPO regulations state that a deposited culture shall be available upon request to any person from the date of publication of the European patent application, and under certain provisions, to any person having the right to inspect the patent files, even prior to that date. Availability of the culture shall be effected by the issue of a culture to the person making the request, but only if the requester has undertaken vis-à-vis the applicant or proprietor of the patent not to make the culture available to a third party and to use the culture for experimental purposes only (Brandon, 1989b).

However, the applicant may choose an option that until the date on which the technical preparations for publication of the application are completed, the availability shall be effected only by the issue of a sample to an independent expert nominated by the requester from a list of experts recognized by the EPO for these purposes. The independent expert would be bound by all of the conditions attached to the original Rule 28 and would therefore not be free to transmit the culture to the third party for whom he was acting. This expert could, nevertheless, perform all the experiments required on behalf of his or her principal to assess the sufficiency of the patent description and to enable the latter to form an opinion of the merit of the invention and its relevance to his or her own activities. This would meet the informational function of the patent disclosure and would enable competitors and others to prepare well in advance for the eventual filing of an opposition to the granted patent.

Interestingly enough, the Netherlands has a dual system of publication, first of the unexamined and then of the accepted application, but only at the later date (the date of the publication for opposition purposes) is it necessary to make the strain available to the public.

The original rule was further modified by extending the scope of the undertakings made by persons obtaining a sample of the deposited organism. These now apply also to cultures "derived" from the deposited culture, defined as those that are derived and that still exhibit the characteristics of the deposited culture essential to carrying out the invention.

An important clarification was made to specify the kind of microorganism to which the rule applies. Originally this was simply defined as a microorganism not already available to the public, but now there has been added the qualification "and which cannot be described in the European Patent application in such a manner as to enable the invention to be carried out by a person skilled in the art." This amendment makes explicit the possibility of avoiding the need to deposit the new microorganism where the applicant can justifiably rely on the reproducibility of a written description of the method of producing and identifying the microorganism.

When cultures are requested in connection with European patent publication, a form for deposits made under the EPC and a form for deposits made under the Budapest Treaty must be completed and sent to the EPO. Once the EPO certifies that the individual has a right to the culture, notification is made available. Individuals making deposits to meet the requirements of the EPO should be aware that a depository may be required to make the culture available before a patent issues (Brandon, 1989b).

If a deposit is deposited to meet the requirements of the EPO and under the provisions of the Budapest Treaty, the provisions of the Budapest Treaty shall prevail in case of conflict.

Countries party to the EPC on January 1, 1987 include the following (Bousfield, 1988).

Austria	Luxembourg
Belgium	Netherlands

Federal Republic of Germany
France
Greece
Italy
Liechtenstein

Spain
Sweden
Switzerland
United Kingdom

VIII. Patent Law in European Patent Convention Countries

The various countries party to the convention have amended their patent regulations to harmonize with the current European practice, but complete unanimity among the contracting states has yet to emerge. France, Italy, Switzerland, and the United Kingdom have specific statutory regulations governing deposit. In Germany the practice has been developed by case law. In the absence of a process to produce the new microorganism, all these countries require a deposit, which must be made not later than the filing date of the national application or the priority date.

The United Kingdom and the Netherlands require the deposit to be maintained at least for the life of the patent. France requires 30 years. In the Federal Republic of Germany it depends on whether the Budapest Treaty is involved, so it may be 20 or 30 years with an additional 5 years from the date of the last request. These countries recognize Budapest Treaty IDAs and other collections that comply with their requirements.

In the Federal Republic of Germany, Switzerland, and the United Kingdom the deposit forms a part of the disclosure and must be available with the publication of the patent application. In the United Kingdom the culture is available subject to EPC Rule 28. German law allows the depositor to place a ban on removing the culture outside the jurisdiction and on distribution to third parties. Switzerland prevents passing the deposit to third parties. France has adopted EPC Rule 28 including the independent expert option.

Sweden, Italy, and Belgium conform to the Budapest Treaty and accept EPC Rule 28. In the Netherlands deposit is necessary if required by the patent office in the specific case and can be deferred until after the application date. It must be available to the public at the date of second publication.

IX. Patent Law in European Countries outside the European Patent Convention

Ireland and Portugal do not have national laws on patent procedures, but they approve and accept procedures generally followed elsewhere and especially that of the Budapest Treaty. Denmark, Finland, and Norway are parties to the Budapest Treaty.

X. Patent Law in Non-European Countries

Canada has no statutory rules or guidelines but will accept a written description or a deposit of a new microorganism in any culture collection available to Canadians. Deposit is recommended for the life of the patent. The culture is freely available after the issue of the patent.

Australia has no statutory requirement but accepts the incorporation of deposit data into the specification. Now a party to the Budapest Treaty, Australia intends to introduce specific biotechnology provisions into new legislation to conform with the general trends in other countries.

New Zealand recommends a deposit of a new strain in a recognized depository before the filing of the application in that country. The identification of a particular strain requires the depository accession number in addition to the morphological description (Crespi, 1985).

In 1965, for the purposes of patent procedures, the JPO required the deposit of a microorganism in a reliable depository and the inclusion of its accession number in the specification. In 1966 the Fermentation Research Institute (FRI) received the first deposit for patent purpose. In 1970 the director general of the JPO designated the FRI as the sole domestic patent microorganism depository, and in 1981 it acquired the status of an IDA under the Budapest Treaty.

In 1971 guidelines for the examination of inventions on applied microbiology were published by the JPO, which have been the basis of examination of biotechnological inventions. The guidelines provided for the descriptions of processes, products, microorganisms, and deposit of microorganisms. In 1984 tentative guidelines for the examination of inventions of genetic engineering were published.

According to these guidelines if the microorganism is not easily available prior to patent application, it should be deposited in a collection designated by the director general of the patent office or an IDA and maintained for 30 years. The release of a deposited microorganism becomes generally available to the public upon the publication after examination of patent application. According to Rule 27 of the patent law in Japan, the release of a deposited microorganism is available to a person who intends to work the invention involving the microorganism for the purposes of tests or experiments, and the released culture shall not be assigned to other persons (Ono, 1988).

Patent law in the People's Republic of China requires a deposit in China before the date of filing, or at the latest, on the date of filing of the application. The application must include relevant information about the culture, the name of the depository, the date of deposit, the accession number, and the receipt from the depository. If the necessary culture is not deposited in time in the patent depository designated by the Chinese Patent Office, either the China Center for Type Culture Collection (CCTCC) in Wuhan or the Center of General Microbiological Culture Collection (CGMCC) in Beijing, it may be rejected by the examiner of the Chinese Patent Office. An import permit must be obtained by the depository before cultures may be shipped to China. After the publication of an application for a patent, any individual can request the microorganism from the patent office by providing his or her name and address, promising not to make it available to a third person, and using it only for experimental purposes until the patent is granted.

On March 28, 1988, the Republic of Korea joined the Budapest Treaty countries. It is no longer a requirement that a deposit be made in Korea if a Korean patent application is filed. A deposit in any IDA will do (Brandon, 1989b). But when an application for patent is filed in a foreign country by any Korean national, he or she must deposit the microorganism at an IDA even though deposited at a domestic institution, since Korean depositary institutions have not been recognized as international institutions (Hwang, 1989). The Korean Federation of Culture Collections (KFCC) and the Korean Collection of Type Cultures (KCTC) at the Korea Advanced Institute of Science and Technology are designated by the administrator of the Office of Patents Administration as official domestic depositories.

XI. Summary of Deposit of Biological Materials

The following will summarize some of the national requirements regarding the deposit and release of microorganisms. Where the names of countries are omitted, there are either no provisions, the provisions are not known, or the information is conflicting (Bousfield, 1988).

A. EUROPEAN PATENT CONVENTION COUNTRIES

Time of deposit: All require deposit by the filing or priority date except the Netherlands, which requires it by the second publication

Earliest release (EPO—first publication):
  First publication: Belgium, France, Federal Republic of Germany, Spain, Sweden, United Kingdom
  Second publication: Netherlands

General availability (EPO—at time of grant):
  First publication: Federal Republic of Germany, Spain, United Kingdom
  Second publication: Netherlands
  Grant: Belgium, France, Sweden

Restrictions on distribution and use (EPO—if applicant chooses, available only to independent expert before grant; must not be passed to third parties before patent expires and used only for experimental purposes):
  Belgium, France, Italy, Sweden: As for EPO
  Federal Republic of Germany: Sample must not be passed to third parties or outside purview of German law until patent expires
  Liechtenstein, Switzerland: Sample must not be passed to third parties
  United Kingdom: Sample not passed to third parties until patent expires and must be used only for experimental purposes
  Netherlands: None

Minimum storage period (EPO—30 years):

  20 Years: Federal Republic of Germany
  30 Years: Belgium, France, Sweden, Switzerland, United Kingdom
  Life of patent: Netherlands

B. European Countries Not in the European Patent Convention

This includes Bulgaria, Denmark, Finland, Hungary, Ireland, and Norway

Time of deposit: All require deposit by the filing or priority date

Earliest release:
  First publication: Denmark, Finland, Hungary, Norway
  Grant: Bulgaria

General availability:
  First publication: Hungary
  Grant: Bulgaria, Denmark, Finland, Norway

Restrictions on distribution and use:
 Bulgaria: As for United Kingdom
 Denmark, Finland, Norway: As for EPO
 Hungary: Sample must not be passed to third party

Minimum storage period:
 20 Years: Hungary
 30 Years: Denmark, Finland, Norway

C. Non-European Countries

This includes Australia, Canada, the People's Republic of China, Japan, Korea, New Zealand, and the United States

Time of deposit: Filing or priority date, later in the United States in certain cases

Earliest release:
 First publication: Australia, China, Korea
 Second publication: Japan
 Grant: Canada, United States

Release to public:
 First publication: Australia
 Second publication: Japan
 Grant: Canada, United States, China, Korea

Restrictions on distribution and use:
 Australia, Korea: As for United Kingdom
 Canada, United States: None
 People's Republic of China: Sample must not be passed to third parties; used only for research before patent granted
 Japan: Sample must not be passed to third parties until patent expires; must be used only for research

Minimum storage period:
 30 Years: Australia, United States
 Life of patent: Canada, Japan

D. Time Scale

The time scale of the patenting procedure with relation to the deposit is shown in the accompanying tabulation. The term of the patent differs country to country. In most European countries it is 20 years from the date of application. In the United States and Canada it is 17 years from the grant.

Time	Invention	Organism
0	(Keep secret/confidential) Basic application (establishes priority)	Deposit of organism: viability testing; preservation
1 Year	Foreign application Official prior art search (European countries)	
18 Months	First publication unexamined application (European countries; China, Japan)	First release of organism (e.g., United Kingdom, China, Germany) Option of release to idependent expert only (EFC)
Variable	Search (in United States, Canada, Japan) Official examination	
Variable	Second publication, accepted text	Release in Japan, Netherlands
Variable	Grant	General release (United States, Canada, China)

The United States and Japanese procedures are the most favorable to the inventor. The EPC plan is midway between the United States and the United Kingdom. Industry generally favors the practice that allows the inventor to have greater control over the new organism in the period before an enforceable right is obtained.

XII. Patent Protection for New Plant Varieties and New Animals

Animal cell lines and plant tissue cultures are generally considered in the same category as microorganisms for patent purposes. However, the patenting of plants and animals is still quite new, and the legislation governing protection has not been universally agreed upon.

A. Protection for New Plant Varieties

There is a sharp contrast between United States patent law and European law in regard to protection for plants produced by either traditional breeding methods or the newer genetic engineering. In the United deposit of animals because it is not practical to maintain or make available whole animals. The deposit of animal embryos, however, does not present the same difficulties. As long as culturing fertilized ova to the blastula stage as an indicator that growth would occur is an acceptable test of viability, and the statistical probability that the ovum–embryo would be capable of implantation and successful gestation is acceptable, it may not be impractical to maintain and make available animal forms. United States patents 4,380,997 and 4,419,986 were issued in 1983 for the process of freezing animal embryos. To date at least 13 species of animal embryos (cattle, mice, rats, rabbits, hamsters, sheep, goats, horses, cats, antelopes, and three species of nonhuman primates) have been frozen and recovered (U.S. Congress, OTA, 1989). In 1988 ATCC accepted its first deposit to support an animal patent application in the form of frozen animal embryos.

XIII. Summary of International Protection for Microorganisms, Plants, and Animals

With the development of biotechnology the question of what constitutes patentable subject matter in the various countries of the world is of increasing concern. The following is a brief review of the protection of inventions where living organisms are involved in countries other than the United States (U.S. Congress, OTA, 1989).

A. EUROPEAN PATENT CONVENTION COUNTRIES

National patent laws complement convention provisions: generally microorganisms are patentable, and animal and plant varieties are not.

Belgium: Microorganisms are patentable; plant and animal species and their varities are not (effective 1986).
Federal Republic of Germany: Microorganisms per se are patentable; plant varieties that are not the subject matter of the specific plant variety law are patentable (effective 1975).
France: Plant varieties that are not the subject matter of the specific plant variety law are patentable (1970 plant variety law).
Switzerland: Product claims to whole plants or their propagating material in which no variety is specified are admitted; the same criteria apply for animals.
United Kingdom: Inventions concerning plants and animals are protectable only at the cellular level (1977 patent act).

B. Eastern European Countries

At the option of the applicant either a patent or an inventor's certificate is granted; for certain categories, only a certificate is obtainable. An invention is state property.

Soviet Union: Patents and inventor's certificates are granted; new strains of microorganisms are recognized as inventions; inventor's certificates are available for new varieties and hybrids of agricultural crops and other cultivated plants and for new breeds of farm animals and poultry, fur-bearing animals, and new species of mulberry silk worms.
Bulgaria: Patent protection is available for animals.
Czechoslovakia: Inventor's certificate are available for inventions relating to medicaments, substances obtained through chemical processes, foodstuffs, and microorganisms used in industrial manufacturing.
German Democratic Republic: Microbiological processes are patentable; solutions for diagnosis, prevention, and treatment of human diseases, plant varieties, animal breeds, and strains of microorganisms are excluded (1984 patent law).
Hungary: New plant varieties and animal breeds are patentable if they are distinguishable, novel, homogeneous, stable, and have been given a variety denomination apt for registration; processes involving the use and preparation of microorganisms are patentable, although the products are not (1983 patent act).
Poland: Neither patents nor inventor's certificates are available for new plant varieties, animal breeds, or processes for curing disease; patents are not available for foodstuffs, pharmaceutical products, or products obtained by chemical processes, although processes for producing the named products are patentable.
Romania: Patent protection is available for animals.
Yugoslavia: Patents for plant and animal varieties and essentially biological processes for the production of plants or animals are excluded.

C. North American Countries

Canada: Essentially the same rules exist as the United States; patents to medically treat humans or animals are not granted; product patents are available (1987 amended patent act).
Mexico: biotechnological products and processes are not patentable.

D. South American Countries

Argentina: Ten- to twenty-year plant variety protection is available for seeds, fruits, bulbs, tubers, buds, and grafts but no patent protection is available for plant varieties; patent protection for new industrial products, new means and new applications of known means, and pharmaceuticals are excluded; no policy exists for genetically engineered animals.

Brazil: Protection for the discovery of varieties or species of microorganisms is excluded; pharmaceuticals and the processes for obtaining them are not patentable; no plant variety or plant patent protection exists.

Chile: Chile is not party to any patent-related bilateral treaty; no protection for biotechnological products is available; trademark protection for seed varieties is available.

E. Australia

Patents are available for any manner of new manufacture, excluding substances capable of being used as food or medicine consisting only of mixtures of known ingredients and the processes for producing them; Australia holds a position similar to the United States regarding the patenting of living organisms; a 20-year certificate for plant variety protection is available.

F. Japan

Japan is similar to the United States regarding biotechnological inventions; no protection is available for inventions producing or utilizing recombinant DNA in higher animals; patent varieties for plants and processes of producing plants is granted; notice of intention to patent nonhuman animals can be granted if they meet the requirements of the patent law.

G. People's Republic of China

Invention–creations are protected; scientific discoveries, rules and methods for mental activities, methods for the diagnosis and treatment of diseases, pharmaceutical products and substances obtained by a chemical process, and animal and plant varieties are precluded (1985 patent law).

XIV. Conclusion

In the early days of biotechnology, trade secrecy was often considered sufficient protection for the inventor, but modern-day biotechnologists are relying more and more on the protection afforded by patents for their inventions. Of the various forms of intellectual property protection, patents are the most difficult to obtain, since strict examination is required. However, once obtained, a patent is generally easy to maintain, requiring only the periodic payment of maintenance fees during the life of the patent. Unfortunately, most patent laws predate modern biotechnology by many years, and inventions that involve the use of living organisms do not fall neatly within their general framework. Whether or not the new technology will merit its own legislation or simply modification of existing laws to deal with special areas of difficulty is yet to be determined (Crespi, 1988).

Since the mid-1970s, the ability to select and manipulate genetic material has generated heightened interest in the commercial uses of living organisms. One result of this development of biotechnology is the creation of inventions that are themselves alive. Since the landmark *Chakrabarty* decision, which allowed the patenting of new life forms such as microorganisms, the patent activity in the biotechnology industry has escalated rapidly. The deposition of living organisms in culture collections has helped in solving the problem of the reproducibility of the written disclosure. With this increased patent activity, culture collections have been recognized as playing a significant role in the development of modern biotechnological patent systems.

References

Auerbach, J. I. (1989). Animal patenting: Ethics, enablement and enforcement. *In* "Biotechnology Patent Conference Workbook," pp. 24–36. Am. Type Cult. Collect., Rockville, Maryland.

Beier, F. K., and Straus, J. (1985). Patents in a time of rapid scientific and technological change: Inventions in biotechnology. *In* "Biotechnology and Patent and Protection: An International Review" (F. K Beier, R. S. Crespi, and J. Straus, eds.), pp. 15–35. OECD, Paris.

Beier, F. K., Crespi, R. S., and Straus, J. (1985). Annex C: Selected court decisions. *In* "Biotechnology and Patent Protection: An Intenational Review," pp. 100–105. OECD, Paris.

Bousfield, I. J. (1988). Patent protection for biotechnological inventions. *In* "Living Resources for Biotechnology: Filamentous Fungi" (D. L. Hawksworth and B. E. Kirsop, eds.), pp. 115–161. Cambridge Univ. Press, London.

Brandon, B. A. (1988). Culture deposits for patent purposes. Rockville, Maryland: *ATCC Q. Newsletter.* **8**, (1), 1–2, 7.

Brandon, B. A. (1989a). ATCC and patenting. In "Biotechnology Patent Conference Workbook," pp. 1–2. Am. Type Cult. Collect., Rockville, Maryland.

Brandon, B. A. (1989b). Deposit requirements for patent purposes. In "Biotechnology Patent Conference Workbook," pp. 82–143. Am. Type Cult. Collect., Rockville, Maryland.

Casella, P. F., and D'Agostino, R. A. (1985). Patenting microorganisms and the role of the International Depository Authority. Am. Biotechnol. Lab. July/Aug. pp. 26–35.

Crespi, R. S. (1985). Patent protection in biotechnology: questions, answers and observations. In "Biotechnology and Patent Protection: An International Review" (F. K. Beier, R. S. Crespi, and J. Straus, eds.), pp. 36–86. OECD, Paris.

Crespi, R.S. (1988). "Patents: A Basic Guide to Patenting in Biotechnology," Cambridge Stud. Biotechnol. Vol. 6. Cambridge Univ. Press, London.

Hoscheit, D. A. (1988). Biotechnology patent law—An overview including recent developments. In "Biotechnology Patent Conference Workbook," pp. 3–13. Am. Type Cult. Collect., Rockville, Maryland.

Hoscheit, D. A. (1989). Biotechnology patent law—An overview including recent developments. In "Biotechnology Patent Conference Workbook," pp. 3–9. Am. Type Cult., Collect., Rockville, Maryland.

Hwang, E. N. (1989). Status of patent protection involving microorganisms in Korea and the requirement for deposit of microorganisms. In "Biotechnology Patent Conference Workbook," pp 75–81. Am. Type Cult. Collect., Rockville, Maryland.

Ono, K. (1988). Biotechnology patenting in Japan. In "Biotechnology Patent Conference Workbook," pp. 62–70. Am. Type Cult. Collect., Rockville, Maryland.

Pridham, T. G., and Hesseltine, C. W. (1975). Culture collections and patent depositions. Adv. Appl. Microbiol. **19** 1–23.

U.S. Congress, Office of Technology Assessment (OTA) (1989). "New Developments in Biotechnology: Patenting Life"–Spec. Rep., OTA-BA-370. U.S. Gov. Print. Off., Washington, D.C.

U.S. Federal Register (1988). Deposit of biological materials for patent purposes: Notice of proposed rule making. Fed. Regist. **53**, No. 194, Oct. 6.

U.S. Federal Register (1989). Deposit of biological materials for patent purposes. Fed. Regist. **54**, No. 161, Aug. 22.

Wahl, R.A. (1971). Deposit of micro-organisms. APLA Bull. pp. 304–305.

INDEX

A

Acetylene, xenobiotics and, 226, 240
Acid
  α-amylase production and
    enzyme characteristics, 14, 29, 30
    modes for economy, 3
    present status, 14, 29, 30
  levan and, 189
Actinomycetes, xenobiotics and, 204, 228, 244
Adhesion
  α-amylase production and, 11
  bacterial gene transfer in soil and, 74
Adsorption, xenobiotics and, 217, 221
AEC, see Anion exchange capacity
Aeration, α-amylase production and, 33, 34
Agriculture Type Culture Collection, patent law and, 257, 271, 286, 288, 289
  Budapest Treaty, 271
  deposits, 262–266
  legal decisions, 280, 281
Aldose, levan and, 175
Algae
  patent law and, 262, 266
  xenobiotics and, 245
    activity, 203, 205
    assessment, 234, 238
    effects, 228, 234
Aluminum, xenobiotics and, 199, 200
Amino acids
  bacterial gene transfer in soil and, 99, 105, 122
  xenobiotics and, 223, 232, 234, 244
Ampicillin, bacterial gene transfer in soil and, 76

α-Amylase, production by solid-state fermentation and, 1–3, 46, 47
  advantages, 8–13
  aeration, 33, 34
  autoclaving, 31
  bacterial cultures, 19–21
  bacterial strains, 7, 8
  clarification, 39, 40
  culture vessels, 30, 31
  economic considerations, 42, 43
  enzyme characteristics, 40–42
  enzyme recovery, 36–39
  enzyme yields, 35, 36
  growth, 34, 35
  history, 18, 19
  incubation, 32, 33
  industrial production, 13–18
  inoculum, 31, 32
  modes for economy, 3–7
  moisture, 26–29
  pH, 29, 30
  research needs, 43–46
  solid substrates, 23, 24
  supplementary nutrients, 25, 26
β-Amylase, solid-state fermentation and, 4
Amyloglucosidase, solid-state fermentation and, 2, 6–8
Amylopectin, solid-state fermentation and, 5
Animals, patent law and, 284, 288–291
Anion exchange capacity, xenobiotics and, 201, 216, 217, 221
Antibiotics
  α-amylase production and, 10
  bacterial gene transfer in soil and, 62, 64
    conjugation, 75, 77, 79, 80, 83

296 INDEX

conjugation study methods, 122–125, 128
environmental factors, 67, 70
recombinants, 140–142
transduction, 102, 137, 138
xenobiotics and, 217
Antibodies, levan and, 180, 190
Antigens
  bacterial gene transfer in soil and, 151
  levan and, 180, 190
Arabinose, levan and, 173
Aspergillus, α-amylase production and, 4
ATCC, *see* Agriculture Type Culture Collection
ATP, xenobiotics and, 234, 244
Autoclaving
  α-amylase production and, 31
  bacterial gene transfer in soil and, 113, 155, 158

B

Bacillus
  α-amylase production and, 7, 8, 10–12
    enzyme characteristics, 40, 41
    enzyme recovery, 39
    enzyme yields, 35, 36
    growth, 34, 35
    incubation, 33
    industrial production, 14, 15, 17
    moisture, 29
    pH, 30
    present status, 18–21, 24
    research needs, 43–45
  bacterial gene transfer in soil and, 67–69, 76, 77, 81
  levan and, 172, 174, 191
    biosynthesis, 174, 176, 177
    chemistry, 180–184
    production, 186
  xenobiotics and, 203, 204
Bacteria
  levan and, 172, 174, 178, 182
  patent law and, 260, 262–264, 266, 286
  xenobiotics and, 245, 246
    activity, 203–205, 208, 209
    assessment, 234, 235, 238, 244
    effects, 224, 226–228, 231–233

Bacterial gene transfer in soil, 60–62, 72, 73
  antimicrobial agents, 158, 159
  biological factors, 62–64
  conjugation, 73, 74, 91–94
    chromosomes, 74, 75
    gene survival, 77–82
    mobilizing elements, 76, 77
    plasmids, 75, 76
    in situ, 82–91
  in soil, 126–132
    spectrum, 77
    in vitro, 115–126
  environmental factors, 64, 65
    electromagnetic radiation, 70
    energy source, 66
    interactions, 71, 72
    ionic composition, 70
    microbial competition, 65, 66
    oxygen, 69
    pH, 68, 69
    surfaces, 70, 71
    temperature, 66–68
    water content, 69
  maintenance, 113, 114
  media composition, 156–158
  microhabitats, 58–60
  preparation, 112, 113
  quality assurance, 156
    analysis, 154, 155
    calibration, 153, 154
    comparability, 153
    internal checks, 155, 156
    sampling, 152, 153
  recombinants, 139, 140
    DNA fingerprinting, 143, 144
    DNA probes, 144–151
    heat induction, 152
    indigenous microbes, 141, 142
    recovery media, 143
    selective media, 140, 141
    serological techniques, 151, 152
    viable bacteria, 142, 143
  sampling, 114, 115
  selection, 111, 112
  storage, 115
  terrestrial microcosms, 105
    complex, 108–111
    simple, 105–108
  transduction, 94–105

INDEX   297

in soil, 134–139
in vitro, 132–134
Bacterial thermostable α-amylase
    production, *see* α-Amylase
Batch systems, xenobiotics and, 236
Bentonite, bacterial gene transfer in soil
    and, 84
Biodegradation, xenobiotics and, 214, 215
Biotin, bacterial gene transfer in soil and,
    95, 146, 147, 149, 151
Bladder, bacterial gene transfer in soil
    and, 92
Bran, *see also* Moldy bran; Wheat bran
    α-amylase production and, 38–40, 44
Budapest Treaty, patent law and,
    269–274
    deposits, 266–268
    European countries, 279, 280
    international agreements, 275, 277, 279
    non-European countries, 281

C

Cadmium, xenobiotics and, 213
    effects, 221–223, 230, 232, 233
    environment, 216
Calcium
    α-amylase production and, 26, 40–42
    bacterial gene transfer in soil and, 133,
        156
Calibration, bacterial gene transfer in soil
    and, 153, 154
Carbohydrate, levan and, 172, 174, 179
Carbon
    α-amylase production and, 15, 23
    bacterial gene transfer in soil and,
        73, 78
    levan and, 181, 182
    xenobiotics and, 245, 247
        activity, 204–209
        assessment, 235, 236, 238, 239, 241,
            243
        classification, 210
        effects, 220–223, 231
        interactions, 215
Carbon dioxide, xenobiotics and, 196,
    245
    activity, 205, 207, 208
    assessment, 236, 238, 239

effects, 220, 234
    soil as microbial habitat, 202
Carotenoids, α-amylase production
    and, 15
Cation exchange capacity, xenobiotics
    and, 199–201, 216
    effects, 221
    environment, 216, 217, 219
CCPA, *see* Court of Customs and Patent
    Appeals
cDNA, bacterial gene transfer in soil and,
    80, 81, 154
CEC, *see* Cation exchange capacity
Cellulose, xenobiotics and, 205, 222, 223,
    239, 241
Centrifugation
    α-amylase production and, 9, 40
    bacterial gene transfer in soil and
        conjugation, 117, 126, 131
        recombinants, 144, 147
        transduction, 133, 135, 138, 139
    levan and, 186
Chloramphenicol, bacterial gene transfer
    in soil and, 78, 116
Chlorination, xenobiotics and, 224, 233,
    234, 246, 248
Chloroform, bacterial gene transfer in soil
    and, 138, 151
Chromatin, xenobiotics and, 234
Chromium, xenobiotics and, 213, 218,
    222, 230, 233
Chromosomes, bacterial gene transfer in
    soil and, 60, 73
    conjugation, 73–75, 77, 81–84, 91,
        93, 94
    study methods, 115–123, 129
    environmental factors, 67–70
    recombinants, 141, 144
    transduction, 94–96, 99, 104, 137
Clarification, α-amylase production and,
    39, 40
Clay
    bacterial gene transfer in soil and,
        58, 59
    conjugation, 82, 84, 91
    environmental factors, 68, 70, 71
    preparation, 112, 113
    quality assurance, 153
    selection, 111, 112
    storage, 115

terrestrial microcosms, 106
transduction, 97, 99, 102, 137–139
xenobiotics and
   assessment, 240
   effects, 228
   environment, 216, 217, 219
   interactions, 214
   soil as microbial habitat, 198–200, 202
Clones, bacterial gene transfer in soil and, 141, 145
Cm, see Chloramphenicol
Cointegrates, bacterial gene transfer in soil and, 76, 77
Colloids, xenobiotics and, 200, 201, 221, 240
Colony-forming units, bacterial gene transfer in soil and
   conjugation, 78, 80, 82, 85, 91, 92
   study methods, 117, 121, 122, 125–127, 130–132
   quality assurance, 153
   recombinants, 142
   transduction, 133–135, 139
Competition
   bacterial gene transfer in soil and, 63–66
   conjugation, 79, 81, 91, 92
   xenobiotics and, 204, 219
Conjugal transfer, bacterial gene transfer in soil and
   conjugation, 82, 93, 94, 118
   environmental factors, 66, 68–71
   transduction, 126, 128
Conjugation, bacterial gene transfer in soil and, 60, 72–74, 91–94
   chromosomes, 74, 75, 126–128
   environmental factors, 71
   mobilization, 76, 77
   plasmids, 75, 76, 128–131
   procedures, 139
   recombinants, 143
   in situ, 82–91
   spectrum, 77
   survival, 77–82
   transduction, 102, 104
   in vitro, 115–126
Contamination, α-amylase production and, 10, 12
Court of Customs and Patent Appeals, 258, 259

Culture
α-amylase production and
   bacteria, 19–21
   genetic improvement, 21–23
   moisture, 29
   present status, 31, 33, 34
   vessels, 30, 31
bacterial gene transfer in soil and
   conjugation study methods, 116–118, 120, 123–125
   quality assurance, 152
   recombinants, 140, 144
   transduction, 133
patent law and, 284, 289, 291
   Budapest Treaty, 270–274
   deposits, 261–266
   European countries, 280
   history, 257, 258
   international agreements, 277–279
   legal decisions, 260, 261
   non-European countries, 280, 281
Cyclic AMP, α-amylase production and, 15
Cyclodextrins, α-amylase production and, 2, 5
Cycloheximide, bacterial gene transfer in soil and, 128
Cytoplasm
bacterial gene transfer in soil and, 74, 133
xenobiotics and, 240

D

DDT, xenobiotics and, 233, 244
Dehydration, α-amylase production and, 45
Dextran, levan and, 171, 172, 191
   analysis, 184
   biosynthesis, 174
   utilization, 189
Dextrins, α-amylase production and, 2, 40, 41, 46
Dextrose, α-amylase production and, 2, 3, 13, 42
DNA
α-amylase production and, 21
bacterial gene transfer in soil and, 59–61
   biological factors, 63, 64

conjugation, 73–76, 84, 94
conjugation study methods, 123, 131
environmental factors, 65, 68, 70
quality assurance, 153, 154
recombinants, 141, 143–152
survival, 77, 79–82
transduction, 94–96, 98, 99, 102, 103, 132
patent law and, 277, 288, 291
xenobiotics and, 244
DNA fingerprinting, bacterial gene transfer in soil and, 143, 144
Downstream processing, α-amylase production and, 40, 43, 44
Drosophila, bacterial gene transfer in soil and, 80

E

Ecological dose, xenobiotics and, 245, 248
Ecology
  bacterial gene transfer in soil and, 61, 62, 73
    environmental factors, 65, 71
    preparation, 112
    recombinants, 150
    selection, 112
    terrestrial microcosms, 105, 107, 108, 110
  xenobiotics and, 197, 246, 247
    assessment, 235–238
    effects, 233
Ecosystems
  bacterial gene transfer in soil and, 96, 108
  xenobiotics and, 197, 210, 236–238
EDTA
  α-amylase production and, 14
  bacterial gene transfer in soil and, 138
E_h, xenobiotics and, 216, 222
Electron microscopy, xenobiotics and, 242
Embryos, patent law and, 288, 289
Endothelium, levan and, 180
Energy
  α-amylase production and, 2–4, 37
  bacterial gene transfer in soil and, 66, 78, 80

levan and, 175
xenobiotics and, 201, 202, 221, 231
Environment
  bacterial gene transfer in soil and, 61–64
  conjugation, 76, 80, 83, 84, 91, 92
  conjugation study methods, 122–124, 128, 131
  environmental factors, 64–72
  recombinants, 140–142, 144, 149–152
  terrestrial microcosms, 105, 108, 110
  transduction, 94, 103, 105
  xenobiotics and, 196–198, 209, 210, 245, 247, 248
    activity, 203, 205, 206, 209
    assessment, 240, 242, 245
    classification, 210, 211
    effects, 221, 222, 226, 227
    interactions, 213–219
Environmental Protection Agency, xenobiotics and, 210, 238, 242
Enzymes, *see also* specific enzyme
  α-amylase production and, *see* α-Amylase
  bacterial gene transfer in soil and, 72, 73
  conjugation, 73, 75, 77, 124
  quality assurance, 154
  recombinants, 145–147, 150, 151
  selection, 112
  terrestrial microcosms, 106, 107
  levan and, 174, 176, 178, 179, 185, 188, 190
  xenobiotics and, 197, 202, 247
    activity, 206, 207
    assessment, 235, 238, 240, 241, 243
    effects, 228, 229, 234
    interactions, 214, 215
EPA, *see* Environmental Protection Agency
EPC, *see* European Patent Convention
EPO, *see* European Patent Office
Erosion, xenobiotics and, 205
Escherichia coli, bacterial gene transfer in soil and, 60–62
  conjugation, 73, 74, 77–85, 91, 91–93
  study methods, 116–119, 121, 124, 125, 128, 129
  environmental factors, 66–70, 72
  recombinants, 146, 148, 152

transduction, 95–99, 102
  study methods, 132–136, 138
Ethanol
  α-amylase production and, 2, 11, 16, 25, 37
  levan and, 184, 188–190
  xenobiotics and, 234
European Patent Convention, 279, 280, 282, 287, 288
  Budapest Treaty, 270, 271
  international agreements, 275–279
European Patent Office, 276–279, 282, 287
Evolution
  bacterial gene transfer in soil and, 73, 94, 106
  xenobiotics and, 236, 239
Exconjugants, bacterial gene transfer in soil and, 130, 131

F

F-factor, bacterial gene transfer in soil and, 73–76, 123, 125
Fermentation
  levan and, 181, 184, 186–189
  liquid surface, see Liquid surface fermentation
  patent law and, 256
  solid-state, see Solid-state fermentation
  submerged, see Submerged fermentation
  xenobiotics and, 203
Fermentation Research Institute, patent law and, 280
Fertility, bacterial gene transfer in soil and, 63, 73
Filtration
  α-amylase production and, 12, 16, 38, 40
  bacterial gene transfer in soil and, 138
  levan and, 184, 190
Fluorescence, bacterial gene transfer in soil and, 82
Fossil fuel, xenobiotics and, 210–213
Fractionation, levan and, 189
Frequency of recombination, bacterial gene transfer in soil and, 122, 123, 126, 128

Fructan, levan and, 172, 174, 179
Fructose
  α-amylase production and, 2, 13
  levan and, 171, 191
    analysis, 184
    biosynthesis, 175–178
    chemistry, 178
    occurrence, 174
    production, 187, 188
    utilization, 190
Fungi
  α-amylase production and, 7, 12
    industrial production, 13
    present status, 25, 30, 31, 34, 45
  bacterial gene transfer in soil and, 59, 71
    conjugation, 128
    recombinants, 142
    transduction, 135, 138
  patent law and, 260, 262–264, 266, 286
  xenobiotics and, 245
    activity, 203, 205, 206, 209
    assessment, 234, 235, 238, 242, 245
    effects, 223, 225, 227, 231–233
Furanose, levan and, 179

G

Galactose, bacterial gene transfer in soil and, 95
Gas, xenobiotics and, 201–203, 223, 226
Gelatinization, α-amylase production and, 3, 4, 13
GEM, see Genetically engineered microorganisms
Genetically engineered microorganisms, bacterial gene transfer in soil and, 60, 61, 73
  conjugation, 84, 91, 92, 94, 131
  environment, 65, 66, 68, 72
  maintenance, 114
  microcosms, 105–108, 110
  preparation, 113
  recombinants, 140, 142–145, 148, 149, 151, 152
  sampling, 114
  selection, 112
  survival, 78, 80, 82
  transduction, 94, 102, 105

Gentamicin, bacterial gene transfer in soil and, 78
Glucoamylase, solid-state fermentation and, 4, 5, 46
Glucose
  α-amylase production and, 2–5, 13–15, 42
  levan and
    analysis, 185
    biosynthesis, 175–178
    chemistry, 181, 183
    occurrence, 173
    production, 187, 188
    xenobiotics and, 245
Glycerol, α-amylase production and, 39
Glycine, α-amylase production and, 15
Glycosides, levan and, 177
Growth factors, bacterial gene transfer in soil and, 122
Gum, levan and, 188, 189

H

Habitat, xenobiotics and
  activity, 203, 206
  phases, 198–203
Heat
  α-amylase production and, 7, 9
  bacterial gene transfer in soil and, 98, 102, 113, 152
  levan and, 189
Herbicides, xenobiotics and, 246
  classification, 210–212
  effects, 223, 227, 229, 232
Hfr cells, bacterial gene transfer in soil and, 74, 116, 122
High-performance liquid chromatography, levan and, 180, 181, 184
Homeostasis, bacterial gene transfer in soil and, 92
Homology, bacterial gene transfer in soil and, 63
  conjugation, 77, 79, 84
  recombinants, 150
  terrestrial microcosms, 108
HPLC, see High-performance liquid chromatography

Hybridization
  bacterial gene transfer in soil and
    conjugation, 81, 82
    recombinants, 144–151
  patent law and, 286, 290
Hybridomas, patent law and, 262, 264, 266
Hydrogen, bacterial gene transfer in soil and, 68
Hydrolases, xenobiotics and, 207
Hydrolysis
  α-amylase production and, 2, 3, 46, 47
    industrial production, 13, 14
    modes for economy, 3–6
    present status, 19, 20, 31, 35, 42, 45
  levan and, 178, 180, 181, 184, 185, 188–191
  xenobiotics and, 214, 215, 244

I

IDA, see International Depository Authority
Immunoglobulins, levan and, 185
Incubation
  α-amylase production and, 32, 33
  bacterial gene transfer in soil and, 119, 121, 129, 135, 136
  xenobiotics and, 223, 240
Inflammation, levan and, 180
Inherent nitrifying capacity, xenobiotics and, 225
Inhibition
  α-amylase production and, 24, 44, 45
  bacterial gene transfer in soil and, 63
    conjugation, 91, 93, 117
    environmental factors, 71
    recombinants, 140, 142
    transduction, 138
  levan and, 177, 180, 188
  xenobiotics and, 220–226, 228, 231–234
Insertion sequences, bacterial gene transfer in soil and, 76, 77
International Depository Authority, patent law and, 267, 270–274, 279–281
International Searching Authority, patent law and, 276

Inulin, levan and, 171, 172, 174
  analysis, 185
  biosynthesis, 178
  chemistry, 178, 180, 182
  utilization, 190
Iodine
  α-amylase production and, 35, 36, 40
  levan and, 180
Ionic composition, bacterial gene transfer in soil and, 70
Iron
  bacterial gene transfer in soil and, 71
  xenobiotics and, 199, 205, 209, 240
Irrigation, bacterial gene transfer in soil and, 69, 110, 114
Isoamylase, solid-state fermentation and, 4, 5
Isopropanol, levan and, 184, 188

J

Japanese Patent Office, 270, 280, 281

K

Ketose, levan and, 175, 184

L

LA, see Luria agar
Lactose
  bacterial gene transfer in soil and, 136, 141
  levan and, 188
LB, see Luria broth
Levan, 171, 172, 191
  analysis, 184–186
  biosynthesis
    hydrolysis, 178
    inhibition, 177, 178
    levansucrase, 174–176
    specificity, 176, 177
  chemistry
    Bacillus polymyxa, 180–184
    properties, 180
    structure, 178–180
  occurrence, 172–174
  production, 186–188
  utilization, 190, 191
    blood plasma extender, 189, 190
    industrial gums, 188, 189
    sweeteners, 190
Levansucrase, 174–176, 185, 186, 188
Levulan, 172, 189
Light microscopy, xenobiotics and, 242
Lignins, xenobiotics and, 205
Liquefication, α-amylase production and, 46
  industrial production, 13, 14
  modes for economy, 5
  present status, 25, 27, 40
Liquid surface fermentation, α-amylase and, 7
  advantages, 11, 12
  industrial production, 14, 16–18
  present status, 18–20, 36
LSF, see Liquid surface fermentation
Luria agar, bacterial gene transfer in soil and, 124, 133, 134, 156, 157
Luria broth, bacterial gene transfer in soil and, 156
  conjugation, 78–81, 85, 91, 93, 124, 125
  transduction, 96, 135
Lysis, bacterial gene transfer in soil and
  recombinants, 148, 151, 152
  transduction, 95, 102, 103, 132–134, 139
Lysogeny, bacterial gene transfer in soil and
  recombinants, 152
  transduction, 96–98, 104, 105
  transduction study methods, 132–134, 136, 138, 139

M

MAC, see MacConkey agar
MacConkey agar, bacterial gene transfer in soil and
  conjugation, 78, 80, 116, 123, 130
  recombinants, 142
  transduction, 98, 102, 134, 136
Magnesium
  α-amylase production and, 42
  bacterial gene transfer in soil and, 133, 156
  xenobiotics and, 199

INDEX

Maltase, α-amylase production and, 5
Maltose
  α-amylase production and, 2, 42
  levan and, 188
Manganese, xenobiotics and, 205, 209, 240
Mating, bacterial gene transfer in soil and, 117–121, 125–128
Mesocosms, xenobiotics and, 236, 237
Methanol, levan and, 189
Microbial levan, see Levan
Microcosms
  bacterial gene transfer in soil and, 105–111
    conjugation, 127, 132
    quality assurance, 155
    sampling, 114, 115
    transduction, 139
  xenobiotics and, 236–238, 244
Microfiltration, α-amylase production and, 8, 9
Microhabitats, bacterial gene transfer in soil and, 58–60, 62–64
  conjugation, 83, 93
  environmental factors, 71
  sampling, 115
  selection, 111
Microorganisms, xenobiotics and, see Xenobiotics, effects on soil microorganisms
Mobilization, bacterial gene transfer in soil and, 75–77, 84
Moisture
  α-amylase production and, 26–30, 39
  bacterial gene transfer in soil and, 113, 127
  xenobiotics and, 205
Moldy bran, α-amylase production and, 8, 9
Monoclonal antibodies, bacterial gene transfer in soil and, 147, 151, 152, 154
Montmorillonite, bacterial gene transfer in soil and, 70, 71
  conjugation, 82, 84, 91
  transduction, 97, 102
Most probable number method
  bacterial gene transfer in soil and, 150
  xenobiotics and, 241, 242

Mutation
  α-amylase production and, 22, 23
  bacterial gene transfer in soil and
    conjugation, 77, 116, 121, 124
    recombinants, 152
    transduction, 96, 132–134, 137
Mycorrhizae, xenobiotics and
  activity, 205, 206, 209
  assessment, 242
  effects, 227, 231–233

N

Nalidixic acid, bacterial gene transfer in soil and, 81, 128, 137, 143
Nitrocellulose, bacterial gene transfer in soil and, 147, 148
Nitrogen
  α-amylase production and, 17, 23, 30
  bacterial gene transfer in soil and, 72, 83, 105, 110
  patent law and, 265
  xenobiotics and, 196, 245–247
    activity, 204–209
    assessment, 234, 235, 238–240, 243
    classification, 213
    effects, 227, 228, 231–233
    soil as microbial habitat, 203
NMR, see Nuclear magnetic resonance
NRRL, patent law and, 257, 258, 271
Nuclear magnetic resonance, levan and, 181, 182, 185, 191
Nucleotides, bacterial gene transfer in soil and, 143, 146, 150
Nutrient broth, bacterial gene transfer in soil and, 124
Nx, see Nalidixic acid

O

Oligosaccharides
  α-amylase production and, 15, 42
  levan and, 175, 178, 189, 190
Oncogenes, patent law and, 261, 262, 288
Oxidation, xenobiotics and, 197, 246
  activity, 208, 209
  assessment, 234
  effects, 221, 224, 227, 233, 234

environment, 218
interactions, 214, 215
Oxygen
  α-amylase production and, 27
  bacterial gene transfer in soil and, 69, 114, 115, 121
  xenobiotics and
    activity, 205
    assessment, 238
    effects, 220, 221, 234
    interactions, 214
    soil as microbial habitat, 199, 202, 203

P

Pancreas, α-amylase production and, 4, 24
Patent Cooperation Treaty, 270, 275, 276
Patent law, 255–257, 289–292
  Budapest Treaty, 269–274
  deposits in culture collection, 261, 262
    availability, 264
    nature, 264, 265
    postal requirements, 262, 263
    release, 265, 266
    security, 265
    time of deposit, 263, 264, 284
  history, 257, 258
  international agreements, 274, 275
    European countries, 279, 280, 282, 283
    European Patent Convention, 276–279, 282
    non-European countries, 280, 281, 283, 284
    Paris Union Convention, 275, 276
    Patent Cooperation Treaty, 276
  legal decisions, 258–261
  new animals, 284, 288, 289
  new plants, 284–287
  regulations, 266–269
Patent Trademark Office, 285, 286, 288, 289
  Budapest Treaty, 271
  deposits, 261, 264–266
  history, 258
  international agreements, 276
  legal decisions, 258–261
  regulations, 266

PCT, see Patent Cooperation Treaty
Percolation, α-amylase production and, 37, 39
Permanent wilting point, bacterial gene transfer in soil and, 112, 114
Pesticides, xenobiotics and, 196
  assessment, 235, 236, 242, 244
  effects, 220, 232, 233
PFU, see Plaque-forming units
pH
  α-amylase production and, 9, 10
    enzyme characteristics, 41, 42
    enzyme recovery, 39
    industrial production, 13–15
    present status, 20, 26, 27, 29, 30
    research needs, 43, 44
  bacterial gene transfer in soil and
    conjugation, 83, 126
    environmental factors, 68, 69, 71
    maintenance, 114
    preparation, 112
    quality assurance, 153
    selection, 112
    terrestrial microcosms, 106
  levan and, 176, 186–189
  xenobiotics and, 196, 246
    activity, 204
    effects, 221, 222
    environment, 216–218
    interactions, 214–216
    soil as microbial habitat, 200
Phenotype, bacterial gene transfer in soil and
  conjugation
    in soil, 127, 128, 131, 132
    in vitro, 116, 117, 121–125
  quality assurance, 152, 154
  recombinants, 140–145, 152
  survival, 81
  transduction study methods, 137, 139
Pheromones, bacterial gene transfer in soil and, 74
Phosphate
  α-amylase production and, 27, 28, 32, 38, 39
  bacterial gene transfer in soil and, 114
  xenobiotics and, 227
Phospholipids, xenobiotics and, 233
Phosphorus, xenobiotics and, 209, 245, 246
  activity, 205, 206, 209

assessment, 227, 228, 238, 240, 243
  effects, 227, 228, 232
Phosphorylation, xenobiotics and, 221
Photosynthesis
  bacterial gene transfer in soil and, 70
  levan and, 174
  xenobiotics and, 197, 246
    activity, 207
    effects, 228, 231, 233, 234
Plant Patent Act, 259, 285, 286
Plant Variety Protection Act, 259, 285, 286
Plants, patent law and, 284–287, 289–291
Plaque-forming units, bacterial gene transfer in soil and, 96, 135, 139
Plasma, levan and, 189, 190
Plasma membrane, xenobiotics and, 234
Plasmids
  α-amylase production and, 21
  bacterial gene transfer in soil and, 61, 63, 72, 73
    conjugation, 74, 75, 77–81, 83–94
    conjugation study methods, 116, 117, 123–126, 128–31
    environmental factors, 66–71
    terrestrial microcosms, 106
    transduction, 93, 101, 102, 104, 136, 137
  patent law and, 259, 261, 262, 264, 266, 288
Pollution, xenobiotics and, 196, 208, 210, 219
Polymerase chain reaction, bacterial gene transfer in soil and, 150
Polymerization
  levan and, 175, 176, 178–180, 191
    analysis, 184
    utilization, 188, 190
  xenobiotics and, 215
Polysaccharides
  α-amylase production and, 39, 44
  levan and, 171–173, 191
    analysis, 185
    biosynthesis, 178
    chemistry, 178–180, 182–184
    production, 186, 188, 189
  xenobiotics and, 240
Population
  bacterial gene transfer in soil and, 61–64, 72

  conjugation, 75, 78, 81, 83, 121, 128, 131
  environmental factors, 65, 66, 71
  preparation, 113
  recombinants, 142, 144, 147, 149
  sampling, 115
  terrestrial microcosms, 106, 109, 110
  transduction, 138
xenobiotics and, 198, 246
  activity, 203–205
  assessment, 235, 238, 241–244
  effects, 220–222, 225, 228, 230–233
  environment, 216, 218
Postal requirements, patent law and, 262, 263
PPA, see Plant Patent Act
Prophages, bacterial gene transfer in soil and, 152
Proteases, α-amylase production and, 8
  industrial production, 13
  present status, 22, 25, 27, 40
  research needs, 43, 44
Protein
  α-amylase production and, 2
    present status, 25, 31, 40
    technique, 10, 12
  bacterial gene transfer in soil and, 63, 70
    conjugation, 73, 74, 80
    quality assurance, 153
    recombinants, 147, 150–152
    terrestrial microcosms, 107
    transduction, 96, 133
  xenobiotics and, 207, 223, 234, 244
Proteolysis
  α-amylase production and, 45
  bacterial gene transfer in soil and, 151
  xenobiotics and, 239
Protozoa
  patent law and, 262, 266
  xenobiotics and, 203, 205, 234, 238, 245
Pseudomonas
  α-amylase production and, 4, 5
  bacterial gene transfer in soil and, 72
    conjugation, 77, 81, 83, 91, 125, 128
    environmental factors, 66, 69
    recombinants, 141, 150
    transduction, 101, 102, 105, 133
  patent law and, 259
PTO, see Patent Trademark Office

Pullulanase, α-amylase production and, 5
Purification
  α-amylase production and, 4, 28, 42, 45
  bacterial gene transfer in soil and, 144, 148
  levan and, 185, 190
PVPA, see Plant Variety Protection Act
PWP, see Permanent wilting point

Q

Quarantine requirements, patent law and, 262, 263

R

Raffinose, levan and, 176, 177, 188
Recombination
  bacterial gene transfer in soil and, 62–64, 66, 72, 139–144
  conjugation, 74, 75, 77, 80, 92, 93
  conjugation study methods, 116, 117, 122, 123, 131, 132
  DNA probes, 144–151
  heat induction, 152
  quality assurance, 153
  serological techniques, 151, 152
  transduction, 103, 105
  patent law and, 288, 291
Redox potential, xenobiotics and, 218
Replica-plating, bacterial gene transfer in soil and, 108, 137, 154
Replication
  bacterial gene transfer in soil and, 61, 63
  conjugation, 73–77, 124
  conjugation study methods, 127, 131
  quality assurance, 154, 155
  recombinants, 152
  sampling, 114
  terrestrial microcosms, 105, 106, 108
  transduction, 94–96, 137
  xenobiotics and, 236, 241
Respiration, xenobiotics and, 246
  assessment, 235, 237–239, 241, 243–245
  effects, 220–222
Restriction endonucleases, bacterial gene transfer in soil and, 63, 143, 144

Rifampicin, bacterial gene transfer in soil and, 81
RNA, patent law and, 277
Roots, xenobiotics and, 205, 206, 231–233, 242

S

Saccharification, α-amylase production and, 1–3, 46
  advantages, 11
  modes for economy, 4–6
  present status, 41, 42, 46
Sampling, bacterial gene transfer in soil and, 153
Selection, bacterial gene transfer in soil and, 141–143, 151
Serological techniques, bacterial gene transfer in soil and, 151, 152, 154
Silicon, xenobiotics and, 199
Simazine, xenobiotics and, 222, 232
Soil
  bacterial gene transfer in, see Bacterial gene transfer in soil
  levan and, 174
  xenobiotics and, see Xenobiotics, effects on soil microorganisms
Solid-state fermentation, α-amylase production by, see α-Amylase
Solid substrates, α-amylase production and, 23, 24
Starch
  α-amylase production and, 1–3, 46, 47
    autoclaving, 31
    clarification, 40
    enzymes, 35, 41, 42
    industrial production, 13, 14, 17
    inoculum, 32
    modes for economy, 3–5
    moisture, 27
    present status, 19, 20, 25, 36
    technique, 11, 12
  levan and, 174
  xenobiotics and, 244
Sterilization
  α-amylase production and, 12, 15, 23
  bacterial gene transfer in soil and, 65, 113, 154, 155

Streptomyces
  bacterial gene transfer in soil and, 73, 76, 77, 84
  patent law and, 257, 258
Streptomycin, bacterial gene transfer in soil and, 78, 122
Submerged fermentation, α-amylase production and, 2, 46, 47
  advantages, 8–13
  aeration, 33
  bacterial strains, 8
  cultures, 22, 23
  economics, 42, 43
  enzyme yields, 36
  incubation, 32
  industrial production, 13–15, 17, 18
  modes for economy, 6, 7
  pH, 30
  present status, 19, 20
  research needs, 43, 44
Sucrose
  levan and, 171, 172, 191
    biosynthesis, 174, 176, 178
    chemistry, 178, 181
    occurrence, 172–174
    production, 186–188
    utilization, 190
  xenobiotics and, 244
Sugar
  α-amylase production and, 2, 13, 35
  levan and, 171, 173, 191
    analysis, 184
    biosynthesis, 176, 177
    production, 188
  xenobiotics and, 232
Sulfur, xenobiotics and, 245, 246
  activity, 205, 206, 209
  assessment, 234, 238, 240, 243
  classification, 213
  effects, 227
Sweeteners, levan and, 171, 190
Symbiosis, xenobiotics and, 226, 227, 232, 240
Synergism, xenobiotics and, 219
Syntrophy, bacterial gene transfer in soil and, 122

T

Temperature
  α-amylase production and, 4, 9, 10
  enzyme characteristics, 42
  industrial production, 13, 15
  present status, 24, 31–33, 38, 39
  research needs, 45
  bacterial gene transfer in soil and
  conjugation, 121, 127
  environmental factors, 66–68, 71
  preparation, 113
  quality assurance, 153, 155
  recombinants, 152
  sampling, 114
  storage, 115
  terrestrial microcosms, 106, 110
  transduction, 95, 96, 132, 133, 135
  levan and, 176, 180, 181, 184, 186
  patent law and, 265
  xenobiotics and, 196, 246
    activity, 205
    effects, 222
    environment, 216, 218
    interactions, 214
Terrestrial microcosms, bacterial gene transfer in soil and, 105–111
Thermostable α-amylase production, *see* α-Amylase
Thymidine kinase, bacterial gene transfer in soil and, 76
Tn, *see* Transposons
Tranduction, bacterial gene transfer in soil and, 60, 72, 94–105
  environmental factors, 65, 68, 69, 71
  recombinants, 148, 152
  study methods, 132–139
Transconjugants, bacterial gene transfer in soil and, 128
Transcription, bacterial gene transfer in soil and, 63
Transferases, xenobiotics and, 207
Transformation, bacterial gene transfer in soil and, 60, 68, 69, 71, 72
Transposons, bacterial gene transfer in soil and, 76, 77, 96
Tryptone glucose yeast extract, bacterial gene transfer in soil and, 124
Tumors, levan and, 180

U

Ultrahigh-fructose glucose syrups, 190
Ultraviolet light
  α-amylase production and, 22, 23

bacterial gene transfer in soil and, 82, 141
Ultraviolet radiation, bacterial gene transfer in soil and, 70
Union for the Protection of New Varieties of Plants, 275, 285–287

V

Viable but nonculturable bacteria, gene transfer in soil and, 142, 143
Virus, patent law and, 263, 266
Viscosity
  α-amylase production and, 3, 12
  levan and, 180, 183, 189, 190

W

Water content
  bacterial gene transfer in soil and, 69, 106, 112, 114
  xenobiotics and, 196, 246
    effects, 222
    environment, 216
    interactions, 214
    soil as microbial habitat, 201, 202
Water-holding capacity
  bacterial gene transfer in soil and, 69, 84, 112, 114
  xenobiotics and, 198, 216
WHC, see Water-holding capacity
Wheat bran, α-amylase production and, 9, 12
  bacterial cultures, 20
  clarification, 39, 40
  inoculum, 32
  present status, 22–29
  recovery of enzyme, 39
  research needs, 44, 45
World Intellectual Property Organization, patent law and, 269–271, 275, 276

X

X-ray
  α-amylase production and, 22
  bacterial gene transfer in soil and, 154
Xenobiotics, bacterial gene transfer in soil and, 108
Xenobiotics, effects on soil microorganisms, 195–198, 209, 210, 245–249
  activity, 203–209
  assessment, 234, 235
  efficacy, 242–245
    laboratory tests, 235–238
    methods, 238–242
  classification, 210–213
  effects, 219, 220
    mode of action, 233, 234
    processes, 220–233
  environment, 216–219
  interactions, 213–216
  soil as microbial habitat, 198
    gaseous phase, 202, 203
    solid phase, 198–201
    water phase, 201, 202

Y

Yeast
  α-amylase production and, 5, 13, 34
  levan and, 178, 186, 187
  patent law and, 260, 262, 266
  xenobiotics and, 205

Z

Zinc, xenobiotics and
  effects, 221–223, 226, 232
  environment, 216, 217
  interactions, 213

CONTENTS OF PREVIOUS VOLUMES

Volume 24

Preservation of Microorganisms
 Robert J. Heckly

Streptococcus mutans Dextransucrase: A Review
 Thomas J. Montville, Charles L. Cooney and Anthony J. Sinskey

Microbiology of Activated Sludge Bulking
 Wesley O. Pipes

Mixed Cultures in Industrial Fermentation Processes
 David E. F. Harrison

Utilization of Methanol by Yeasts
 Yoshiki Tani, Nobuo Kato, and Hideaki Yamada

Recent Chemical Studies on Peptide Antibiotics
 Jun'ichi Shoji

The CBS Fungus Collection
 J. A. Von Arx and M. A. A. Schipper

Microbiology and Biochemistry of Oil-Palm Wine
 Nduka Okafor

Bacterial-Amylases
 M. B. Ingle and R. J. Erickson

SUBJECT INDEX

Volume 25

Introduction to Extracellular Enzymes: From the Ribosome to the Market Place
 Rudy J. Wodzinski

Applications of Microbial Enzymes in Food Systems and in Biotechnology
 Matthew J. Taylor and Tom Richardson

Molecular Biology of Extracellular Enzymes
 Robert F. Ramaley

Increasing Yields of Extracellular Enzymes
 Douglas E. Eveleigh and Bland S. Montenecourt

Regulation of Chorismate-Derived Antibiotic Production
 Vedpal S. Malik

Structure-Activity Relationships in Fusidic Acid-Type Antibiotics
 W. von Daehne, W. O. Godtfredsen, and P. R. Rasmussen

Antibiotic Tolerance in Producer Organisms
 Leo C. Vining

Microbial Models for Drug Metabolism
 John P. Rosazza and Robert V. Smith

Plant Cell Cultures, a Potential Source of Pharmaceuticals
 W. G. W. Kurz and F. Constabel

Bacteriophages of the Genus Clostridium
 Seiya Ogata and Motoyoshi Hongo

SUBJECT INDEX

Volume 26

Microbial Oxidation of Gaseous Hydrocarbons
 Ching-Tsang Hou

Ecology and Diversity of Methylotrophic Organisms
 R. S. Hanson

Epoxidation and Ketone Formation by C_1-Utilizing Microbes
Ching-Tsang Hou, Ramesh N. Patel, and Allen I. Laskin

Oxidation of Hydrocarbons by Methane Monooxygenases from a Variety of Microbes
Howard Dalton

Propane Utilization of Microorganisms
Jerome J. Perry

Production of Intracellular and Extracellular Protein from n-Butane by Pseudomonas butanovora sp. nov.
Joji Takahashi

Effects of Microwave Irradiation on Microorganisms
John R. Chipley

Ethanol Production by Fermentation: An Alternative Liquid Fuel
N. Kosaric, D. C. M. Ng, I. Russell, and G. C. Stewart

Surface-Active Compounds from Microorganisms
D. G. Cooper and J. E. Zajic

INDEX

Volume 27

Recombinant DNA Technology
Vedpal Singh Malik

Nisin
A. Hurst

The Coumermycins: Developments in the Late 1970s
John C. Godfrey

Instrumentation for Process Control in Cell Culture
Robert J. Fleischaker, James C. Weaver, and Anthony J. Sinskey

Rapid Counting Methods for Coliform Bacteria
A. M. Cundell

Training in Microbiology at Indiana University-Bloomington
L. S. McClung

INDEX

Volume 28

Immobilized Plant Cells
P. Brodelius and K. Mosbach

Genetics and Biochemistry of Secondary Metabolism
Vedpal Singh Malik

Partition Affinity Ligand Assay (PALA): Applicatons in the Analysis of Haptens, Macromolecules, and Cells
Bo Mattiasson, Matts Ramstorp, and Torbjörn G. I. Ling

Accumulation, Metabolism, and Effects of Organophosphorus Insecticides on Microorganisms
Rup Lal

Solid Substrate Fermentations
K. E. Aidoo, R. Hendry, and B. J. B. Wood

Microbiology and Biochemistry of Miso (Soy Paste) Fermentation
Sumbo H. Abiose, M. C. Allan, and B. J. B. Wood

INDEX

Volume 29

Stabilization of Enzymes against Thermal Inactivation
Alexander M. Klibanov

Production of Flavor Compounds by Microorganisms
G. M. Kempler

New Perspectives on Aflatoxin Biosynthesis
J. W. Bennett and Siegfried B. Christensen

Biofilms and Microbial Fouling
W. G. Characklis and K. E. Cooksey

Microbial Inulinases: Fermentation Process, Properties, and Applications
Erick J. Vandamme and Dirk G. Derycke

Enumeration of Indicator Bacteria Exposed to Chlorine
Gordon A. McFeters and Anne K. Camper

Toxicity of Nickel to Microbes: Environmental Aspects
H. Babich and G. Stotzky

INDEX

Volume 30

Interactions of Bacteriophages with Lactic Streptococci
Todd R. Klaenhammer

Microbial Metabolism of Polycyclic Aromatic Hydrocarbons
Carl E. Cerniglia

Microbiology of Potable Water
Betty H. Olson and Laslo A. Nagy

Applied and Theoretical Aspects of Virus Adsorption to Surfaces
Charles P. Gerba

Computer Applications in Applied Genetic Engineering
Joseph L. Modelevsky

Reduction of Fading of Fluorescent Reaction Product for Microphotometric Quantitation
G. L. Picciolo and D. S. Kaplan

INDEX

Volume 31

Genetics and Biochemistry of Clostridium Relevant to Development of Fermentation Processes
Palmer Rogers

The Acetone Butanol Fermentation
B. McNeil and B. Kristiansen

Survival of, and Genetic Transfer by, Genetically Engineered Bacteria in Natural Environments
G. Stotzky and H. Babich

Apparatus and Methodology for Microcarrier Cell Culture
S. Reuveny and R. W. Thoma

Naturally Occurring Monobactams
William L. Parker, Joseph O'Sullivan, and Richard B. Sykes

New Frontiers in Applied Sediment Microbiology
Douglas Gunnison

Ecology and Metabolism of Thermothrix thiopara
Daniel K. Brannan and Douglas E. Caldwell

Enzyme-Linked Immunoassays for the Detection of Microbial Antigens and Their Antibodies
John E. Herrmann

The Identification of Gram-Negative, Nonfermentative Bacteria from Water: Problems and Alternative Approaches to Identification
N. Robert Ward, Roy L. Wolfe, Carol A. Justice, and Betty H. Olson

INDEX

Volume 32

Microbial Corrosion of Metals
Warren P. Iverson

Economics of the Bioconversion of Biomass to Methane and Other Vendable Products
Rudy J. Wodzinski, Robert N. Gennaro, and Michael H. Scholla

The Microbial Production of 2,3-Butanediol
Robert J. Magee and Naim Kosaric

Microbial Sucrose Phosphorylase: Fermentation process, Properties, and Biotechnical Applications
Erick J. Vandamme, Jan Van Loo, Lieve Machtelinckx, and Andre De Laports

Antitumor Anthracyclines Produced by Streptomyces peucetius
A. Grein

INDEX

Volume 33

The Cellulosome of Clostridium thermocellum
Raphael Lamed and Edward A. Bayer

Clonal Populations with Special Reference to Bacillus sphaericus
Samuel Singer

Molecular Mechanisms of Viral Inactivation by Water Disinfectants
R. B. Thurman and C. P. Gerba

Microbial Ecology of the Terrestrial Subsurface
William C. Ghiorse and John T. Wilson

Foam Control in Submerged Fermentation: State of the Art
N. P. Ghildyal, B. K. Lonsane, and N. G. Karanth

Applications and Mode of Action of Formaldehyde Condensate Biocides
H. W. Rossmoore and M. Sondossi

Occurrence and Mechanisms of Microbial Oxidation of Manganese
Kenneth H. Nealson, Bradley M. Tebo, and Reinhardt A. Rosson

Recovery of Bioproducts in China: A General Review
Xiong Zhenping

INDEX

Volume 34

What's in a Name?—Microbial Secondary Metabolism
J. W. Bennett and Ronald Bentley

Microbial Production of Gibberellins: State of the Art
P. K. R. Kumar and B. K. Lonsane

Microbial Dehydrogenations of Monosaccharides
Miloš Kulhánek

Antitumor and Antiviral Substances from Fungi
Shung-Chang Jong and Richard Donovick

Biotechnology—The Golden Age
V. S. Malik

INDEX

AUG 2 4 1990